Student Solutions

Wendy Pell
University of Ottawa
John Carran
Queen's University
Vladimir Kitaev
Wilfrid Laurier University
Rabin Bissessur
University of Prince Edward Island
John Sorensen
University of Manitoba
Rashmi Venkateswaran
University of Ottawa

CHEMISTRY

Second Canadian Edition

JOHN OLMSTED
California State University, Fullerton

GREG WILLIAMS
University of Oregon

ROBERT C. BURK
Carleton University

John Wiley & Sons Canada, Ltd.
6045 Freemont Blvd.
Mississauga, Ontario L5R 4J3

www.wiley.ca

Table of Contents

Chapter 1 Fundamental Concepts of Chemistry

Solutions to Problems in Chapter 1

1.1 You must commit to memory the correspondence between names and symbols of various elements, but remembering them is simplified by the fact that most English names and elemental symbols are related: (a) H; (b) He; (c) Hf; (d) N; (e) Ne; and (f) Nb.

1.3 Associating an element's name with its symbol requires memorization of both names and symbols. The examples in this problem all begin with the letter "A" but their names do not necessarily begin with "A" too: (a) arsenic; (b) argon; (c) aluminum; (d) americium; (e) silver; (f) gold; (g) astatine; and (h) actinium.

1.5 To convert a molecular picture into a molecular formula, count the atoms of each type and consult (or recall) the colour scheme used for the elements. See Figure 1-3 of your textbook for the colour scheme used in this and many other texts: (a) Br_2; (b) HCl; (c) C_2H_5I; and (d) C_3H_6O.

1.7 In writing a chemical formula, remember to use elemental symbols and subscripts for the number of atoms: (a) CCl_4; (b) H_2O_2; (c) P_4O_{10}; and (d) Fe_2S_3.

1.9 Scientific notation expresses any number as a value between 1 and 10 times a power of ten. Trailing zeros are retained only when they are significant: (a) 1.00000×10^5; (b) 1.0×10^4; (c) 4.00×10^{-4}; (d) 3×10^{-4}; and (e) 2.753×10^2.

1.11 To do unit conversions, multiply by a ratio that cancels the unwanted unit(s). Refer to Table 1-2 for the SI base units:

(a) $432\ kg = 4.32 \times 10^2\ kg$

(b) $624\ ps \left(\dfrac{10^{-12}\ s}{1\ ps} \right) = 6.24 \times 10^{-10}\ s$

(c) $1024\ ng \left(\dfrac{10^{-9}g}{1\ ng} \right) \left(\dfrac{10^{-3}\ kg}{1\ g} \right) = 1.024 \times 10^{-9}\ kg$

(d) $93\ 000\ km \left(\dfrac{10^3\ m}{1\ km} \right) = 9.300 \times 10^7\ m$

(e) $1\ day \left(\dfrac{24\ h}{1\ day} \right) \left(\dfrac{60\ min}{1\ h} \right) \left(\dfrac{60\ s}{1\ min} \right) =$ exactly 8.64×10^4 s (assuming exactly 1 day)

(f) 0.0426 in $\left(\dfrac{2.54 \text{ cm}}{1 \text{ in}}\right)\left(\dfrac{1 \text{ m}}{100 \text{ cm}}\right) = 1.08 \times 10^{-3}$ m

1.13 This is a unit-conversion problem involving summation and unusual units. First convert all masses into kg, then put them into the same power of ten and add the masses:

$$m_{\text{diamonds}} = 5.0 \times 10^{-1} \text{ carat} \left(\dfrac{3.168 \text{ grains}}{1 \text{ carat}}\right)\left(\dfrac{1 \text{ g}}{15.4 \text{ grains}}\right)\left(\dfrac{10^{-3} \text{ kg}}{1 \text{ g}}\right) = 1.0 \times 10^{-4} \text{ kg}$$

$$m_{\text{gold}} = 7.00 \text{ g} \left(\dfrac{10^{-3} \text{ kg}}{1 \text{ g}}\right) = 7.00 \times 10^{-3} \text{ kg}$$

$$m_{\text{total}} = (7.00 \times 10^{-3} \text{ kg}) + (0.10 \times 10^{-3} \text{ kg}) = 7.10 \times 10^{-3} \text{ kg}$$

1.15 The question asks for the density of water expressed in SI units (kg/m^3). Begin by analyzing the given information. The mass of the container is given before and after the water has been added. Thus, the mass of water can be obtained from the difference between the masses of the filled and empty container:

$m = 270.064 \text{ g} - 93.054 \text{ g} = 177.010 \text{ g } H_2O$

In SI units, $m = 177.010 \text{ g}\left(\dfrac{10^{-3} \text{ kg}}{1 \text{ g}}\right) = 0.177010 \text{ kg}$

Convert the units from inches to metres before computing the volume:

$r = 2.22 \text{ cm}\left(\dfrac{1 \text{ m}}{100 \text{ cm}}\right) = 0.0222 \text{ m}$

$h = 11.43 \text{ cm}\left(\dfrac{1 \text{ m}}{100 \text{ cm}}\right) = 0.1143 \text{ m}$

$V = \pi r^2 h = (3.1416)(0.0222 \text{ m})^2(0.1143 \text{ m}) = 1.77 \times 10^{-4} \text{ m}^3$

Calculate the density by dividing the mass of water by the volume:

$\rho = \dfrac{m}{V} = \dfrac{0.177010 \text{ kg}}{1.77 \times 10^{-4} \text{ m}^3} = 9.98 \times 10^2 \text{ kg/m}^3$ or 0.998 g/cm^3 (Round to three significant figures because the radius of the cylinder is known to only three significant figures.)

1.17 The question asks for a comparison of the masses of two objects of different densities and shapes. For each object, $m = \rho V$:

Au sphere: $\rho = 19.3$ g/cm^3, $V = \left(\dfrac{4}{3}\right)\pi r^3$ and $r = $ diameter/2 $= 1.00$ cm

$$V = \frac{4\pi(1.00 \text{ cm})^3}{3} = 4.19 \text{ cm}^3 \text{ and } m_{\text{gold}} = 4.19 \text{ cm}^3\left(\frac{19.3 \text{ g}}{1 \text{ cm}^3}\right) = 80.9 \text{ g}$$

Ag cube: $\rho = 10.50$ g/cm^3, $V = lwh$, and $l = w = h = 2.00$ cm

$$V = (2.00 \text{ cm})^3 = 8.00 \text{ cm}^3 \text{ and } m_{\text{silver}} = 8.00 \text{ cm}^3\left(\frac{10.50 \text{ g}}{1 \text{ cm}^3}\right) = 84.0 \text{ g}$$

The silver cube has more mass than the gold sphere.

1.19 Volume and mass are related through density:

$$V = \frac{m}{\rho} = \left(\frac{36.5 \text{ g}}{3.12 \text{ g/mL}}\right) = 11.7 \text{ mL}$$

1.21 The molar mass of a naturally occurring element can be calculated by summing the product of the fractional abundance of each isotope times its isotopic molar mass:

^{36}Ar: 35.96755 g/mol $\left(\dfrac{0.337\%}{100\%}\right) = 0.121$ g/mol

^{38}Ar: 37.96272 g/mol $\left(\dfrac{0.063\%}{100\%}\right) = 0.024$ g/mol

^{40}Ar: 39.96238 g/mol $\left(\dfrac{99.600\%}{100\%}\right) = 39.803$ g/mol

Elemental molar mass $= 0.121$ g/mol $+ 0.024$ g/mol $+ 39.803$ g/mol $= 39.948$ g/mol

1.23 Mass–mole conversions require the use of masses in grams and molar masses in grams per mole. To determine the number of moles, convert the mass into grams and divide by the molar mass:

$$n = \frac{m}{M}$$

(a) $n = 7.85 \text{ g}\left(\dfrac{1 \text{ mol}}{55.85 \text{ g}}\right) = 0.141 \text{ mol}$

(b) $n = 65.5 \ \mu g \left(\dfrac{10^{-6} \ g}{1 \ \mu g} \right) \left(\dfrac{1 \ mol}{12.01 \ g} \right) = 5.45 \times 10^{-6} \ mol$

(c) $n = 4.68 \ mg \left(\dfrac{10^{-3} \ g}{1 \ mg} \right) \left(\dfrac{1 \ mol}{28.09 \ g} \right) = 1.67 \times 10^{-4} \ mol$

(d) $n = 1.46 \ ton \left(\dfrac{10^{6} \ g}{1 \ ton} \right) \left(\dfrac{1 \ mol}{26.98 \ g} \right) = 5.41 \times 10^{4} \ mol$

1.25 To calculate the number of atoms in a mass, convert the mass to grams, divide by molar mass to obtain moles and multiply by Avogadro's number to obtain the number of atoms:

(a) $\# = 5.86 \ mg \left(\dfrac{10^{-3} \ g}{1 \ mg} \right) \left(\dfrac{1 \ mol}{9.012 \ g} \right) \left(\dfrac{6.022 \times 10^{23} \ atoms}{1 \ mol} \right) = 3.92 \times 10^{20} \ atoms$

(b) $\# = 5.86 \ mg \left(\dfrac{10^{-3} \ g}{1 \ mg} \right) \left(\dfrac{1 \ mol}{30.97 \ g} \right) \left(\dfrac{6.022 \times 10^{23} \ atoms}{1 \ mol} \right) = 1.14 \times 10^{20} \ atoms$

(c) $\# = 5.86 \ mg \left(\dfrac{10^{-3} \ g}{1 \ mg} \right) \left(\dfrac{1 \ mol}{91.22 \ g} \right) \left(\dfrac{6.022 \times 10^{23} \ atoms}{1 \ mol} \right) = 3.87 \times 10^{19} \ atoms$

(d) $\# = 5.86 \ mg \left(\dfrac{10^{-3} \ g}{1 \ mg} \right) \left(\dfrac{1 \ mol}{238.0 \ g} \right) \left(\dfrac{6.022 \times 10^{23} \ atoms}{1 \ mol} \right) = 1.48 \times 10^{19} \ atoms$

1.27 To calculate the molar mass of a compound, multiply each elemental molar mass by the number of atoms in the formula and sum over the elements:

(a) CCl_4: $M = 12.01$ g/mol C + 4(35.45 g/mol Cl) = 153.81 g/mol

(b) K_2S: $M = 2(39.10$ g/mol K) + 32.07 g/mol S = 110.27 g/mol

(c) O_3: $M = 3(16.00$ g/mol O) = 48.00 g/mol

(d) LiBr: $M = 6.94$ g/mol Li + 79.90 g/mol Br = 86.84 g/mol

(e) GaAs: $M = 69.72$ g/mol Ga + 74.92 g/mol As = 144.64 g/mol

(f) $AgNO_3$: $M = 107.87$ g/mol Ag + 14.01 g/mol N + 3(16.00 g/mol O) = 169.88 g/mol

1.29 Determine the molecular formula from the line drawing, taking into account the "missing" carbon atoms at the ends and vertices of lines and the "missing" hydrogen atoms attached to carbon atoms. To calculate the molar mass, multiply each elemental molar mass by the number of atoms in the formula and sum over the elements:

(a) Tyrosine: 9 C atoms, 7 missing H atoms, 4 shown H atoms, 1 N atom, 3 O atoms.

$C_9H_{11}NO_3$: $M = 9(12.01 \text{ g/mol}) + 11(1.008 \text{ g/mol}) + 1(14.01 \text{ g/mol}) + 3(16.00 \text{ g/mol})$
$= 181.19 \text{ g/mol}$

(b) Tryptophan: 11 C atoms, 8 missing H atoms, 4 shown H atoms, 2 N atoms, 2 O atoms.

$C_{11}H_{12}N_2O_2$: $M = 11(12.01 \text{ g/mol}) + 12(1.008 \text{ g/mol}) + 2(14.01 \text{ g/mol}) + 2(16.00 \text{ g/mol}) = 204.23 \text{ g/mol}$

(c) Glutamic acid: 5 C atoms, 9 H atoms, 1 N atom, 4 O atoms.;

$C_5H_9NO_4$: $M = 5(12.01 \text{ g/mol}) + 9(1.008 \text{ g/mol}) + 1(14.01 \text{ g/mol}) + 4(16.00 \text{ g/mol}) = 147.13 \text{ g/mol}$

(d) Lysine: 6 C atoms, 14 shown H atoms, 2 N atoms, 2 O atoms.

$C_6H_{14}N_2O_2$: $M = 6(12.01 \text{ g/mol}) + 14(1.008 \text{ g/mol}) + 2(14.01 \text{ g/mol}) + 2(16.00 \text{ g/mol}) = 146.19 \text{ g/mol}$

1.31 To calculate the number of atoms in a mass, convert the mass to grams, divide by molar mass to obtain moles, and multiply by Avogadro's number to obtain the number of atoms:

(a) CH_4: $M = 1(12.01 \text{ g/mol}) + 4(1.008 \text{ g/mol}) = 16.04 \text{ g/mol}$

$$\# = 5.86 \text{ mg} \left(\frac{10^{-3}\text{g}}{1 \text{ mg}}\right)\left(\frac{1 \text{ mol}}{16.04 \text{ g}}\right)\left(\frac{6.022\times10^{23} \text{ molecules}}{1 \text{ mol}}\right) = 2.20\times10^{20} \text{ molecules}$$

(b) Phosphorus trichloride is PCl_3:

$M = 1(30.974 \text{ g/mol}) + 3(35.453 \text{ g/mol}) = 137.33 \text{ g/mol}$

$$\# = 5.86 \text{ mg} \left(\frac{10^{-3}\text{g}}{1 \text{ mg}}\right)\left(\frac{1 \text{ mol}}{137.33 \text{ g}}\right)\left(\frac{6.022\times10^{23} \text{ molecules}}{1 \text{ mol}}\right) = 2.57\times10^{19} \text{ molecules}$$

(c) C_2H_6O: $M = 2(12.01 \text{ g/mol}) + 6(1.008 \text{ g/mol}) + 1(15.999 \text{ g/mol}) = 46.07 \text{ g/mol}$

$$\# = 5.86 \text{ mg} \left(\frac{10^{-3}\text{g}}{1 \text{ mg}}\right)\left(\frac{1 \text{ mol}}{46.07 \text{ g}}\right)\left(\frac{6.022\times10^{23} \text{ molecules}}{1 \text{ mol}}\right) = 7.66\times10^{19} \text{ molecules}$$

(d) Uranium hexafluoride is UF_6:

$$M = 1(238.03 \text{ g/mol}) + 6(18.998 \text{ g/mol}) = 352.02 \text{ g/mol}$$

$$\# = 5.86 \text{ mg} \left(\frac{10^{-3}\text{g}}{1 \text{ mg}}\right)\left(\frac{1 \text{ mol}}{352.02 \text{ g}}\right)\left(\frac{6.022\times10^{23} \text{ molecules}}{1 \text{ mol}}\right) = 1.00\times10^{19} \text{ molecules}$$

1.33 To calculate the mass of some number of molecules of a substance, divide by Avogadro's number to obtain moles and multiply by molar mass to obtain grams:

(a) CH_4: $M = 1(12.01 \text{ g/mol}) + 4(1.008 \text{ g/mol}) = 16.04 \text{ g/mol}$

$$m = 3.75\times10^5 \text{ molecules}\left(\frac{1 \text{ mol}}{6.022\times10^{23} \text{ molecules}}\right)\left(\frac{16.04 \text{ g}}{1 \text{ mol}}\right) = 9.99\times10^{-18} \text{ g}$$

(b) $C_9H_{13}NO_3$:

$$M = 9(12.01 \text{ g/mol}) + 13(1.008 \text{ g/mol}) + 1(14.01 \text{ g/mol}) + 3(16.00\text{g/mol}) = 183.2 \text{ g/mol}$$

$$m = 2.5\times10^9 \text{ molecules}\left(\frac{1 \text{ mol}}{6.022\times10^{23} \text{ molecules}}\right)\left(\frac{183.2 \text{ g}}{1 \text{ mol}}\right) = 7.6\times10^{-13} \text{ g}$$

(c) $C_{55}H_{72}MgN_4O_5$:

$$M = 55(12.01 \text{ g/mol}) + 72(1.008 \text{ g/mol}) + 1(24.305 \text{ g/mol}) + 4(14.01 \text{ g/mol}) + 5(16.00 \text{ g/mol}) = 893.5 \text{ g/mol}$$

$$m = 1 \text{ molecules}\left(\frac{1 \text{ mol}}{6.022\times10^{23} \text{ molecules}}\right)\left(\frac{893.5 \text{ g}}{1 \text{ mol}}\right) = 1.484\times10^{-21} \text{ g}$$

1.35 All parts of this question involve mass–mole–number conversions. Moles and mass in grams are related through the equation, $n = \dfrac{m}{M}$. Use Avogadro's number to convert between number and moles. When masses are not given in grams, unit conversions must be made. The chemical formula states the number of atoms of each element per molecule of substance:

$$M_{\text{vitamin A}} = 20(12.01 \text{ g/mol}) + 30(1.008 \text{ g/mol}) + 1(16.00 \text{ g/mol}) = 286.44 \text{ g/mol}$$

$$m = 0.75 \text{ mg}\left(\frac{10^{-3} \text{ g}}{1 \text{ mg}}\right)\left(\frac{1 \text{ mol}}{286.44 \text{ g}}\right) = 2.6\times10^{-6} \text{ mol vitamin A}$$

$$\# = nN_A = (2.6 \times 10^{-6} \text{ mol}) \left(\frac{6.022 \times 10^{23} \text{ molecules}}{1 \text{ mol}} \right) = 1.6 \times 10^{18} \text{ molecules}$$

There are 30 H atoms for every molecule of vitamin A:

atoms = (atoms/molecule)(# molecules)

$$\# \text{ H} = 1.6 \times 10^{18} \text{ molecules} \left(\frac{30 \text{ atoms H}}{1 \text{ molecule}} \right) = 4.8 \times 10^{19} \text{ atoms of H in 0.75 mg of vitamin A}$$

$$m_H = \frac{\#\text{H}}{N_A} M = 4.8 \times 10^{19} \text{ atoms} \left(\frac{1 \text{ mol}}{6.022 \times 10^{23} \text{ atoms}} \right) \left(\frac{1.008 \text{ g}}{1 \text{ mol}} \right) = 8.0 \times 10^{-5} \text{ g}$$

1.37 The solution process is $MgCl_2 (s) \rightarrow Mg^{2+} (aq) + 2 Cl^- (aq)$. Each mole of solid generates 1 mole of magnesium cations and 2 moles of chloride anions:

(a) Molarity is found using the equations, $c = \dfrac{n}{V}$ and $n = \dfrac{m}{M}$

$M = 1(24.31 \text{ g/mol}) + 2(35.45 \text{ g/mol}) = 95.21 \text{ g/mol}$

$$n_{Mg^{2+}} = 4.68 \text{ g } MgCl_2 \left(\frac{1 \text{ mol}}{95.21 \text{ g}} \right) \left(\frac{1 \text{ mol } Mg^{2+}}{1 \text{ mol } MgCl_2} \right) = 4.915 \times 10^{-2} \text{ mol } Mg^{2+}$$

$$n_{Cl^-} = 2\, n_{Mg^{2+}} = 9.831 \times 10^{-2} \text{ mol } Cl^-$$

Divide by the volume in litres to obtain the concentrations:

$$V = 1.50 \times 10^2 \text{ mL} \left(\frac{10^{-3} \text{ L}}{1 \text{ mL}} \right) = 0.150 \text{ L}$$

$$c_{Mg^{2+}} = \frac{4.915 \times 10^{-2} \text{ mol}}{0.150 \text{ L}} = 0.328 \text{ M}$$

$$c_{Cl^-} = 2\, c_{Mg^{2+}} = 0.656 \text{ M}$$

(Answers have three significant figures because the mass and volume are known to three significant figures.)

(b) Molecular pictures of solutions must illustrate the relative numbers of ions of each type present in the solution. Your picture must show twice as many chloride ions as magnesium ions. There are many more molecules of water than of either of the ions:

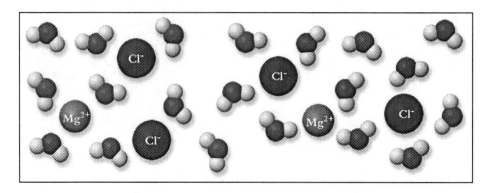

1.39 The solution process is KOH $(s) \rightarrow K^+ (aq) + OH^- (aq)$. Each mole of solid generates 1 mole of each ion:

(a) Molarity is found using the equations, $c = \dfrac{n}{V}$ and $n = \dfrac{m}{M}$

M = 39.098 g/mol + 15.999 g/mol + 1.0079 g/mol = 56.11 g/mol

$$c = \frac{n}{V} = \left(\frac{4.75 \text{ g}}{275 \text{ mL}} \right) \left(\frac{1 \text{ mol}}{56.11 \text{ g}} \right) \left(\frac{10^3 \text{ mL}}{1 \text{ L}} \right) = 0.308 \text{ M} = c_{K^+} = c_{OH^-}$$

(b) In a dilution, the number of moles of solute remains constant whereas volume increases, so $c_f V_f = c_i V_i$. The starting volume is 25.00 mL and the final volume is 100.00 mL:

$$c_f = \frac{c_i V_i}{V_f} = \frac{(0.308 \text{ M})(25.00 \text{ mL})}{100. \text{ mL}} = 0.0770 \text{ M} = c_{K^+} = c_{OH^-}$$

(c) Molecular pictures of solutions must illustrate the relative numbers of ions of each type present in the solution. Because of the four-fold dilution, the solution in (b) is 1/4 as concentrated as the solution in (a). We omit solvent molecules for clarity, but remember that there are many more molecules of water than of either of the ions:

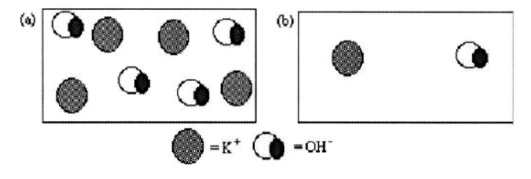

1.41 This is a dilution type problem. Rearrange the dilution equation to give an expression for the initial volume:

$$c_i V_i = c_f V_f \qquad V_i = \frac{c_f V_f}{c_i}$$

$$V_i = \frac{(0.125 \text{ M})(0.500 \text{ L})}{12.1 \text{ M}} = 0.00517 \text{ L or } 5.17 \text{ mL}$$

1.43 In each of the following, remember that concentration is $c = \dfrac{n}{V}$. Begin each part by identifying the major ionic species present in solution.

(a) Na_2CO_3 contains a Group 1 metal ion, Na^+, and a polyatomic anion, CO_3^{2-}, which are the major ionic species. Determine the number of moles in 4.55 g of Na_2CO_3 and use stoichiometric ratios to determine the number of moles of each ion:

$$M_{Na_2CO_3} = 2(22.99 \text{g/mol}) + 1(12.01 \text{ g/mol}) + 3(16.00 \text{g/mol}) = 105.99 \text{ g/mol}$$

$$n_{Na_2CO_3} = 4.55 \text{ g}\left(\frac{1 \text{ mol}}{105.99 \text{ g}}\right) = 0.0429 \text{ mol } Na_2CO_3$$

$$n_{CO_3^{2-}} = 0.0429 \text{ mol } Na_2CO_3 \left(\frac{1 \text{ mol } CO_3^{2-}}{1 \text{ mol } Na_2CO_3}\right) = 0.0429 \text{ mol } CO_3^{2-}$$

There are 2 moles of Na^+ per mole of Na_2CO_3:

$$n_{Na^+} = 0.0429 \text{ mol } Na_2CO_3 \left(\frac{2 \text{ mol } Na^+}{1 \text{ mol } Na_2CO_3}\right) = 0.0858 \text{ mol } Na^+$$

Now determine the molarities by dividing the moles by the volume of the solution:

$$V = 245 \text{ ml} \left(\frac{10^{-3} \text{ L}}{1 \text{ mL}}\right) = 0.245 \text{ L}$$

$$c_{CO_3^{2-}} = \left(\frac{0.0429 \text{ mol}}{0.245 \text{ L}}\right) = 0.175 \text{ M}$$

$$c_{Na^+} = 2\, c_{CO_3^{2-}} = 0.350 \text{ M}$$

(b) NH_4Cl contains a polyatomic ion and is therefore an ionic compound with two major ionic species, NH_4^+ and Cl^-. Determine the number of moles in 27.45 mg of the salt. The stoichiometric ratio is 1:1, so the number of moles of each ion is equal to the number of moles of the salt.

$$M_{NH_4Cl} = 1(14.01 \text{ g/mol}) + 4(1.008 \text{ g/mol}) + 1(35.45 \text{ g/mol}) = 53.49 \text{ g/mol}$$

$$n_{NH_4Cl} = 27.45 \text{ mg}\left(\frac{10^{-3} \text{ g}}{1 \text{ mg}}\right)\left(\frac{1 \text{ mol}}{53.49 \text{ g}}\right) = 5.132\times10^{-4} \text{ mol} = n_{NH_4^+} = n_{Cl^-}$$

Determine the molarities of each ion by dividing moles by the volume of the solution:

$$c_{NH_4^+} = c_{Cl^-} = \left(\frac{5.132\times10^{-4} \text{ mol}}{1.55\times10^{-2} \text{ L}}\right) = 0.033 \text{ M}$$

(c) Potassium sulfate (K_2SO_4) contains a Group 1 metal ion, K^+, and a polyatomic anion, SO_4^{2-}, which are the major ionic species. Determine the number of moles in 1.85 kg of the salt and use stoichiometric ratios to determine the number of moles of each ion:

$$M_{K_2SO_4} = 2(39.01 \text{ g/mol}) + 1(32.06 \text{ g/mol}) + 4(16.00 \text{ g/mol}) = 174.08 \text{ g/mol}$$

$$n_{K_2SO_4} = 1.85 \text{ kg}\left(\frac{10^3 \text{ g}}{1 \text{ kg}}\right)\left(\frac{1 \text{ mol}}{174.08 \text{ g}}\right) = 10.6 \text{ mol } K_2SO_4$$

$$n_{K_2SO_4^{2-}} = 10.6 \text{ mol } K_2SO_4\left(\frac{1 \text{ mol } SO_4^{2-}}{1 \text{ mol } K_2SO_4}\right) = 10.6 \text{ mol } SO_4^{2-}$$

$$n_{K^+} = 2n_{SO_4^{2-}} = 21.2 \text{ mol } K^+$$

Obtain the molarities by dividing the moles of each ion by the volume of the solution:

$$c_{SO_4^{2-}} = \frac{10.6 \text{ mol}}{5.75\times10^3 \text{ L}} = 1.84\times10^{-3} \text{ M}$$

Similarly to part (a), K^+ has twice the concentration as the anion:

$$c_{K^+} = \frac{21.2 \text{ mol}}{5.75\times10^3 \text{ L}} = 3.69\times10^{-3} \text{ M}$$

1.45 A balanced chemical equation must have equal numbers of atoms of each element on each side of the arrow. Balance each element in turn, beginning with those that appear in only one reactant and product, by adjusting stoichiometric coefficients.

Generally, H and O are balanced last. In each case, we start by determining the number of atoms on each side of the chemical equation:

(a) $NH_4NO_3 \rightarrow N_2O + H_2O$

$2\,N + 3\,O + 4\,H \rightarrow 2\,N + 2\,O + 2\,H$

There are two nitrogen atoms in the reactant and two in the products, thus nitrogen is already balanced. Balance H by making the stoichiometric coefficient of water 2:

$NH_4NO_3 \rightarrow N_2O + \mathbf{2}\,H_2O$

$2\,N + 3\,O + 4\,H \rightarrow 2\,N + 3\,O + 4\,H$ (all elements balanced)

(b) $P_4O_{10} + H_2O \rightarrow H_3PO_4$

$4\,P + 11\,O + 2\,H \rightarrow P + 4\,O + 3\,H$

Start by balancing P, which occurs in only one reactant and one product. There are 4 P on the reactant side and 1 P on the product side. Hence, balance P by giving H_3PO_4 a coefficient of 4:

$$P_4O_{10} + H_2O \rightarrow \mathbf{4}\,H_3PO_4$$

$$4\,P + 11\,O + 2\,H \rightarrow 4\,P + 16\,O + 12\,H$$

Next, balance H by giving H_2O a coefficient of 6, which also balances O:

$$P_4O_{10} + \mathbf{6}\,H_2O \rightarrow 4\,H_3PO_4$$

$$4\,P + 16\,O + 12\,H \rightarrow 4\,P + 16\,O + 12\,H \text{ (all elements balanced)}$$

(c) $HIO_3 \rightarrow I_2O_5 + H_2O$

$$I + H + 3\,O \rightarrow 2\,I + 2\,H + 6\,O$$

Notice that there are half as many of each atom on the reactant side as on the product side. Therefore, this equation can be balanced by simply increasing the HIO_3 coefficient from 1 to 2.

$$\mathbf{2}\,HIO_3 \rightarrow I_2O_5 + H_2O$$

$$2\,I + 2\,H + 6\,O \rightarrow 2\,I + 2\,H + 6\,O \text{ (all elements balanced)}$$

(d) $As + Cl_2 \rightarrow AsCl_5$

$$1\,As + 2\,Cl \rightarrow 1\,As + 5\,Cl$$

The As is already balanced. There are 2 Cl on the reactant side and 5 on the product. To balance Cl we need to change the Cl_2 coefficient from 1 to $\dfrac{5}{2}$:

$$As + \frac{5}{2}\,Cl_2 \rightarrow AsCl_5$$

$$1\,As + 5\,Cl \rightarrow 1\,As + 5\,Cl$$

Notice that both As and Cl are now balanced. However, we do not want fractions in a chemical equation. Therefore, multiply **all** coefficients by 2:

$$\mathbf{2}(As + \frac{5}{2}Cl_2 \rightarrow AsCl_5)$$

$$\mathbf{2}\ As + \mathbf{5}\ Cl_2 \rightarrow \mathbf{2}\ AsCl_5$$

$$2\ As + 10\ Cl \rightarrow 2\ As + 10\ Cl\ \text{(all elements balanced)}$$

1.47 Molecular pictures must show the correct number of molecules undergoing the reaction. In Problem 1.45(d), two atoms of As react with five molecules of Cl_2 to form two molecules of $AsCl_5$. Remember that when drawing molecular pictures you must differentiate between the different atom types by colour, labelling, or shading (here we use shading):

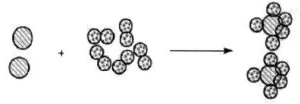

1.49 A balanced chemical equation must have equal numbers of atoms of each element on each side of the arrow. Balance each element in turn, beginning with those that appear in only one reactant and product, by adjusting stoichiometric coefficients. Generally, H and O are balanced last. When balancing an equation, start by determining the number of atoms on each side of the chemical equation:

(a) Start by determining the chemical formula of each compound. Molecular hydrogen: H_2; carbon monoxide: CO; methanol: CH_3OH

The chemical equation (unbalanced) is

$$H_2 + CO \rightarrow CH_3OH$$

$$2\ H + 1\ C + 1\ O \rightarrow 4\ H + 1\ C + 1\ O$$

Both carbon and oxygen are balanced. There are 4 H on the product side and 2 H on the reactant. Change the coefficient of H_2 from 1 to 2:

$$\mathbf{2}\ H_2 + CO \rightarrow CH_3OH$$

$$4\ H + 1\ C + 1\ O \rightarrow 4\ H + 1\ C + 1\ O\ \text{(all elements balanced)}$$

(b) $CaO + C \rightarrow CO + CaC_2$

$$1\ Ca + 1\ O + 1\ C \rightarrow 1\ Ca + 1\ O + 3\ C$$

Note that both Ca and O are balanced. Multiply the coefficient of C by 3:

$$CaO + \mathbf{3}\ C \rightarrow CO + CaC_2$$

$$1\ Ca + 1\ O + 3\ C \rightarrow 1\ Ca + 1\ O + 3\ C\ \text{(all elements balanced)}$$

(c) $C_2H_4 + O_2 + HCl \rightarrow C_2H_4Cl_2 + H_2O$

$$2\,C + 5\,H + 2\,O + Cl \rightarrow 2\,C + 6\,H + 1\,O + 2\,Cl$$

Carbon is already balanced. Both O and Cl are found in only one reactant and one product, so we can start with either. We choose Cl. Since there is one Cl on the reactant side and 2 on the product side, change the HCl coefficient to 2:

$$C_2H_4 + O_2 + \textbf{2}\ HCl \rightarrow C_2H_4Cl_2 + H_2O$$

$$2\,C + 6\,H + 2\,O + 2\,Cl \rightarrow 2\,C + 6\,H + 1\,O + 2\,Cl$$

This balances both Cl and H while leaving C balanced. To balance O, change the coefficient on O_2 from 1 to $\dfrac{1}{2}$:

$$C_2H_4 + \frac{1}{2}\,O_2 + 2\ HCl \rightarrow C_2H_4Cl_2 + H_2O$$

$$2\,C + 6\,H + 1\,O + 2\,Cl \rightarrow 2\,C + 6\,H + 1\,O + 2\,Cl$$

The reaction is now balanced. Multiply **all** coefficients by 2 to eliminate fractions from the chemical equation:

$$\textbf{2}(C_2H_4 + \frac{1}{2}\,O_2 + 2\ HCl \rightarrow C_2H_4Cl_2 + H_2O)$$

$$\textbf{2}\ C_2H_4 + O_2 + \textbf{4}\ HCl \rightarrow \textbf{2}\ C_2H_4Cl_2 + \textbf{2}\ H_2O$$

1.51 Molecular pictures must show the correct number of molecules undergoing the reaction. In Problem 1.49(a), two molecules of hydrogen react with a molecule of CO to form one molecule of CH_3OH.

1.53 A balanced chemical equation must have equal numbers of atoms of each element on each side of the arrow. Balance each element in turn, beginning with those that appear in only one reactant and product, by adjusting stoichiometric coefficients. Start by determining the number of atoms on each side of the chemical equation.

(a) $Ca(OH)_2 + H_3PO_4 \rightarrow H_2O + Ca_3(PO_4)_2$

     $1\,Ca + 6\,O + 5\,H + 1\,P \rightarrow 3\,Ca + 9\,O + 2\,H + 2\,P$

Start by balancing Ca by changing the coefficient of $Ca(OH)_2$ from 1 to 3:

     $\textbf{3}\ Ca(OH)_2 + H_3PO_4 \rightarrow H_2O + Ca_3(PO_4)_2$

     $3\,Ca + 10\,O + 9\,H + 1\,P \rightarrow 3\,Ca + 9\,O + 2\,H + 2\,P$

Next, balance P by multiplying the H_3PO_4 coefficient by 2:

     $3\ Ca(OH)_2 + \textbf{2}\ H_3PO_4 \rightarrow H_2O + Ca_3(PO_4)_2$

$$3\,Ca + 14\,O + 12\,H + 2\,P \rightarrow 3\,Ca + 9\,O + 2\,H + 2\,P$$

Now balance H by multiplying the water coefficient by 6

$$3\,Ca(OH)_2 + 2\,H_3PO_4 \rightarrow \mathbf{6}\,H_2O + Ca_3(PO_4)_2$$

$3\,Ca + 14\,O + 12\,H + 2\,P \rightarrow 3\,Ca + 14\,O + 12\,H + 2\,P$ (all elements balanced)

(b) $Na_2O_2 + H_2O \rightarrow NaOH + H_2O_2$

$$2\,Na + 3\,O + 2\,H \rightarrow 1\,Na + 3\,O + 3\,H$$

Start by balancing Na by changing the NaOH coefficient from 1 to 2:

$$Na_2O_2 + H_2O \rightarrow \mathbf{2}\,NaOH + H_2O_2$$

$$2\,Na + 3\,O + 2\,H \rightarrow 2\,Na + 4\,O + 4\,H$$

Since H occurs in only one reactant (O occurs in both), balance H next. Balance H by multiplying the water coefficient by 2:

$$Na_2O_2 + \mathbf{2}\,H_2O \rightarrow 2\,NaOH + H_2O_2$$

$2\,Na + 4\,O + 4\,H \rightarrow 2\,Na + 4\,O + 4\,H$ (all elements balanced)

(c) $BF_3 + H_2O \rightarrow HF + H_3BO_3$

$$1\,B + 3\,F + 2\,H + 1\,O \rightarrow 1\,B + 1\,F + 4\,H + 3\,O$$

Boron is already balanced. Balance F by multiplying the HF coefficient by 3:

$$BF_3 + H_2O \rightarrow \mathbf{3}\,HF + H_3BO_3$$

$$1\,B + 3\,F + 2\,H + 1\,O \rightarrow 1\,B + 3\,F + 6\,H + 3\,O$$

Next, balance H by multiplying the water coefficient by 3:

$$BF_3 + \mathbf{3}\,H_2O \rightarrow 3\,HF + H_3BO_3$$

$1\,B + 3\,F + 6\,H + 3\,O \rightarrow 1\,B + 3\,F + 6\,H + 3\,O$ (all elements balanced)

(d) $NH_3 + CuO \rightarrow Cu + N_2 + H_2O$

$$1\,N + 3\,H + 1\,Cu + 1\,O \rightarrow 2\,N + 2\,H + 1\,Cu + 1\,O$$

Cu is balanced. To balance H, multiply the NH_3 coefficient by 2 and H_2O by 3:

$$\mathbf{2}\,NH_3 + CuO \rightarrow Cu + N_2 + \mathbf{3}\,H_2O$$

$$2\,N + 6\,H + 1\,Cu + 1\,O \rightarrow 2\,N + 6\,H + 1\,Cu + 3\,O$$

This also balances N. Finally, balance O without unbalancing Cu by giving *both* CuO *and* Cu coefficients of 3:

$$2\,NH_3 + \mathbf{3}\,CuO \rightarrow \mathbf{3}\,Cu + N_2 + 3\,H_2O$$

$$2\,N + 6\,H + 3\,Cu + 3\,O \rightarrow 2\,N + 6\,H + 3\,Cu + 3\,O$$

1.55 We must calculate the mass of the second reactant that will completely react with 5.00 g of the first. Remember that calculations of amounts in chemistry always centre on the mole. Thus, we need first to determine how many moles there are in the first reactant and then determine how many moles of the second are required to react completely:

(a) Balanced equation: $2\ H_2 + CO \rightarrow CH_3OH$

$$5.00\ g\ H_2 \left(\frac{1\ mol\ H_2}{2.016\ g\ H_2}\right)\left(\frac{1\ mol\ CO}{2\ mol\ H_2}\right)\left(\frac{28.01\ g\ CO}{1\ mol\ CO}\right) = 34.7\ g\ CO$$

(b) Balanced equation: $CaO + 3\ C \rightarrow CO + CaC_2$

$$5.00\ g\ CaO \left(\frac{1\ mol\ CaO}{56.08\ g\ CaO}\right)\left(\frac{3\ mol\ C}{1\ mol\ CaO}\right)\left(\frac{12.01\ g\ C}{1\ mol\ C}\right) = 3.21\ g\ C$$

(c) Balanced equation: $2\ C_2H_4 + O_2 + 4\ HCl \rightarrow 2\ C_2H_4Cl_2 + 2\ H_2O$

$$5.00\ g\ C_2H_4 \left(\frac{1\ mol\ C_2H_4}{28.05\ g\ C_2H_4}\right)\left(\frac{1\ mol\ O_2}{2\ mol\ C_2H_4}\right)\left(\frac{32.00\ g\ O_2}{1\ mol\ O_2}\right) = 2.85\ g\ O_2$$

1.57 You are asked to calculate the mass of sodium iodide required to produce 1.50 kg of iodine. The balanced equation is given in the problem. Remember to convert the mass into grams before dividing by the molar mass:

$$1.50\ kg\ I_2 \left(\frac{1000\ g}{1\ kg}\right)\left(\frac{1\ mol}{253.8\ g}\right)\left(\frac{2\ mol\ NaI}{1\ mol\ I_2}\right)\left(\frac{149.89\ g}{1\ mol}\right) = 1.77 \times 10^3\ g\ NaI$$

1.59 This problem tells you how much of the reactant you have (in kg) and asks you to calculate the amount of product formed. Remember that calculations of amounts in chemistry always centre on the mole:

$$1.00\ kg\ sugar \left(\frac{10^3\ g}{1\ kg}\right)\left(\frac{1\ mol}{180.2\ g}\right)\left(\frac{2\ mol\ C_2H_5OH}{1\ mol\ sugar}\right)\left(\frac{46.07\ g}{1\ mol}\right) = 511\ g\ C_2H_5OH$$

1.61 To calculate masses for a chemical reaction, balance the equation and then do appropriate mass–mole–mass conversions. The reaction is

$$CCl_4 + HF \rightarrow CCl_2F_2 + HCl$$

$$1\ C + 4\ Cl + 1\ H + 1\ F \rightarrow 1\ C + 3\ Cl + 2\ F + 1\ H$$

Carbon is balanced. Fluorine occurs in only 1 reactant and 1 product, and therefore, balance it next. Fluorine can be balanced by giving HF a coefficient of 2:

$$CCl_4 + \mathbf{2}\ HF \rightarrow CCl_2F_2 + HCl$$

$$1\ C + 4\ Cl + 2\ H + 2\ F \rightarrow 1\ C + 3\ Cl + 2\ F + 1\ H$$

Finally, balance H and Cl by giving HCl a coefficient of 2:

$$CCl_4 + 2\ HF \rightarrow CCl_2F_2 + \mathbf{2}\ HCl$$

$$1\ C + 4\ Cl + 2\ H + 2\ F \rightarrow 1\ C + 4\ Cl + 2\ F + 2\ H\ \text{(all elements balanced)}$$

To determine the amount of HF required to completely react with the CCl_4, convert to moles and do the appropriate conversions:

$$175 \text{ kg } CCl_4 \left(\frac{10^3 \text{ g}}{1 \text{ kg}}\right)\left(\frac{1 \text{ mol}}{153.8 \text{ g}}\right)\left(\frac{2 \text{ mol HF}}{1 \text{ mol } CCl_4}\right)\left(\frac{20.01 \text{ g}}{1 \text{ mol}}\right)\left(\frac{1 \text{ kg}}{10^3 \text{ g}}\right) = 45.5 \text{ kg HF}$$

Since this reaction is 100% efficient, all of the reactants will be converted to products. Mass of products formed:

$$175 \text{ kg } CCl_4 \left(\frac{10^3 \text{ g}}{1 \text{ kg}}\right)\left(\frac{1 \text{ mol}}{153.8 \text{ g}}\right)\left(\frac{1 \text{ mol } CCl_2F_2}{1 \text{ mol } CCl_4}\right)\left(\frac{120.9 \text{ g}}{1 \text{ mol}}\right)\left(\frac{1 \text{ kg}}{10^3 \text{ g}}\right) = 138 \text{ kg } CCl_2F_2$$

$$175 \text{ kg } CCl_4 \left(\frac{10^3 \text{ g}}{1 \text{ kg}}\right)\left(\frac{1 \text{ mol}}{153.8 \text{ g}}\right)\left(\frac{2 \text{ mol HCl}}{1 \text{ mol } CCl_4}\right)\left(\frac{36.46 \text{ g}}{1 \text{ mol}}\right)\left(\frac{1 \text{ kg}}{10^3 \text{ g}}\right) = 83.0 \text{ kg } HCl_2$$

1.63 In a yield problem, we compare actual amounts with theoretical amounts. Carry out standard mole–mass conversions to find the theoretical amount.

The reaction (balanced) is

$$Ca_5(PO_4)_3F + 5 H_2SO_4 + 10 H_2O \rightarrow 3 H_3PO_4 + 5 CaSO_4 \cdot 2H_2O + HF$$

$$\% \text{ Yield} = (100\%)\left(\frac{\text{Actual yield}}{\text{Theoretical yield}}\right)$$

The problem gives the actual yield (400 g), but we need to calculate the theoretical yield. Use stoichiometry to calculate the theoretical yield (FA = fluoroapatite):

$$1.00 \text{ kg FA}\left(\frac{10^3 \text{ g}}{1 \text{ kg}}\right)\left(\frac{1 \text{ mol}}{504.3 \text{ g}}\right)\left(\frac{3 \text{ mol } H_3PO_4}{1 \text{ mol FA}}\right)\left(\frac{97.99 \text{ g}}{1 \text{ mol}}\right) = 583 \text{ g } H_3PO_4$$

Now divide the actual yield by the theoretical yield to obtain the percent yield:

$$\% \text{ Yield} = (100\%)\left(\frac{\text{Actual yield}}{\text{Theoretical yield}}\right) = (100\%)\left(\frac{400 \text{ g}}{583 \text{ g}}\right) = 68.6\%$$

1.65 In a multiple-step synthesis, the yield of each step must be multiplied to determine the overall yield. In this case, there are eight steps, each with a yield of 88%, so the overall fractional yield is $(0.88)(0.88)(0.88)(0.88)(0.88)(0.88)(0.88)(0.88) = (0.88)^8 = 0.36$. Use this value and the desired amount of phenobarbital in moles to determine how many moles of toluene are required. Do the usual mole–mass conversions:

For phenobarbital, $M = 12(12.01 \text{ g/mol}) + 12(1.01 \text{ g/mol}) + 2(14.01 \text{ g/mol}) + 3(16.00 \text{ g/mol}) = 232.3 \text{ g/mol}$

$$n_{\text{toluene}} = 25 \text{ kg}\left(\frac{10^3 \text{ g}}{1 \text{ kg}}\right)\left(\frac{1 \text{ mol phenolbarbital}}{232.2 \text{ g}}\right)\left(\frac{1 \text{ mol toluene}}{1 \text{ mol phenobarbital}}\right) = 107.7 \text{ mol toluene}$$

$$\text{Moles of toluene required} = \frac{107.7 \text{ mol}}{0.36} = 299 \text{ mol}$$

Mass required $= 299 \text{ mol } (92.13 \text{ g/mol})(10^{-3} \text{ kg/g}) = 28 \text{ kg}$

(The result has two significant figures because the mass and yield are only known to two significant figures.)

1.67 The calculation of a yield requires a theoretical amount and an actual amount. The theoretical amount is calculated as in Problem 1.61:

$$175 \text{ kg CCl}_4 \left(\frac{10^3 \text{ g}}{1 \text{ kg}}\right)\left(\frac{1 \text{ mol}}{153.8 \text{ g}}\right)\left(\frac{1 \text{ mol CCl}_2\text{F}_2}{1 \text{ mol CCl}_4}\right)\left(\frac{120.9 \text{ g}}{1 \text{ mol}}\right)\left(\frac{10^{-3} \text{ kg}}{1 \text{ g}}\right) = 137.6 \text{ kg CCl}_2\text{F}_2$$

(Carry an additional significant figure until the calculation is complete.) The actual amount (given) is 105 kg.

$$\% \text{ Yield} = (100\%) \left(\frac{105 \text{ kg}}{137.6 \text{ kg}}\right) = 76.3\%$$

To find how much of each reactant is required, divide the desired amount by the percent yield to obtain the theoretical yield and then do the usual stoichiometric calculations:

$$\text{Theoretical yield} = 155 \text{ kg} \left(\frac{100\%}{76.3\%}\right) = 203 \text{ kg CCl}_2\text{F}_2$$

$$203 \text{ kg CCl}_2\text{F}_2 \left(\frac{10^3 \text{ g}}{1 \text{ kg}}\right)\left(\frac{1 \text{ mol}}{120.9 \text{ g}}\right)\left(\frac{1 \text{ mol CCl}_4}{1 \text{ mol CCl}_2\text{F}_2}\right)\left(\frac{153.8 \text{ g}}{1 \text{ mol}}\right)\left(\frac{10^{-3} \text{ kg}}{1 \text{ g}}\right) = 258 \text{ kg CCl}_4$$

$$203 \text{ kg CCl}_2\text{F}_2 \left(\frac{10^3 \text{ g}}{1 \text{ kg}}\right)\left(\frac{1 \text{ mol}}{120.9 \text{ g}}\right)\left(\frac{2 \text{ mol HF}}{1 \text{ mol CCl}_2\text{F}_2}\right)\left(\frac{20.01 \text{ g}}{1 \text{ mol}}\right)\left(\frac{10^{-3} \text{ kg}}{1 \text{ g}}\right) = 67.2 \text{ kg HF}$$

1.69 To determine which ingredient will run out first, calculate how many cheeseburgers could be made with each ingredient:

Ingredient	Inventory	Amount/Burger	# Burgers
Roll	(12 dozen)(12/dozen)	1	144
Beef	40 lb	2(1/4 lb)	80
Cheese	(2 pkg)(65/pkg)	1	130
Tomato	40	1/4	160
Lettuce	(1 kg)(10^3 g/kg)	15 g	66.7

The lettuce will run out first, after 66 burgers have been made.

1.71 The problem gives information about the amounts of both starting materials, so this is a limiting reactant situation. We must calculate the number of moles of each species, construct a table of amounts, and use the results to determine the mass of the product formed. Starting amounts are in kilograms (1 kg = 10^3 g), so it will be convenient to work with 10^3 mol amounts. The balanced equation is given in the problem.

Begin by calculating the initial amounts:

$$75.0 \text{ kg N}_2\left(\frac{10^3 \text{ g}}{1 \text{ kg}}\right)\left(\frac{1 \text{ mol}}{28.02 \text{ g}}\right) = 2.68 \times 10^3 \text{ mol}$$

$$75.0 \text{ kg H}_2\left(\frac{10^3 \text{ g}}{1 \text{ kg}}\right)\left(\frac{1 \text{ mol}}{2.02 \text{ g}}\right) = 37.1 \times 10^3 \text{ mol}$$

Next, using the balanced chemical equation, construct an amounts table:

Reaction	$N_2 +$	$3 H_2 \rightarrow$	$2 NH_3$
Initial amount (10^3 mol)	2.68	37.1	0
10^3 mol/coeff	2.68 (LR)	12.4	
Change in amount (10^3 mol)	−2.68	−8.04	+5.36
Final amount (10^3 mol)	0	29.06	5.36

Do a mole–mass conversion to determine the mass of ammonia that could be produced:

$$5.36 \times 10^3 \text{ mol}\left(\frac{17.04 \text{ g}}{1 \text{ mol}}\right)\left(\frac{10^{-3} \text{ kg}}{1 \text{ g}}\right) = 91.3 \text{ kg}$$

1.73 The problem gives information about the amounts of starting materials, so this is a limiting reactant situation. We must calculate the number of moles of each species, construct a table of amounts, and use the results to determine the final product masses. Starting amounts are in metric tons (1 metric ton = 10^6 g), so it will be convenient to work with 10^6 mol amounts. See the answers to Problem 1.49 for the balanced equations:

(a) Calculate the initial amounts:

$$1000 \text{ kg CO}\left(\frac{10^3 \text{ g}}{1 \text{ kg}}\right)\left(\frac{1 \text{ mol}}{28.01 \text{ g}}\right) = 0.0357 \times 10^6 \text{ mol CO}$$

$$1000 \text{ kg H}_2\left(\frac{10^3 \text{ g}}{1 \text{ kg}}\right)\left(\frac{1 \text{ mol}}{2.02 \text{ g}}\right) = 0.495 \times 10^6 \text{ mol H}_2$$

Now set up an amounts table:

Reaction	$2 H_2 +$	$CO \rightarrow$	CH_3OH
Initial amount (10^6 mol)	0.495	0.0357	0
10^6 mol/coeff	0.248	0.0357 (LR)	
Change (10^6 mol)	−2(0.0357)	−0.0357	+0.0357
Final amount (10^6 mol)	0.424	0.00	0.0357

The mass that could be produced is

$$0.0357 \times 10^6 \text{ mol} \left(\frac{32.04 \text{ g}}{1 \text{ mol}} \right) \left(\frac{1 \text{ ton}}{10^6 \text{ g}} \right) = 1.14 \text{ metric ton CH}_3\text{OH}$$

(b) Calculate the initial amounts:

$$1000 \text{ kg CaO} \left(\frac{10^3 \text{ g}}{1 \text{ kg}} \right) \left(\frac{1 \text{ mol}}{56.08 \text{ g}} \right) = 0.0178 \times 10^6 \text{ mol CaO}$$

$$1000 \text{ kg C} \left(\frac{10^3 \text{ g}}{1 \text{ kg}} \right) \left(\frac{1 \text{ mol}}{12.01 \text{ g}} \right) = 0.0833 \times 10^6 \text{ mol C}$$

Now set up an amounts table:

Reaction	CaO +	3 C →	CO +	CaC$_2$
Initial amount (10^6 mol)	0.0178	0.0833	0	0
10^6 mol/coeff	0.0178 (LR)	0.0278		
Change (10^6 mol)	−0.0178	−3(0.0178)	+0.0178	+0.0178
Final amount (10^6 mol)	0	0.0299	+0.0178	+0.0178

The masses that could be produced are

$$\text{CO: } 0.0178 \times 10^6 \text{ mol} \left(\frac{28.01 \text{ g}}{1 \text{ mol}} \right) \left(\frac{1 \text{ ton}}{10^6 \text{ g}} \right) = 0.499 \text{ metric ton}$$

$$\text{CaC}_2 = 0.0178 \times 10^6 \text{ mol} \left(\frac{64.10 \text{ g}}{1 \text{ mol}} \right) \left(\frac{1 \text{ ton}}{10^6 \text{ g}} \right) = 1.14 \text{ metric ton}$$

(c) Calculate the initial amounts:

$$1000 \text{ kg C}_2\text{H}_4 \left(\frac{10^3 \text{ g}}{1 \text{ kg}} \right) \left(\frac{1 \text{ mol}}{28.05 \text{ g}} \right) = 0.0357 \times 10^6 \text{ mol C}_2\text{H}_4$$

$$1000 \text{ kg O}_2 \left(\frac{10^3 \text{ g}}{1 \text{ kg}} \right) \left(\frac{1 \text{ mol}}{32.00 \text{ g}} \right) = 0.0313 \times 10^6 \text{ mol O}_2$$

$$1000 \text{ kg HCl} \left(\frac{10^3 \text{ g}}{1 \text{ kg}} \right) \left(\frac{1 \text{ mol}}{36.46 \text{ g}} \right) = 0.0274 \times 10^6 \text{ mol HCl}$$

Now set up an amounts table:

Reaction	2 C$_2$H$_4$ +	O$_2$ +	4 HCl →	2 C$_2$H$_4$Cl$_2$ +	2 H$_2$O
Initial amount (10^6 mol)	0.0357	0.0313	0.0274	0	0
10^6 mol/coeff	0.0179	0.0313	0.00685 (LR)		

Change (10^6 mol)	$-\dfrac{2}{4}(0.0274)$	$-\dfrac{1}{4}(0.0274)$	-0.0274	$+\dfrac{2}{4}(0.0274)$	$+\dfrac{2}{4}(0.0274)$
Final amount (10^6 mol)	0.0220	0.0245	0	0.0137	0.0137

The masses that could be produced are

$$C_2H_4Cl_2\text{: } 0.0137\times10^6 \text{ mol}\left(\frac{98.96 \text{ g}}{1 \text{ mol}}\right)\left(\frac{1 \text{ ton}}{10^6 \text{ g}}\right) = 1.36 \text{ metric ton}$$

$$H_2O\text{: } 0.0137\times10^6 \text{ mol}\left(\frac{18.02 \text{ g}}{1 \text{ mol}}\right)\left(\frac{1 \text{ ton}}{10^6 \text{ g}}\right) = 0.247 \text{ metric ton}$$

$$\text{Mass NO: } 0.187\times10^6 \text{ mol}\left(\frac{30.01 \text{ g}}{1 \text{ mol}}\right)\left(\frac{1 \text{ ton}}{10^6 \text{ g}}\right) = 5.61 \text{ metric ton}$$

$$\text{Mass } H_2O\text{: } 0.281\times10^6 \text{ mol}\left(\frac{18.02 \text{ g}}{1 \text{ mol}}\right)\left(\frac{1 \text{ ton}}{10^6 \text{ g}}\right) = 5.06 \text{ metric ton}$$

1.75 The problem gives information about the amounts of both starting materials, so this is a limiting reactant situation. We must calculate the number of moles of each species, construct a table of amounts, and use the results to determine the masses of the product formed and the remaining reactant.

Begin by determining the balanced chemical equation:

$$P_4 + O_2 \rightarrow P_4O_{10}$$

P is already balanced. To balance O, give O_2 a coefficient of 5:

$$P_4 + \mathbf{5}\, O_2 \rightarrow P_4O_{10}$$

Next, calculate the initial amounts:

$$3.75 \text{ g } P_4\left(\frac{1 \text{ mol}}{123.9 \text{ g}}\right) = 0.0303 \text{ mol} \qquad 6.55 \text{ g } O_2\left(\frac{1 \text{ mol}}{32.00 \text{ g}}\right) = 0.205 \text{ mol}$$

Construct an amounts table:

Reaction	$P_4 +$	$5\, O_2 \rightarrow$	P_4O_{10}
Initial amount (mol)	0.0303	0.205	0
mol/coeff	0.0303 (LR)	0.0410	
Change (mol)	−0.0303	−5(0.0303)	+0.0303
Final amount (mol)	0	0.0535	0.0303

Now obtain the mass of P_4O_{10} produced, using the information from the amounts table and the molar mass: M = 4(30.97 g/mol) + 10(16.00 g/mol) = 283.9 g/mol

The mass that could be produced is $0.0303 \text{ mol}\left(\dfrac{283.9}{1 \text{ mol}}\right) = 8.60 \text{ g } P_4O_{10}$

Finally, determine the mass of O_2 left over:

There would be $0.0535 \text{ mol}\left(\dfrac{32.00 \text{ g}}{1 \text{ mol}}\right) = 1.71 \text{ g of } O_2$ left unreacted.

1.77 Both molecular fluorine and molecular chlorine are diatomic (two atoms per molecule) species. To have a total of 20 molecules in a 3:1 ratio there should be 5 molecules of chlorine (dark) and 15 molecules of fluorine (light).

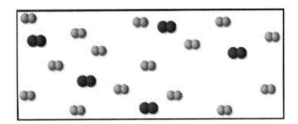

1.79 Kelvin–Celsius conversions involve addition or subtraction rather than multiplication or division, because the two scales have the same unit size but different zero points:

$$T(\text{K}) = T(^{\circ}\text{C}) + 273.15 \quad T = -11.5 \,^{\circ}\text{C} + 273.15 = 261.7 \text{ K}$$

1.81 The number of significant figures in a result of multiplication/division is determined by the number that contains the least number of significant figures:

(a) $\dfrac{\left(6.531\times10^{13}\right)\left(6.02\times10^{23}\right)}{(435)(2.000)} = 4.52\times10^{34}$

(b) $\dfrac{4.476+(3.44)(5.6223)+5.666}{(4.3)\left(7\times10^4\right)} = 1\times10^{-4}$

1.83 To draw molecular models, make use of the colour-coded, scaled atoms shown in Figure 1-4 in your textbook. Here, we use shading to indicate different elements:

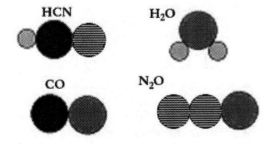

1.85 "Same number of atoms" also means "same number of moles," so work with moles:

$$n_{Li} = n_{Pt} \text{ and } n = \frac{m}{M}$$

$$m_{Li} = 5.75 \text{ g Pt} \left(\frac{1 \text{ mol Pt}}{195.08 \text{ g Pt}} \right) \left(\frac{1 \text{ mol Li}}{1 \text{ mol Pt}} \right) \left(\frac{6.941 \text{ g Li}}{1 \text{ mol Li}} \right) = 0.205 \text{ g}$$

1.87 The chemical formula of a substance provides all the information needed to compute its molar characteristics:

(a) $M = 2(26.98 \text{ g/mol}) + 3[32.07 \text{ g/mol} + 4(16.00 \text{ g/mol})] = 342.17 \text{ g/mol}$

(b) $n = \dfrac{m}{M} = 25.0 \text{ g} \left(\dfrac{1 \text{ mol}}{342.17 \text{ g}} \right) = 7.31 \times 10^{-2} \text{ mol}$

(c) To determine the percent composition, work with 1 mole of substance. Take the ratio of the mass of each element that 1 mole contains to the mass of 1 mole (molar mass):

$$\%Al = (100\%) \left(\frac{2(26.98 \text{ g/mol})}{342.17 \text{ g/mol}} \right) = 15.77\%$$

$$\%S = (100\%) \left(\frac{3(32.07 \text{ g/mol})}{342.17 \text{ g/mol}} \right) = 28.12\%$$

$$\%O = (100\%) \left(\frac{12(16.00 \text{ g/mol})}{342.17 \text{ g/mol}} \right) = 56.11\%$$

(d) $1.00 \text{ mol O} \left(\dfrac{1 \text{ mol Al}_2(SO_4)_3}{12 \text{ mol O}} \right) \left(\dfrac{342.17 \text{ g}}{1 \text{ mol}} \right) = 28.5 \text{ g}$

1.89 Convert from mass to moles to determine molarity:

$$c = \frac{n}{V} = \frac{m}{MV}$$

$$c = 8.3 \text{ mg/L} \left(\frac{10^{-3} \text{ g}}{1 \text{ mg}} \right) \left(\frac{1 \text{ mol}}{32.00 \text{ g}} \right) = 2.6 \times 10^{-4} \text{ mol/L}$$

1.91 The question asks about mole–mass–number quantities. Although there is much interesting biomedical information provided, the only relevant data are the chemical formula of verapamil, $C_{27}H_{38}O_4N_2$, and the amount of verapamil per tablet, 120.0 mg:

(a) $M = 27(12.01 \text{ g/mol}) + 38(1.008 \text{ g/mol}) + 4(16.00 \text{ g/mol}) + 2(14.01 \text{ g/mol}) = 454.59 \text{ g/mol}$

(b) $n = \dfrac{m}{M} = 120.0 \text{ mg} \left(\dfrac{10^{-3} \text{ g}}{1 \text{ mg}} \right) \left(\dfrac{1 \text{ mol}}{454.59 \text{ g}} \right) = 2.640 \times 10^{-4} \text{ mol}$

(c) $\text{\# atoms} = nN_A = 2.640 \times 10^{-4} \text{ mol} \left(\dfrac{6.022 \times 10^{23} \text{ molec. verap.}}{1 \text{ mol}} \right) \left(\dfrac{2 \text{ atoms N}}{1 \text{ molec. verap.}} \right)$

$$= 3.179 \times 10^{20} \text{ atoms N}$$

1.93 This problem describes a redox reaction. We are asked to determine the percent yield for the reaction. The starting materials are Xe and F_2, and the product is XeF_4:

$$Xe + 2 F_2 \rightarrow XeF_4$$

The problem gives information about the amount of one of the starting materials (the other is in excess), so this is a simple mass–mole–mass problem. We must calculate the theoretical yield, determine the percent yield, and use the actual yield to determine the final amount of Xe left over.

Calculations of theoretical yield:

$$M_{XeF_4} = 131.3 \text{ g/mol} + 4(19.0 \text{ g/mol}) = 207.3 \text{ g/mol}$$

The theoretical yield is determined from the starting amount of Xe:

$$n = 5.00 \text{ g} \left(\dfrac{1 \text{ mol}}{131.3 \text{ g}} \right) \left(\dfrac{1 \text{ mol XeF}_4}{1 \text{ mol Xe}} \right) = 3.81 \times 10^{-2} \text{ mol XeF}_4$$

The actual yield of XeF_4 is $4.00 \text{ g} \left(\dfrac{1 \text{ mol}}{207.3 \text{ g}} \right) = 1.93 \times 10^{-2} \text{ mol}$

Now determine the percent yields using the actual yield and the theoretical yield:

$$\% \text{ Yield} = (100\%) \left(\dfrac{1.93 \times 10^{-2} \text{ mol}}{3.81 \times 10^{-2} \text{ mol}} \right) = 50.7\%$$

The mass of Xe left unreacted is the difference between the starting mass and the amount converted into XeF_4:

$$m = 5.00 \text{ g} - \left(1.93 \times 10^{-2} \text{ mol} \right) \left(\dfrac{131.3 \text{ g}}{1 \text{ mol}} \right) = 2.47 \text{ g}$$

1.95 Examine the molecular picture to see what chemical reaction occurs. The reactants are atomic X and atomic Y and the product of the reaction is YX_2 (there are atoms of X left over). Thus, the reaction is $Y + 2X \rightarrow YX_2$.

1.97 This problem describes a combustion reaction. Begin by analyzing the chemistry. The problem asks about the amount of a product that forms from a given amount of reactant. First balance the chemical equation. The reaction is

$$C_2H_5OH + O_2 \rightarrow CO_2 + H_2O$$

$$2\,C + 6\,H + 3\,O \rightarrow 1\,C + 2\,H + 3\,O$$

Give CO_2 a coefficient of 2 and H_2O a coefficient of 3 to balance C and H:

$$C_2H_5OH + O_2 \rightarrow \mathbf{2}\,CO_2 + \mathbf{3}\,H_2O$$

$$2\,C + 6\,H + 3\,O \rightarrow 2\,C + 6\,H + 7\,O$$

There are 7 O on the product side, so give O_2 a coefficient of 3 to balance O:

$$C_2H_5OH + 3\,O_2 \rightarrow 2\,CO_2 + 3\,H_2O$$

Use the balanced equation to do the appropriate mass–mole–number calculations:

(a) $n_{H_2O} = 4.6\ \text{g}\left(\dfrac{1\ \text{mol}}{46.07\ \text{g}}\right)\left(\dfrac{3\ \text{mol}\ H_2O}{1\ \text{mol}\ C_2H_5OH}\right) = 0.30\ \text{mol}\ H_2O$

(b) $\# = nN_A = 0.30\ \text{mol}\left(\dfrac{6.022\times10^{23}\ \text{molecules}}{1\ \text{mol}}\right) = 1.8\times10^{23}\ \text{molecules}$

(c) $m = nM = 0.30\ \text{mol}\left(\dfrac{18.02\ \text{g}}{1\ \text{mol}}\right) = 5.4\ \text{g}$

1.99 There is much information about propylene oxide provided in this problem, but all that is needed for the calculations are the balanced chemical equation, the amount of starting material, and molar masses. The stoichiometry of this reaction is 1:1 in all reagents:

$M_{\text{propylene oxide}} = 3(12.01\ \text{g/mol}) + 6(1.01\ \text{g/mol}) + 1(16.00\ \text{g/mol}) = 58.1\ \text{g/mol}$

$M_{\text{propene}} = 3(12.01\ \text{g/mol}) + 6(1.01\ \text{g/mol}) = 42.1\ \text{g/mol}$

$M_{\text{hydroperoxide}} = 4(12.01\ \text{g/mol}) + 10(1.01\ \text{g/mol}) + 2(16.00\ \text{g/mol}) = 90.1\ \text{g/mol}$

(a) $m_{\text{propylene oxide}} = 75\ \text{kg}\left(\dfrac{1\ \text{kmol}}{90.1\ \text{kg}}\right)\left(\dfrac{1\ \text{kmol propylene oxide}}{1\ \text{kmol hydroperoxide}}\right)\left(\dfrac{58.1\ \text{kg}}{1\ \text{kmol}}\right) = 48\ \text{kg}$

(b) $m_{\text{propene}} = 75\ \text{kg}\left(\dfrac{1\ \text{kmol}}{90.1\ \text{kg}}\right)\left(\dfrac{1\ \text{kmol propene}}{1\ \text{kmol hydroperoxide}}\right)\left(\dfrac{42.1\ \text{kg}}{1\ \text{kmol}}\right) = 35\ \text{kg}$

1.101 Each atom contributes a distance equal to its diameter, so the number of atoms is the total distance divided by the diameter of an atom. A unit conversion is required between centimetres and picometres:

$$\text{\# atoms } = 2.54 \text{ cm} \left(\frac{1 \text{ m}}{100 \text{ cm}} \right)\left(\frac{10^{12} \text{ pm}}{1 \text{ m}} \right)\left(\frac{1 \text{ atom}}{200 \text{ pm}} \right) = 1.27 \times 10^8 \text{ atoms}$$

1.103 To determine how many atoms of an isotope are in a sample of an element, we must first determine how many atoms of that element are in the sample. Then, we can use the percentage composition to calculate how many atoms of one particular isotope there are. Use the molar mass of chromium to determine the number of moles, and then multiply by Avogadro's number to convert to number of atoms:

$$n = \frac{m}{M} = \frac{0.123 \text{ g}}{51.996 \text{ g/mol}} = 2.366 \times 10^{-3} \text{ mol}$$

$$\# = nN_A = (2.366 \times 10^{-3} \text{ mol})(6.022 \times 10^{23} \text{ atoms/mol}) = 1.425 \times 10^{21} \text{ atoms of Cr}$$

If 9.5% of the Cr is chromium-53:
$$\# = (1.425 \times 10^{21} \text{ atoms})(0.095) = 1.35 \times 10^{20} \text{ atoms of } ^{53}\text{Cr}$$

1.105 Moles and mass in grams are related through the equation, $n = \dfrac{m}{M}$. When masses are not given in grams, unit conversions must be made (1 lb = 453.6 g):

$$M_{\text{H}_3\text{PO}_4} = 3(1.008 \text{ g/mol}) + 1(30.97 \text{ g/mol}) + 4(16.00 \text{ g/mol}) = 97.99 \text{ g/mol}$$

$$n = 2.619 \times 10^8 \text{ lb} \left(\frac{453.6 \text{ g}}{1 \text{ lb}} \right)\left(\frac{1 \text{ mol}}{97.99 \text{ g}} \right) = 1.212 \times 10^9 \text{ mol H}_3\text{PO}_4$$

The chemical formula shows that there is 1 mole of P in every mole of H_3PO_4. The problem states that 15% of the annual production of H_3PO_4 comes from elemental P:

$$n_\text{P} = (1.212 \times 10^9 \text{ mol})\left(\frac{15\%}{100\%} \right) = 1.8 \times 10^8 \text{ mol}$$

$$m_\text{P} = nM = 1.8 \times 10^8 \text{ mol}\left(\frac{30.97 \text{ g}}{1 \text{ mol}} \right)\left(\frac{10^{-3} \text{ kg}}{1 \text{ g}} \right) = 5.6 \times 10^6 \text{ kg of P consumed}$$

1.107 The problem asks about yields and provides information about the amounts of both starting materials, so it is both a limiting reagent and a yield problem. Use a table of amounts to determine the theoretical yield.

Calculations of initial amounts:

$$N_2:\ 84.0\ kg\left(\frac{1\ g}{10^{-3}\ kg}\right)\left(\frac{1\ mol}{28.02\ g}\right) = 3.00 \times 10^3\ mol$$

$$H_2:\ 24.0\ kg\left(\frac{1\ g}{10^{-3}\ kg}\right)\left(\frac{1\ mol}{2.016\ g}\right) = 1.19 \times 10^4\ mol$$

Reaction	$N_2 +$	$3\ H_2 \rightarrow$	$2\ NH_3$
Initial amount (10^3 mol)	3.00	11.9	0
mol/coeff	3.00 (LR)	3.97	
Change (10^3 mol)	−3.00	−9.00	+6.00
Final amount (10^3 mol)	0	2.9	6.00

Use the table to obtain the theoretical yield of NH_3:

$$6.00 \times 10^3\ mol\left(\frac{17.04\ g}{1\ mol}\right)\left(\frac{10^{-3}\ kg}{1\ g}\right) = 102\ kg$$

The theoretical yield is 102 kg, and the actual yield is 68 kg:

$$\%\ Yield = (100\%)\left(\frac{68\ kg}{102\ kg}\right) = 67\%$$

Multiply the theoretical change by the percent yield to get the actual change:

Actual change = (67%)(6.00 × 10^3 mol) = 4.00 × 10^3 mol

Repeat the table of amounts using the actual change rather than the theoretical change:

Reaction	$N_2 +$	$3\ H_2 \rightarrow$	$2\ NH_3$
Amount (10^3 mol)	3.00	11.9	0
Change (10^3 mol)	−2.00	−6.00	+4.00
Final (10^3 mol)	1.00	5.9	4.00

Do mole–mass conversions to determine the masses of leftover reactants:

$$N_2:\ 1.00 \times 10^3\ mol\left(\frac{28.02\ g}{1\ mol}\right)\left(\frac{1\ kg}{10^3\ g}\right) = 28\ kg$$

$$H_2:\ 5.9 \times 10^3\ mol\left(\frac{2.016\ g}{1\ mol}\right)\left(\frac{1\ kg}{10^3\ g}\right) = 12\ kg$$

1.109 This problem asks about a mixture of two salts, $CaSO_4$ and $Ca(H_2PO_4)_2$:

   (a) The 2:1 mole ratio indicates the coefficients for the balanced equation. There is no water present during this reaction, so we need not ask about species present:

$$2\ H_2SO_4 + Ca_3(PO_4)_2 \rightarrow 2\ CaSO_4 + Ca(H_2PO_4)_2$$

(b) Because the mixture has fixed stoichiometric proportions of 2:1, it has a combined molar mass of $2\ M_{CaSO_4} + 1\ M_{Ca(H_2PO_4)_2}$:

$$M_{combination} = 2(136.1\ g/mol) + 1(234.1\ g/mol) = 506.3\ g/mol$$

$$n_{combination} = 50.0\ kg \left(\frac{10^3\ g}{1\ kg} \right) \left(\frac{1\ mol}{506.3\ g} \right) = 98.76\ mol$$

$$n_{acid} = 2\ n_{combination} \qquad\qquad n_{Ca_3(PO_4)_2} = n_{combination}$$

$$m_{acid} = 98.76\ mol \left(\frac{2\ mol\ H_2SO_4}{1\ mol\ combination} \right) \left(\frac{98.09\ g}{1\ mol} \right) \left(\frac{10^{-3}\ kg}{1\ g} \right) = 19.4\ kg$$

$$m_{Ca_3(PO_4)_2} = 98.76\ mol \left(\frac{1\ mol\ Ca_3(PO_4)_2}{1\ mol\ combination} \right) \left(\frac{310.2\ g}{1\ mol} \right) \left(\frac{10^{-3}\ kg}{1\ g} \right) = 30.6\ kg$$

(c) Each mole of $Ca_3(PO_4)_2$ yields 2 moles of phosphate ion, so the amount of phosphate ion available is $2\ n_{Ca_3(PO_4)_2} = 2(98.76\ mol) = 198\ mol$

1.111 This is a density problem. The data provided are sufficient to determine the number density of red blood cells. The number of red blood cells ($\#_{rbc}$) in an adult human is then the volume of blood multiplied by the number density:

$$\#_{rbc} = \rho V \quad \text{and} \quad V = 5\ L$$

$$\rho = \left(\frac{6.0 \times 10^3\ rbc}{(0.1\ mm)^3} \right) \left(\frac{10\ mm}{1\ cm} \right)^3 \left(\frac{1000\ cm^3}{1\ L} \right) = 6 \times 10^{12}\ rbc/L$$

$$\#_{rbc} = (6 \times 10^{12}\ rbc/L)(5\ L) = 3 \times 10^{13}\ rbc\ \text{in a typical adult}$$

1.113 Make use of tabulated data in the *CRC Handbook of Chemistry and Physics* to identify the Z-values of elements falling in each category, and use the periodic table to determine the symbol for each:

Elements with only one stable isotope, with their Z-values, are the following: 4, Be; 9, F; 11, Na; 13, Al; 15, P; 21, Sc; 23, V; 25, Mn; 27, Co; 33, As; 39, Y; 41, Nb; 45, Rh; 53, I; 55, Cs; 59, Pr; 65, Tb; 67, Ho; 69, Tm; 79, Au; and 83, Bi. Elements with four or more stable isotopes, with their Z-values, are the following: 16, S; 20, Ca; 22, Ti; 24, Cr; 26, Fe; 28, Ni; 30, Zn; 32, Ge; 34, Se; 36, Kr; 38, Sr; 40, Zr; 42, Mo; 44, Ru; 46, Pd; 48, Cd; 50, Sn; 52, Te; 54, Xe; 56, Ba; 58, Ce; 60, Nd; 62, Sm; 64, Gd; 66, Dy; 68, Er; 70, Yb; 72, Hf; 74, W; 76, Os; 78, Pt; and 82, Pb. Apart from Be ($Z = 4$), all the elements with just one stable isotope have an odd number of protons (odd Z-value). In sharp contrast, all the elements having four or more stable isotopes have an even number of protons (even Z-value). In addition, five light elements ($Z < 16$) have just one stable isotope, whereas none of the light elements has four or more.

1.115 (a) Molar mass of HFC-134a (CF_3CH_2F)
$$= 2(12.011 \text{ g/mol}) + 4(18.998 \text{ g/mol}) + 2(1.008 \text{ g/mol})$$
$$= 102.03 \text{ g/mol}$$

Molar mass of trichloroethylene (CCl_2CHCl)
$$= 2(12.011 \text{ g/mol}) + 3(35.453 \text{ g/mol}) + 1.008 \text{ g/mol}$$
$$= 131.389 \text{ g/mol}$$

Moles of HFC-134a produced $= (65 \times 10^6 \text{ kg})(1000 \text{ g/kg})/(102.03 \text{ g/mol})$
$$= 6.37 \times 10^8 \text{ moles}$$
The reaction yield is 47%; therefore, to produce this amount of HFC-134a, you would require enough raw materials to make
$$\frac{6.37 \times 10^8 \text{ mol}}{0.47} = 1.36 \times 10^9 \text{ moles of HFC-134a.}$$

$$n_{\text{trichloroethylene}} = n_{\text{HFC-134a}}$$

$$m_{\text{trichloroethylene}} = nM = (1.36 \times 10^9 \text{ mol})(131.389 \text{ g/mol})$$
$$= 1.78 \times 10^{11} \text{ g}$$
$$= 178 \text{ million kg}$$

(b) If the process were 53% efficient (rather than 47%), then the 178 million kg (1.36×10^9 moles) of trichloroethylene would produce:

$$(1.36 \times 10^9 \text{ mol})(0.53)\left(\frac{1 \text{ mol HFC-134a}}{1 \text{ mol trichloroethylene}}\right) = 7.21 \times 10^8 \text{ mol HFC-134a}$$

$$m_{\text{HFC-134a}} = nM = (7.21 \times 10^8 \text{ mol})(102.03 \text{ g/mol}) = 7.35 \times 10^{10} \text{ g}$$
$$= 74 \text{ million kg}$$

(c) Excess production $= (74 - 65)$ million kg $= 9$ million kg

$$\text{Excess } \$ = (9 \times 10^6 \text{ kg})(\$5.50/\text{kg}) = \$50 \times 10^6$$

1.117 To solve this problem requires molecular weights of the starting material and of the product, but not of the two intermediate molecules. The molecular formula of isobutylbenzene is $C_{10}H_{14}$, which has a molecular weight of 134.2 g/mol, and that of ibuprofen is $C_{13}H_{18}O_2$, which has a molecular weight of 206.3 g/mol.

If the yields of all three reactions were 100%, the theoretical mass of the product

would be limited by the mass of the reactant:

$$100 \text{ kg C}_{10}\text{H}_{14} \times \frac{1000 \text{ g}}{1 \text{ kg}} \times \frac{1 \text{ mol C}_6\text{H}_{14}}{134.2 \text{ g C}_6\text{H}_{14}} \times \frac{1 \text{ mol C}_{13}\text{H}_{18}\text{O}_2}{1 \text{ mol C}_6\text{H}_{14}} \times \frac{206.3 \text{ g C}_{13}\text{H}_{18}\text{O}_2}{1 \text{ mol C}_{13}\text{H}_{18}\text{O}_2}$$

$$= 153{,}700 \text{ g C}_{13}\text{H}_{18}\text{O}_2$$

$$= 153.7 \text{ kg C}_{13}\text{H}_{18}\text{O}_2$$

But the actual overall yield will be a function of the product of the yields of the three

reactions, namely 91% x 90% x 85% = 69.6%, so the actual mass of ibuprofen

produced will be $\dfrac{69.6\%}{100\%}(153.7 \text{kg}) = 107.0 \text{ kg}$

Chapter 2 The Behaviour of Gases

Solutions to Problems in Chapter 2

2.1 A pinhole in the top of the tube would let air leak into the space until the internal pressure matched the external, atmospheric pressure. The barometer would then indicate zero pressure, because the height of the mercury column is determined by the pressure *difference* between inside and outside, and this difference would be zero in the presence of a pinhole.

2.3 When air is pumped into a flat tire, the tire expands and becomes hard. Both these observations are the result of the pressure exerted by the gas molecules making up air.

2.5 Useful pressure conversion factors are 1 atm = 1.01325×10^5 Pa, 1 atm = 760 Torr, and 1 bar = 10^5 pascals (Pa).

(a) $455 \text{ Torr} \left(\dfrac{1 \text{ atm}}{760 \text{ Torr}} \right) \left(\dfrac{1.01325 \times 10^5 \text{ Pa}}{1 \text{ atm}} \right) = 6.07 \times 10^4 \text{ Pa} = 0.607 \text{ bar}$

(b) $2.45 \text{ atm} \left(\dfrac{1.01325 \times 10^5 \text{ Pa}}{1 \text{ atm}} \right) = 2.48 \times 10^5 \text{ Pa} = 2.48 \text{ bar}$

(c) $0.46 \text{ Torr} \left(\dfrac{1 \text{ atm}}{760 \text{ atm}} \right) \left(\dfrac{1.01325 \times 10^5 \text{ Pa}}{1 \text{ atm}} \right) = 61 \text{ Pa} = 6.1 \times 10^{-4} \text{ bar}$

(d) $1.33 \times 10^{-3} \text{ atm} \left(\dfrac{1.01325 \times 10^5 \text{ Pa}}{1 \text{ atm}} \right) = 1.35 \times 10^2 \text{ Pa} = 1.35 \times 10^{-3} \text{ bar}$

2.7 The first part of this question asks about number of moles. Rearrange the ideal gas equation to solve for *n*, and then substitute the appropriate values and do the calculation:

$$n = \frac{pV}{RT} = \frac{(5.00 \text{ bar})(20.0 \text{ L})}{(0.08314 \text{ L bar mol}^{-1} \text{ K}^{-1})(298 \text{ K})} = 4.04 \text{ mol}$$

The second part of the question asks about the volume if the pressure was different. We can calculate this volume using *n* and the ideal gas equation, but we can also do the calculation by using proportionalities.

$$p_i V_i = p_f V_f, \text{from which } V_i = \frac{p_f V_f}{p_i} = \frac{(5.00 \text{ bar})(20.0 \text{ L})}{1.00 \text{ bar}} = 100. \text{ L}$$

2.9 When some variables are held fixed, rearrange the ideal gas equation, $pV = nRT$, to collect fixed values on the right.

(a) n, R, V are constant: $\dfrac{p}{T} = \dfrac{nR}{V} = \text{constant};\ \dfrac{p_i}{T_i} = \dfrac{p_f}{T_f}$

(b) $V = \dfrac{nRT}{p}$

(c) n, R, T are constant: $pV = nRT = \text{constant};\ p_iV_i = p_fV_f$

2.11 When some conditions change but others remain fixed, rearrange the ideal gas equation so the constant terms are grouped on the right. In this problem, n and p are fixed:

$$\frac{V}{T} = \frac{nR}{p} = \text{constant, so } \frac{V_i}{T_i} = \frac{V_f}{T_f}$$

$$V_i = 0.255\ \text{L}$$

Convert temperature to kelvins:

$$T_i = 25 + 273.15 = 298\ \text{K} \quad T_f = -15 + 273.15 = 258\ \text{K}$$

$$V_f = \frac{V_iT_f}{T_i} = \frac{(0.255\ \text{L})(258\ \text{K})}{(298\ \text{K})} = 0.221\ \text{L}$$

2.13 The equation is valid only for a gas for which n and T are fixed. (a) n and T are fixed, so the equation is valid. (b) n can change, so the equation is not valid. (c) T changes, so the equation is not valid. (d) The equation is not valid for liquids.

2.15 According to Dalton's law of partial pressures, each gas exerts a pressure equal to the total pressure times its mole fraction. Mole fraction can be found from concentration in ppm:

$$X = \left(\frac{\text{ppm}}{10^6}\right)$$

$$p_{NO_2} = 758.4\ \text{Torr} \left(\frac{1\ \text{atm}}{760\ \text{Torr}}\right)\left(\frac{0.78\ \text{molecules } NO_2}{10^6\ \text{molecules of air}}\right) = 7.8 \times 10^{-7}\ \text{atm}$$

2.17 According to Dalton's law of partial pressures, each gas exerts a pressure equal to its mole fraction times the total pressure: $p_i = X_i p_{\text{total}}$. Standard atmospheric conditions correspond to $p_{\text{total}} = 1$ atm.

$$p_{N_2} = (0.7808)(1\ \text{atm})\left(\frac{760\ \text{Torr}}{1\ \text{atm}}\right) = 593.4\ \text{Torr}$$

$$p_{O_2} = (0.2095)(1\ \text{atm})\left(\frac{760\ \text{Torr}}{1\ \text{atm}}\right) = 159.2\ \text{Torr}$$

$$p_{Ar} = (9.34 \times 10^{-3})(1 \text{ atm}) \left(\frac{760 \text{ Torr}}{1 \text{ atm}} \right) = 7.10 \text{ Torr}$$

$$p_{CO_2} = (3.25 \times 10^{-4})(1 \text{ atm}) \left(\frac{760 \text{ Torr}}{1 \text{ atm}} \right) = 2.47 \times 10^{-1} \text{ Torr}$$

2.19 Molecular pictures show the relative numbers of molecules of various substances, which also represent the relative numbers of moles of those substances. The figure contains 8 He atoms and 4 O_2 molecules.

(a) and (b) There are more He atoms, so the pressure due to He and the mole fraction of He are higher.

(c) $X_{He} = \dfrac{n_{He}}{n_{total}} = \dfrac{8}{12} = 0.67$

2.21 To calculate partial pressures from a total pressure, mole fractions are required: $p_i = X_i\, p_{total}$. Mole fractions can be calculated from the analytical data for the gas sample. The equations needed to solve this problem are

$$n = \frac{m}{M} \qquad X_i = \frac{n_i}{n_{total}} \qquad p_i = \frac{n_i R T}{V}$$

$$n_{CH_4} = 1.57 \text{ g} \left(\frac{1 \text{ mol}}{16.04 \text{ g}} \right) = 9.79 \times 10^{-2} \text{ mol}$$

$$n_{C_2H_6} = 0.41 \text{ g} \left(\frac{1 \text{ mol}}{30.07 \text{ g}} \right) = 1.36 \times 10^{-2} \text{ mol}$$

$$n_{C_3H_8} = 0.020 \text{ g} \left(\frac{1 \text{ mol}}{44.09 \text{ g}} \right) = 4.54 \times 10^{-4} \text{ mol}$$

$n_{total} = 9.79 \times 10^{-2} \text{ mol} + 1.36 \times 10^{-2} \text{ mol} + 4.54 \times 10^{-4} \text{ mol} = 1.120 \times 10^{-1} \text{ mol}$. Determine the mole fraction by dividing the moles of each component by the total number of moles:

$$X_{CH_4} = \frac{9.79 \times 10^{-2} \text{ mol}}{1.120 \times 10^{-1} \text{ mol}} = 0.874 \qquad X_{C_2H_6} = \frac{1.36 \times 10^{-2} \text{ mol}}{1.120 \times 10^{-1} \text{ mol}} = 0.121$$

$$X_{C_3H_8} = \frac{4.54 \times 10^{-4} \text{ mol}}{1.120 \times 10^{-1} \text{ mol}} = 4.06 \times 10^{-3}$$

Multiply the mole fractions by the total pressure to obtain the partial pressure (round the final values for ethane and propane to two significant figures, because their masses are only known to two significant figures):

$$p_{CH_4} = (0.874)(2.35 \text{ bar}) = 2.05 \text{ bar}$$

$$p_{C_2H_6} = (0.121)(2.35 \text{ bar}) = 0.28 \text{ bar}$$

$$p_{C_3H_8} = (4.06 \times 10^{-3})(2.35 \text{ bar}) = 9.5 \times 10^{-3} \text{ bar}$$

2.23 Stoichiometric calculations require moles and a balanced chemical equation. Balance the equation, determine the number of moles of CO_2 produced, and apply the ideal gas equation to calculate V:

The unbalanced chemical equation is

$$C_6H_{12}O_6 + O_2 \rightarrow CO_2 + H_2O$$

$$6 \text{ C} + 12 \text{ H} + 8 \text{ O} \rightarrow 1 \text{ C} + 2 \text{ H} + 3 \text{ O}$$

Give both CO_2 and H_2O a coefficient of 6 to balance C and H:

$$C_6H_{12}O_6 + O_2 \rightarrow 6 \text{ CO}_2 + 6 \text{ H}_2O;$$

$$6 \text{ C} + 12 \text{ H} + 8 \text{ O} \rightarrow 6 \text{ C} + 12 \text{ H} + 18 \text{ O}$$

This leaves 18 O on the product side. 6 O are from glucose, so O_2 needs a coefficient of 6 to balance O and obtain the balanced chemical equation:

$$C_6H_{12}O_6 \, (g) + 6 \text{ O}_2 \, (g) \rightarrow 6 \text{ CO}_2 \, (g) + 6 \text{ H}_2O \, (l)$$

$M_{glucose} = 6(12.01 \text{ g/mol}) + 12(1.01 \text{ g/mol}) + 6(16.00 \text{ g/mol}) = 180.18 \text{ g/mol}$. Use mass–mole conversions to determine the moles of CO_2 formed:

$$n_{CO_2} = 4.65 \text{ g C}_6H_{12}O_6 \left(\frac{1 \text{ mol}}{180.18 \text{ g}} \right) \left(\frac{6 \text{ mol CO}_2}{1 \text{ mol C}_6H_{12}O_6} \right) = 1.548 \times 10^{-1} \text{ mol}$$

$$V = \frac{nRT}{p} = \frac{(1.548 \times 10^{-1} \text{ mol})(8.314 \text{ L kPa mol}^{-1} \text{ K}^{-1})(310 \text{ K})}{101.325 \text{ kPa}} = 3.94 \text{ L}$$

2.25 This is a stoichiometry problem that involves gases. We are asked to determine the final pressure of the container (which is the pressure of chlorine gas). Begin by analyzing the chemistry. The starting materials are Na metal and Cl_2, a gas. The product is NaCl. The balanced chemical reaction is $2 \text{ Na} \, (s) + Cl_2 \, (g) \rightarrow 2 \text{ NaCl} \, (s)$.

The problem gives information about the amounts of both starting materials, so this is a limiting reactant situation. We must calculate the number of moles of each species, construct a table of amounts, and use the results to determine the final pressure. Calculations of initial amounts:

$$n_{Na} = 6.90 \text{ g} \left(\frac{1 \text{ mol}}{22.99 \text{ g}} \right) = 0.300 \text{ mol}$$

For chlorine gas, use the ideal gas equation to do pressure–mole conversions. The data must have the same units as those for R:

$$p_{Cl_2} = 1.25 \times 10^3 \text{ Torr} \left(\frac{1 \text{ bar}}{750.06 \text{ Torr}} \right) = 1.67 \text{ bar}$$

$$n_{Cl_2} = \frac{pV}{RT} = \frac{(1.67 \text{ bar})(3.00 \text{ L})}{(0.08314 \text{ L bar mol}^{-1} \text{ K}^{-1})(300 \text{ K})} = 0.200 \text{ mol}$$

Divide each initial amount by its coefficient to determine the limiting reactant:

$$Cl_2: 0.200 \text{ mol} \qquad Na: \frac{0.300 \text{ mol}}{2} = 0.150 \text{ mol (LR)}$$

Use the initial amounts and the balanced equation to construct a table of amounts:

Reaction:	2 Na (s) +	Cl$_2$ (g) →	2 NaCl (s)
Initial amount (mol)	0.300	0.200	0.000
Change (mol)	−0.300	−0.150	+0.300
Final amount (mol)	0.000	0.050	0.300

Use the final amount of Cl_2 and the new temperature (47 °C) to calculate the pressure:

$$p = \frac{nRT}{V} = \frac{(5.0 \times 10^{-2} \text{ mol})(0.08314 \text{ L bar mol}^{-1} \text{ K}^{-1})(320 \text{ K})}{3.00 \text{ L}} = 0.443 \text{ bar}$$

2.27 This is a stoichiometry problem that involves gases. We are asked to determine the mass of product produced. Begin by analyzing the chemistry. The starting materials are N_2 and H_2, both gases. The product of the reaction is NH_3.

The balanced chemical reaction is $N_2 + 3 H_2 \rightarrow 2 NH_3$.

The problem gives information about the amounts of both starting materials, so this is a limiting reactant situation. We must calculate the number of moles of each species, construct a table of amounts, and use the results to determine the partial pressure. Use the ideal gas equation to determine the initial amounts of each gas. The reactor initially contains the gases in 1:1 mole ratio, so

$$p_i = \frac{275 \text{ bar}}{2} = 137.5 \text{ bar for each gas}$$

$$n_i = \frac{p_i V}{RT} = \frac{(137.5 \text{ bar})(8.75 \times 10^3 \text{ L})}{(0.08314 \text{ L bar mol}^{-1} \text{ K}^{-1})(455 + 273.15 \text{ K})} = 1.99 \times 10^4 \text{ mol}$$

Since both gases have the same initial amount, the limiting reactant will be the one with the larger stoichiometric coefficient: H_2 is limiting. Here is the complete amounts table:

Reaction:	N_2 (g)	+ 3 H_2 (g)	→ 2 NH_3 (g)
Initial amount (10^4 mol)	1.99	1.99	0.00
Change (10^4 mol)	$-\frac{1}{3}(1.99)$	−1.99	$+\frac{2}{3}(1.99)$
Final amount (10^4 mol)	1.33	0.00	1.33

Obtain the mass of product formed from the final amount in the table:

$$m_{NH_3} = nM = 1.33 \times 10^4 \text{ mol} \left(\frac{17.04 \text{ g}}{1 \text{ mol}} \right) = 2.27 \times 10^5 \text{ g}$$

2.29 The yield in a reaction is the ratio of the actual amount produced to the theoretical amount. The calculation in Problem 2.27 gives the theoretical amount:

$$\text{Actual amount} = \text{theoretical amount} \left(\frac{\% \text{ yield}}{100\%} \right)$$

$$\text{Actual amount} = 2.27 \times 10^5 \text{ g} \left(\frac{13\%}{100\%} \right) = 2.9 \times 10^4 \text{ g}$$

2.31 Molecular speed is related to temperature by the equations, $v = \sqrt{\dfrac{2E_{\text{kinetic}}}{m}}$,

and $\overline{E}_{\text{kinetic}} = \dfrac{3RT}{2N_A}$, so $\overline{v} = \sqrt{\dfrac{3RT}{mN_A}}$

. Thus, molecular speed increases with the square root of the absolute temperature, so molecules will reach the detector sooner at higher temperature, because they have greater speed.

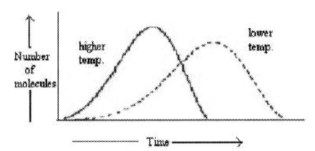

2.33 Molecular speed is related to temperature by two equations: $v = \sqrt{\dfrac{2E_{\text{kinetic}}}{m}}$,

$\overline{E}_{\text{kinetic}} = \dfrac{3RT}{2N_A}$, so $\overline{v} = \sqrt{\dfrac{3RT}{mN_A}}$. Each distribution of molecular speeds will have the

same general shape, with the position of the peak depending on $\sqrt{\dfrac{T}{m}}$:

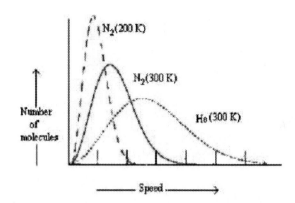

2.35 Average molecular kinetic energy is directly proportional to temperature and is independent of molecular mass: $\bar{E}_{\text{kinetic}} = \dfrac{3RT}{2N_A}$. N_2 and He at 300 K have identical kinetic energy distributions, whereas the distribution for N_2 at 200 K is shifted to lower energy: The shape of the curve is slightly different for energies than for speeds.

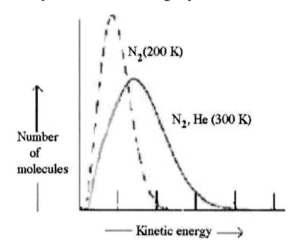

2.37 The most probable speed at any temperature is related to the most probable kinetic energy by the equation $E_{\text{kinetic}} = \dfrac{1}{2}mv^2$, or $v_{\text{mp}} = \sqrt{\dfrac{2E_{\text{kinetic, mp}}}{m}}$. According to your textbook, the most probable kinetic energy at 300 K is 4.13×10^{-21} J/molecule. Because energy is proportional to temperature, the most probable kinetic energy at any other temperature can be calculated using proportions: $\dfrac{E_{\text{kinetic, 2}}}{E_{\text{kinetic, 1}}} = \dfrac{T_2}{T_1}$. Average kinetic energy, which is not the same as most probable kinetic energy because the distribution of energies is asymmetric, is calculated using $\bar{E}_{\text{kinetic, molar}} = \dfrac{3}{2}RT$ (T in kelvins; 1 J = 1 kg m^2 s^{-2}):

(a) He: $m = \left(\dfrac{4.00 \text{ g}}{1 \text{ mol}}\right)\left(\dfrac{10^{-3} \text{ kg}}{1 \text{ g}}\right)\left(\dfrac{1 \text{ mol}}{6.022 \times 10^{23} \text{ atoms}}\right) = 6.64 \times 10^{-27}$ kg

$$627\,^{\circ}\text{C} + 273 = 900\text{ K}, \quad E_{\text{kinetic, mp}} = 4.13 \times 10^{-21}\text{ J}\left(\frac{900\text{ K}}{300\text{ K}}\right) = 1.24 \times 10^{-20}\text{ J}$$

$$v_{\text{mp}} = \sqrt{\frac{2(1.24 \times 10^{-20}\text{ J})}{6.64 \times 10^{-27}\text{ kg}}} = 1.93 \times 10^{3}\text{ m/s}$$

$$\overline{E}_{\text{kinetic, molar}} = \frac{3}{2}(8.314\text{ J mol}^{-1}\text{ K}^{-1})(900\text{ K}) = 1.12 \times 10^{4}\text{ J/mol}$$

(b) O_2: $m = \left(\dfrac{32.00\text{ g}}{1\text{ mol}}\right)\left(\dfrac{10^{-3}\text{ kg}}{1\text{ g}}\right)\left(\dfrac{1\text{ mol}}{6.022 \times 10^{23}\text{ atoms}}\right) = 5.31 \times 10^{-26}\text{ kg}$

$$27\,^{\circ}\text{C} + 273 = 300\text{ K}, \quad E_{\text{kinetic, mp}} = 4.13 \times 10^{-21}\text{ J}$$

$$v_{\text{mp}} = \sqrt{\frac{2(4.13 \times 10^{-21}\text{ J})}{5.31 \times 10^{-26}\text{ kg}}} = 3.94 \times 10^{2}\text{ m/s}$$

$$\overline{E}_{\text{kinetic, molar}} = \frac{3}{2}(8.314\text{ J mol}^{-1}\text{ K}^{-1})(300\text{K}) = 3.74 \times 10^{3}\text{ J/mol}$$

(c) SF_6: $M = 32.066\text{ g/mol} + 6(18.998\text{ g/mol}) = 146.1\text{ g/mol}$

$$m = \left(\frac{146.1\text{ g}}{1\text{ mol}}\right)\left(\frac{10^{-3}\text{ kg}}{1\text{ g}}\right)\left(\frac{1\text{ mol}}{6.022 \times 10^{23}\text{ atoms}}\right) = 2.43 \times 10^{-25}\text{ kg}$$

$$627\,^{\circ}\text{C} + 273 = 900\text{ K}, \quad E_{\text{kinetic, mp}} = 4.13 \times 10^{-21}\text{ J}\left(\frac{900\text{K}}{300\text{K}}\right) = 1.24 \times 10^{-20}\text{J}$$

$$v_{\text{mp}} = \sqrt{\frac{2(1.24 \times 10^{-20}\text{ J})}{2.43 \times 10^{-25}\text{ kg}}} = 3.20 \times 10^{2}\text{ m/s}$$

$$\overline{E}_{\text{kinetic, molar}} = \frac{3}{2}(8.314\text{ J mol}^{-1}\text{ K}^{-1})(900\text{ K}) = 1.12 \times 10^{4}\text{ J/mol}$$

2.39 The ideal gas is defined by the conditions that molecular volumes and intermolecular forces both are negligible.

(a) At very high pressure, molecules are very close together, so their volumes are significant compared with the volume of their container; because the first condition is not met, the gas is not ideal. Furthermore, intermolecular forces are significant when the molecules are close together, which influences molecular motion.

(b) At very low temperature, molecules move very slowly, so the forces between molecules, even though small, are sufficient to influence molecular motion; because the second condition is not met, the gas is not ideal.

2.41 (a) If the piston is stationary and there is no friction, the forces and pressures on each side must be equal, so the internal pressure is also 1 bar. This pressure is generated by gas molecules colliding with the face of the piston.

(b) When temperature doubles, the increase in average molecular speed results in a doubling of the pressure. The piston will move outward, causing the concentration of gas molecules to decrease. This will lead to a lower frequency of collisions with the wall, reducing the pressure. The piston will stop when the internal pressure is once again 1 bar.

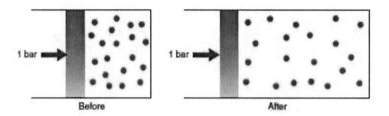

2.43 As a gas cools, its molecules move more slowly, so they impart smaller impulses on the walls of their container. This reduces the internal pressure, so the balloon collapses until the increase in gas density inside the balloon brings the internal pressure back up to the external pressure of 1.000 bar.

2.45 The ideal gas equation, $pV = nRT$, can be used to calculate moles using p–V–T data. Molar mass is related to moles through $n = \dfrac{m}{M}$:

$$n = \frac{pV}{RT} = \frac{m}{M} \qquad M = \frac{mRT}{pV}$$

Begin by converting the initial data into the units of R:

$$T = 25.0 + 273.15 = 298. \text{ K}$$

$$p = 262 \text{ Torr} \left(\frac{1 \text{ bar}}{750.06 \text{ Torr}} \right) = 0.349 \text{ bar}$$

Use the modified ideal gas equation to obtain the molar mass:

$$M = \frac{(2.55 \text{ g})(0.08314 \text{ L bar mol}^{-1} \text{ K}^{-1})(298 \text{ K})}{(0.349 \text{ bar})(1.50 \text{ L})} = 121 \text{ g/mol}$$

The compound contains only C (12.0 g/mol), F (19.0 g/mol), and Cl (35.5 g/mol). The formula can be determined by trial and error. The combination of 1 C, 2 F, and 2 Cl has

$$M = 1(12.0 \text{ g/mol}) + 2(19.0 \text{ g/mol}) + 2(35.5 \text{ g/mol}) = 121 \text{ g/mol}$$

This matches the experimental value. The formula is CF_2Cl_2.

2.47 Density can be calculated from the ideal gas equation and mole–mass conversions:

$$n \ = \ \frac{m}{M} = \frac{pV}{RT} \qquad \frac{m}{V} = \ \frac{pM}{RT}$$

$$p = 755 \text{ Torr} \left(\frac{1 \text{ bar}}{750.06 \text{ Torr}} \right) = 1.007 \text{ bar}$$

$$\frac{m}{V} = \ \frac{pM}{RT} = \frac{(1.007 \text{ bar})(146.05 \text{ g/mol})}{(0.08314 \text{ L bar mol}^{-1} \text{ K}^{-1})(27 + 273 \text{ K})} = 5.90 \text{ g/L}$$

2.49 Molecular speed can be calculated from temperature and molar mass:

$$\bar{E}_{\text{kinetic}} = \frac{3RT}{2N_A} \qquad \qquad \bar{E}_{\text{kinetic}} = \frac{1}{2} m \bar{v}^2_{\text{rms}}$$

$$\bar{v}^2 = \frac{3RT}{mN_A} = \frac{3RT}{M} \qquad \bar{v} \ = \left(\frac{3RT}{M} \right)^{1/2}$$

$T = 27\ °C + 273 = 300.\ \text{K}$; $M = 32.06 \text{ g/mol} + 6(18.998 \text{ g/mol}) = 146.0 \text{ g/mol}$. The mass must be converted to kg/mol for mass units to cancel $(1 \text{ J} = 1 \text{ kg m}^2 \text{ s}^{-2})$: $M = (146.0 \text{ g/mol})(10^{-3} \text{ kg/g}) = 1.460 \times 10^{-1} \text{ kg/mol}$

$$\bar{v}_{\text{rms}} \ = \left(\frac{3(8.314 \text{ J mol}^{-1} \text{ K}^{-1})(1 \text{ kg m}^2\text{s}^{-2} \text{ /J})(300.\ \text{K})}{1.460 \times 10^{-1} \text{ kg/mol}} \right)^{1/2} = 226 \text{ m/s}$$

2.51 Rates of diffusion and effusion depend on the molar masses of the gases because average speeds determine these rates, and these are given by Equation 2-5:

$\bar{v}_{\text{rms}} = \left(\dfrac{3RT}{M} \right)^{1/2}$. For two gases at the same temperature, the one with the smaller molar mass has the faster molecular speed and will diffuse faster. Thus, CH_4 ($M = 16$ g/mol) diffuses faster through the atmosphere than C_2H_6 ($M = 30$ g/mol).

2.53 $\dfrac{\text{effusion rate}_A}{\text{effusion rate}_B} = \sqrt{\dfrac{M_B}{M_A}}$ (by Graham's Law)

$$M_{^{14}CO_2} = 14 + 2(16) = 46 \text{ g/mol}$$

$$M_{^{12}CO_2} = 12 + 2(16) = 44 \text{ g/mol}$$

$$\text{thus, } \frac{\text{effusion rate}_{^{14}CO_2}}{\text{effusion rate}_{^{12}CO_2}} = \sqrt{\frac{44 \text{ g/mol}}{46 \text{ g/mol}}} = 0.98$$

$$M_{^{14}CO} = 14 + 16 = 30 \text{ g/mol}$$

$$M_{^{12}CO} = 12 + 16 = 28 \text{ g/mol}$$

$$\text{thus, } \frac{\text{effusion rate}_{^{14}CO}}{\text{effusion rate}_{^{12}CO}} = \sqrt{\frac{28 \text{ g/mol}}{30 \text{ g/mol}}} = 0.97$$

It should be easier to separate ^{14}C from ^{12}C in the form of the monoxide rather than the dioxide, but the values are very close.

2.55 Ideal pressure for F_2 *(g)*:

$$p = \frac{nRT}{V} = \frac{(1.00 \text{ mol})(0.08314 \text{ L bar K}^{-1}\text{mol}^{-1})(298 \text{ K})}{30.0 \text{ L}} = 0.826 \text{ bar}$$

Real pressure for F_2 *(g)*:

$$p = \frac{nRT}{V - nb} - a\left(\frac{n}{V}\right)^2$$

$$= \frac{(1.00 \text{ mol})(8.314 \text{ L kPa K}^{-1}\text{mol}^{-1})(298 \text{ K})}{30.0 \text{ L} - (1.00 \text{ mol})(0.0290 \text{ L/mol})} - (117.1 \text{ kPa L}^2 \text{ mol}^{-2})\left(\frac{1.00 \text{ mol}}{30.0 \text{ L}}\right)^2$$

$$= 82.7 \text{ kPa} - 0.13 \text{ kPa}$$

$$= 82.6 \text{ kPa}$$

$$= 0.826 \text{ bar}$$

The two values are similar because the pressure is low enough that the molecules are far apart and intermolecular forces are small. Also, the temperature is high enough that the kinetic energy of the molecules is high enough to overcome what intermolecular forces do exist.

2.57 Real pressure of H_2 *(g)*:

$$p = \frac{nRT}{V - nb} - a\left(\frac{n}{V}\right)^2$$

$$= \frac{(100 \text{ mol})(8.314 \text{ L kPa K}^{-1} \text{mol}^{-1})(298 \text{ K})}{30.0 \text{ L} - (100 \text{ mol})(0.0266 \text{ L/mol})} - (24.7 \text{ kPa L}^2 \text{ mol}^{-2})\left(\frac{100 \text{ mol}}{30.0 \text{ L}}\right)^2$$

$$= 9062 \text{ kPa} - 274 \text{ kPa}$$

$$= 8788 \text{ kPa}$$

$$= 87.8 \text{ bar}$$

This value is higher than that calculated for fluorine in the previous problem because the intermolecular forces in hydrogen are smaller, because hydrogen is a much smaller molecule than fluorine.

2.59 Real pressure of O_2 *(g)*:

$$p = \frac{nRT}{V - nb} - a\left(\frac{n}{V}\right)^2$$

$$= \frac{(75 \text{ mol})(8.314 \text{ L kPa K}^{-1} \text{mol}^{-1})(298 \text{ K})}{15.0 \text{ L} - (75 \text{ mol})(0.0318 \text{ L/mol})} - (137.8 \text{ kPa L}^2 \text{ mol}^{-2})\left(\frac{75 \text{ mol}}{15.0 \text{ L}}\right)^2$$

$$= 14\,730 \text{ kPa} - 3445 \text{ kPa}$$

$$= 11\,285 \text{ kPa}$$

$$= 112.8 \text{ bar}$$

2.61 Table 2-4 gives the mole fractions of the components in dry air (0% humidity), from which $X_{O_2} = 0.2095$. Table 2-5 gives the vapour pressure of water at 20 °C, 17.535 Torr; this is the partial pressure at 100% humidity for that temperature. To find the mole fraction of O_2 at 100% humidity, first determine the partial pressure of O_2. We can assume air is composed only of oxygen, nitrogen, and water:

$$p_{H_2O} = 17.535 \text{ Torr}\left(\frac{1 \text{ bar}}{750.06 \text{ Torr}}\right) = 2.338 \times 10^{-2} \text{ bar}$$

$$p_{N_2 + O_2} = (1.000 \text{ bar}) - (2.338 \times 10^{-2} \text{ bar}) = 0.9766 \text{ bar}$$

$$p_{O_2} = (0.2095)(0.9766 \text{ bar}) = 0.2046 \text{ bar}$$

$$X_{O_2} = \frac{p_{O_2}}{P_{\text{total}}} = \frac{0.2046 \text{ bar}}{1.000 \text{ bar}} = 0.2046 \qquad \text{Change} = 0.2095 - 0.2046 = 0.0049$$

In other words, the mole fraction of O_2 in the atmosphere increases by 0.0049 when the relative humidity decreases from 100% to 0%.

2.63 The dew point is defined as the temperature at which the partial pressure of water vapour in the atmosphere matches the vapour pressure of water. Refer to Table 2-5 for vapour pressures: at 25 °C, $p_{vap} = 23.756$ Torr

$$p_{vap} = 23.756 \text{ Torr} \left(\frac{78\%}{100\%}\right) = 18.53 \text{ Torr}$$

This vapour pressure corresponds to a dew point between 20 and 25 °C. Assuming a linear variation in vapour pressure between these two temperatures (which is not exactly true, but is close enough over this small range of temperature):

$$\text{Dew point} = 20.0 \text{ °C} + \left(\frac{18.53-17.535}{23.756-17.535}\right)(5 \text{ °C}) = 20.0+0.8 = 20.8 \text{ °C}$$

2.65 We must calculate the volume of dry air that contains 10 mg of neon. Use moles, mole fractions from Table 2-4, and the ideal gas equation:

$$n_{Ne} = \frac{m}{M} = 10 \text{ mg}\left(\frac{10^{-3} \text{ g}}{1 \text{ mg}}\right)\left(\frac{1 \text{ mol}}{20.180 \text{ g}}\right) = 5.0 \times 10^{-4} \text{ mol}$$

From Table 2-4, $X_{Ne} = 1.82 \times 10^{-5} = \dfrac{n_{Ne}}{n_{air}}$

Thus, the number of moles of air containing 10 mg of neon is

$$n_{air} = \frac{n_{Ne}}{X_{Ne}} = \frac{5.0 \times 10^{-4} \text{mol}}{1.82 \times 10^{-5}} = 27.5 \text{ mol}$$

The volume occupied by this amount of air can be calculated from the ideal gas equation:

$$V = \frac{nRT}{p} = \frac{(27.5 \text{ mol})(0.08314 \text{ L bar mol}^{-1} \text{ K}^{-1})(298 \text{ K})}{1.00 \text{ bar}} = 6.8 \times 10^2 \text{ L}$$

(There are two significant figures because the moles of air should only have two significant figures.)

2.67 Use the ideal gas equation to determine moles, and then do mole–mass–number conversions:

$$n = \frac{pV}{RT} = \left(\frac{(10.0 \text{ bar})(725 \text{ mL})}{(0.08314 \text{ L bar mol}^{-1} \text{ K}^{-1})(925 + 273 \text{ K})}\right)\left(\frac{10^{-3} \text{L}}{1 \text{ mL}}\right) = 7.28 \times 10^{-2} \text{ mol}$$

$$m = nM = (7.28 \times 10^{-2} \text{ mol})(83.80 \text{ g/mol}) = 6.10 \text{ g}$$

$$\# \text{ atoms} = nN_A = (7.28 \times 10^{-2} \text{ mol})(6.022 \times 10^{23} \text{ atoms/mol}) = 4.38 \times 10^{22} \text{ atoms}$$

2.69 Molecular pictures show the relative numbers of molecules of various substances, which also represent the relative numbers of moles of those substances. Chamber A contains 6 atoms, B contains 12 atoms, and C contains 9 atoms, for a total of 27 atoms, and the chambers have equal volumes and are at the same temperature.

(a) Chamber B contains the most atoms, so it exhibits the highest pressure.

(b) The pressures in the chambers are proportional to the number of atoms:

$$p_A = p_B \left(\frac{\#_A}{\#_B} \right) = 1.0 \text{ bar} \left(\frac{6}{12} \right) = 0.50 \text{ bar}$$

(c) The pressure will increase in proportion to the number of atoms:

$$p_{new} = p_{old} \left(\frac{\#_{new}}{\#_{old}} \right) = 1.0 \text{ bar} \left(\frac{27}{6} \right) = 4.5 \text{ bar}$$

(d) When the valves are opened, the number of atoms in each chamber equalizes:

$$\frac{27}{3} = 9$$

$$p_{new} = p_{old} \left(\frac{\#_{new}}{\#_{old}} \right) = 0.50 \text{ bar} \left(\frac{9}{12} \right) = 0.38 \text{ bar}$$

2.71 Much of the information in this problem is not needed, because molecular speed can be calculated from temperature and molar mass:

$$\bar{E}_{kinetic} = \frac{3RT}{2N_A} \qquad \bar{E}_{kinetic} = \frac{1}{2} m \bar{v}^2$$

$$\bar{v}^2 = \frac{3RT}{mN_A} = \frac{3RT}{M} \qquad \bar{v} = \left(\frac{3RT}{M} \right)^{1/2}$$

The mass must be converted to kg/mol for mass units to cancel (1 J = 1 kg m^2 s^{-2}):

$$M = 3 \left(\frac{16.00 \text{ g}}{1 \text{ mol}} \right) \left(\frac{1 \text{ kg}}{1000 \text{ g}} \right) = 48.00 \times 10^{-3} \text{ kg/mol}$$

$$\bar{v} = \left(\frac{3(8.314 \text{ J mol}^{-1} \text{ K}^{-1})(-25 + 273 \text{ K})}{48.00 \times 10^{-3} \text{ kg/mol}} \right)^{1/2} = 359 \text{ m/s}$$

2.73 Use the ideal gas equation to find molar density and Avogadro's number to convert to molecular density:

$$\frac{n}{V} = \frac{p}{RT} = \frac{10^{-10} \text{ bar}}{(0.08314 \text{ L bar mol}^{-1} \text{ K}^{-1})(310 \text{ K})} = 4 \times 10^{-12} \text{ mol/L}$$

$$\frac{\#}{V} = N_A \left(\frac{n}{V}\right) = \left(\frac{6.022 \times 10^{23} \text{ molecules}}{1 \text{ mol}}\right) \left(\frac{4 \times 10^{-12} \text{ mol}}{1 \text{ L}}\right) = 2 \times 10^{12} \text{ molecules/L}$$

2.75 To determine the dew point, first calculate the partial pressure of water vapour at the existing temperature and then estimate the temperature at which this partial pressure equals the vapour pressure of water, assuming a linear variation of vapour pressure with temperature. Use data in Table 2-5:

$$p_{H_2O} = \frac{(p_{vap, H_2O})(\% \text{ relative humidity})}{100\%}$$

(a) $p_{H_2O} = 42.175 \text{ Torr} \left(\frac{80\%}{100\%}\right) = 33.74 \text{ Torr}$

Dew point $= 30.0 \text{ °C} + \left(\frac{33.74 - 31.824}{42.175 - 31.824}\right)(5 \text{ °C}) = 30.0 + 0.9 = 30.9 \text{ °C}$

(b) $p_{H_2O} = 12.788 \text{ Torr} \left(\frac{50\%}{100\%}\right) = 6.394 \text{ Torr}$

Dew point $= 0 \text{ °C} + \left(\frac{6.394 - 4.579}{6.543 - 4.579}\right)(5 \text{ °C}) = 0.0 + 4.6 = 4.6 \text{ °C}$

(c) $p_{H_2O} = 23.756 \text{ Torr} \left(\frac{30\%}{100\%}\right) = 7.1268 \text{ Torr}$

Dew point $= 5 \text{ °C} + \left(\frac{7.1268 - 6.543}{9.209 - 6.543}\right)(5 \text{ °C}) = 5.0 + 1.1 = 6.1 \text{ °C}$

2.77 The molar mass of gaseous oxygen can be determined using the ideal gas law:

$$M = \frac{mRT}{pV}$$

Pump out a bulb of known volume, weigh the empty bulb, fill the bulb with oxygen at a measured pressure, and weigh again. For monatomic oxygen, this experiment would give $M = 16.00 \text{ g/mol}$, whereas it would give $M = 32.00 \text{ g/mol}$ for the diatomic gas.

2.79 Use Dalton's law of partial pressures to determine partial pressures, and then apply the ideal gas equation to calculate masses.

$$X_{C_3H_6} = \frac{1.00}{5.00} = 0.200 \qquad p_{C_3H_6} = 0.200 \text{ bar}$$

$$X_{O_2} = 1.000 - 0.200 = 0.800 \quad p_{O_2} = 0.800 \text{ bar}$$

$$m_{C_3H_6} = \frac{pVM}{RT} = \frac{(0.200 \text{ bar})(2.00 \text{ L})(42.078 \text{ g/mol})}{(0.08314 \text{ L bar mol}^{-1} \text{ K}^{-1})(23.5 + 273.15 \text{ K})} = 0.682 \text{ g}$$

$$m_{O_2} = \frac{pVM}{RT} = \frac{(0.800 \text{ bar})(2.00 \text{ L})(32.00 \text{ g/mol})}{(0.08314 \text{ L bar mol}^{-1} \text{ K}^{-1})(23.5 + 273.15 \text{ K})} = 2.08 \text{ g}$$

2.81 All conditions except the volumes are different for the two samples.

(a) The bulb with the larger number of moles will contain more molecules:

$$n_{H_2} = \frac{pV}{RT} = \frac{(2 \text{ bar})V}{(273 \text{ K})R} = 7 \times 10^{-3} \left(\frac{V}{R}\right) \text{ mol}$$

$$n_{O_2} = \frac{pV}{RT} = \frac{(1 \text{ bar})V}{(298 \text{ K})R} = 3 \times 10^{-3} \left(\frac{V}{R}\right) \text{ mol}$$

The bulb of hydrogen contains more molecules.

(b) $m_{H_2} = (7 \times 10^{-3}) \left(\frac{V}{R}\right) \text{ mol} \left(\frac{2.01 \text{ g}}{1 \text{ mol}}\right) = 1 \times 10^{-2} \left(\frac{V}{R}\right) \text{ g}$

$m_{O_2} = (3 \times 10^{-3}) \left(\frac{V}{R}\right) \text{ mol} \left(\frac{32.00 \text{ g}}{1 \text{ mol}}\right) = 1 \times 10^{-1} \left(\frac{V}{R}\right) \text{ g}$

The bulb of oxygen contains more mass.

(c) The average kinetic energy of molecules depends only on the temperature of the gas, so the oxygen molecules, being at higher temperature, have greater average kinetic energy.

(d) The average molecular speed depends on $\sqrt{\dfrac{T}{m}}$ because $\bar{v} = \sqrt{\dfrac{3RT}{mN_A}}$. The temperature ratio of H_2:O_2 is $\left(\dfrac{273}{298}\right) = 0.916$, whereas the mass ratio is

$\left(\dfrac{2}{32}\right) = 0.0625$ Thus, $\sqrt{\dfrac{T}{m}} = \sqrt{\dfrac{0.916}{0.0625}} = 3.83$, so the hydrogen molecules have greater average speed by this factor.

2.83 (a) The most probable kinetic energy at 300 K is 4×10^{-21} J.

(b) $E_{\text{kinetic (most probable)}} = \dfrac{m(v_{\text{most probable}})^2}{2}$ (Remember $1\,J = 1\,kg\,m^2\,s^{-2}$)

$$v_{\text{most probable}} = \sqrt{\frac{2(4 \times 10^{-21}\,J)(6.022 \times 10^{23}\,\text{molecules/mol})}{32.00 \times 10^{-3}\,\text{kg/mol}}} = 4 \times 10^2\,\text{m/s}$$

2.85 Your explanation should include the fact that sulphur in coal combines with oxygen from the air to form SO_2 and then SO_3, and that when SO_3 reacts with water, sulphuric acid is an end product. Chemical equations:

S (s) + O_2 (g) → SO_2 (g)

2 SO_2 (g) + O_2 (g) → 2 SO_3 (g)

SO_3 (g) + H_2O (l) → H_2SO_4 (aq)

2.87 Each part of this problem represents a change of one or more conditions, for which a rearranged version of the ideal gas law can be used.

(a) T and n are fixed, so $pV = nRT$ = constant and $V_f = \dfrac{p_i V_i}{p_f}$

$$V_f = \frac{(525\,\text{Torr})(3.00\,L)}{755\,\text{Torr}} = 2.09\,L$$

(b) T and n are fixed, so $pV = nRT$ = constant and $p_f = \dfrac{p_i V_i}{V_f}$

$$p_f = \frac{(525\,\text{Torr})(3.00\,L)}{2.00\,L} = 788\,\text{Torr}$$

(c) V and n are fixed, so $\dfrac{p}{T}$ = constant and $p_f = \dfrac{p_i T_f}{T_i}$

$$p_f = \frac{(525\,\text{Torr})(50.0 + 273.15\,K)}{273.15\,K} = 621\,\text{Torr}$$

2.89 The figure illustrates atomic density and atomic motion of a sample of helium gas. The figure contains eight atoms:

(a) Your drawing should show the same atomic density as the original figure but with longer "tails" on the atoms, indicating that they are moving at higher speeds:

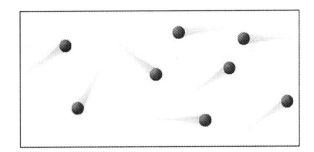

(b) Your drawing should show the same lengths of "tails" as in the original figure, but there should be two fewer atoms:

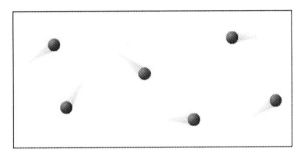

(c) Your drawing should be the same as the original figure except that half the atoms should be replaced with diatomic molecules:

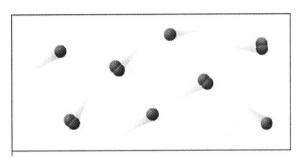

2.91 Use the ideal gas equation, $pV = nRT$, to determine which statements are correct:

(a) At constant T and V, $\dfrac{p}{n} = \dfrac{RT}{V}$ = constant. Thus, p is directly proportional to number of moles of gas. The statement is false.

(b) At constant V and n, $\dfrac{p}{T} = \dfrac{nR}{V}$ = constant ; $\dfrac{p}{T}$ is constant. The statement is true.

(c) At fixed V and n, $\dfrac{p}{T} = \dfrac{nR}{V}$ = constant ; $\dfrac{p}{T}$ is constant; pT is not. The statement is false.

2.93 This is an application of the ideal gas equation to changing gas conditions. The only constant quantity is n, so $\dfrac{pV}{T} = nR = $ constant. Therefore, $\dfrac{p_iV_i}{T_i} = \dfrac{p_fV_f}{T_f}$.

$p_i = 0.969$ bar $V_i = 0.963$ L $p_f = 1.00$ bar

$T_i = 22 + 273.15 = 295$ K $T_f = 15 + 273.15 = 288$ K

$$V_f = \left(\frac{p_iV_i}{T_i}\right)\left(\frac{T_f}{p_f}\right) = \frac{(0.969 \text{ bar})(0.963 \text{ L})(288 \text{ K})}{(295 \text{ K})(1.00 \text{ bar})} = 0.911 \text{ L}$$

2.95 Concentrations in parts per billion can be converted to mole fractions, and the partial pressure can then be calculated using $p_i = X_i p_{total}$. To determine molecular concentration, first use the ideal gas equation to calculate n in moles, and then multiply by $\dfrac{N_A}{V}$:

$X_{SF_6} = (1.0 \text{ ppb})(10^{-9}/\text{ppb}) = 1.0 \times 10^{-9}$ $p_{SF_6} = (1.0 \times 10^{-9})(1 \text{ bar}) = 1.0 \times 10^{-9}$ bar

$$V = 1.0 \text{ cm}^3 \left(\frac{10^{-3} \text{ L}}{1 \text{ cm}^3}\right) = 1.0 \times 10^{-3} \text{ L}$$

$$n = \frac{pV}{RT} = \frac{(1.0 \times 10^{-9} \text{ bar})(1.0 \times 10^{-3} \text{ L})}{(0.08314 \text{ L bar mol}^{-1} \text{ K}^{-1})(21 + 273.15 \text{ K})} = 4.1 \times 10^{-14} \text{ mol}$$

$$\frac{\#}{V} = \frac{(4.1 \times 10^{-14} \text{ mol})(6.022 \times 10^{23} \text{ molecules/mol})}{1 \text{ L}} = 2.5 \times 10^{10} \text{ molecules/L}$$

2.97 This is a P–V–T problem. First calculate moles of CO_2, and then use the ideal gas equation to calculate the final pressure:

$$n = \frac{m}{M} = \frac{15.00 \text{ g}}{44.01 \text{ g/mol}} = 0.3408 \text{ mol}$$

$$p_{CO_2} = \frac{(0.3408 \text{ mol})(0.08314 \text{ L bar mol}^{-1} \text{ K}^{-1})(273.15 \text{ K})}{0.750 \text{ L}} = 10.3 \text{ bar}$$

2.99 Density can be calculated from the ideal gas equation and mole–mass conversions:

$$n = \frac{m}{M} = \frac{pV}{RT}, \text{ from which } \frac{m}{V} = \frac{pM}{RT}$$

First, calculate the average molar masses of dry air and of moist air:

$M_{air} = \Sigma_i(X_i M_i)$

$M_{dry\ air} = (0.7808)(28.01 \text{ g/mol}) + (0.2095)(32.00 \text{ g/mol}) + (9.34 \times 10^{-3})(39.948$

g/mol) = 28.95 g/mol

For moist air at 298 K:

$$p_{H_2O} = 23.756 \text{ Torr} \left(\frac{1 \text{ bar}}{750.06 \text{ Torr}} \right) = 3.17 \times 10^{-2} \text{ bar} \qquad X_{H_2O} = 0.0317$$

The total mole fraction of the dry air components is reduced from 1 to $(1 - .0317) =$ 0.9683. Each dry air mole fraction must be multiplied by this factor to obtain the mole fractions in moist air. Thus,

$M_{moist\ air} = (0.0317)(18.01 \text{ g/mol}) + (0.9683)(28.95 \text{ g/mol}) = 28.60 \text{ g/mol}$

$$\left(\frac{m}{V} \right)_{dry\ air} = \frac{(1 \text{ bar})(28.95 \text{ g/mol})}{(0.08314 \text{ L bar mol}^{-1} \text{ K}^{-1})(298 \text{ K})} = 1.17 \text{ g/L}$$

$$\left(\frac{m}{V} \right)_{moist\ air} = \frac{(1 \text{ bar})(28.60 \text{ g/mol})}{(0.08314 \text{ L bar mol}^{-1} \text{ K}^{-1})(298 \text{ K})} = 1.15 \text{ g/L}$$

2.101 First, determine moles of liquid oxygen in 150 L. Then, calculate how many moles of dry air contain this amount of oxygen. Finally, use the ideal gas equation to calculate the volume:

$$n = \frac{\rho V}{M} = 150 \text{ L} \left(\frac{10^3 \text{ mL}}{1 \text{ L}} \right) \left(\frac{1.14 \text{ g}}{1 \text{ mL}} \right) \left(\frac{1 \text{ mol}}{32.00 \text{ g}} \right) = 5.34 \times 10^3 \text{ mol}$$

$$n_{air} = \frac{n_{O_2}}{X_{O_2}} = \frac{5.34 \times 10^3 \text{ mol}}{0.2095} = 2.55 \times 10^4 \text{ mol}$$

$$p = 750 \text{ Torr} \left(\frac{1 \text{ bar}}{750.06 \text{ Torr}} \right) = 1.000 \text{ bar}$$

$$V = \frac{nRT}{p} = \frac{(2.55 \times 10^4 \text{ mol})(0.08314 \text{ L bar mol}^{-1} \text{ K}^{-1})(298 \text{ K})}{1.000 \text{ bar}} = 6.32 \times 10^5 \text{ L}$$

2.103 In addition to a yield problem, information is given about both starting materials, so this is a limiting reactant problem that involves gases. We must calculate the number of moles of each species, construct a table of amounts, and use the results to determine the theoretical yield. Water can be ignored, because CH_4 is the product of interest. All of the reagents of interest are gases, so pressures can be used as the measures of amounts:

Reaction:	3 H₂ (g)	+	CO (g) →	CH₄ (g)
Initial pressure (bar)	20.0		10.0	0.0
Change (bar)	−20.0		−6.67	+6.67
Final pressure (bar)	0.0		3.33	6.67

Use the ideal gas equation and mole–mass conversion to obtain the theoretical yield in grams:

$$n = \frac{pV}{RT} = \frac{(6.67 \text{ bar})(100 \text{ L})}{(0.08314 \text{ L bar mol}^{-1} \text{ K}^{-1})(575 \text{ K})} = 14.0 \text{ mol}$$

Theoretical amount $= nM = (14.0 \text{ mol})(16.04 \text{ g/mol}) = 225 \text{ g}$

Obtain the percent yield by dividing the actual yield by the theoretical yield:

$$\% \text{ Yield} = 100\% \left(\frac{145 \text{ g}}{225 \text{ g}}\right) = 64.4\%$$

2.105 The chemical reaction is $N_2 + 3 H_2 \rightarrow 2 NH_3$. The molecular picture shows 4 moles of N_2 and 12 moles of H_2, which represents stoichiometric proportions. If the reaction goes to completion, all molecules will react to form 8 moles of NH_3:

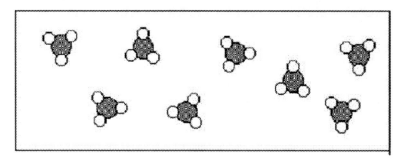

2.107 The question looks like an empirical formula problem, but the only data provided are p–V–T data for the gaseous substance, which suggests that an ideal gas law calculation can be done. Use the gas data to calculate n, and then use $n = \dfrac{m}{M}$ to determine M and subsequently determine the formula of the compound:

$$p = 552 \text{ Torr} \left(\frac{1 \text{ bar}}{750.06 \text{ Torr}}\right) = 0.736 \text{ bar}$$

$$n = \frac{pV}{RT} = \frac{(0.736 \text{ bar})(0.100 \text{ L})}{(0.08314 \text{ L bar mol}^{-1} \text{ K}^{-1})(30 + 273 \text{ K})} = 2.92 \times 10^{-3} \text{ mol}$$

$$M = \frac{m}{n} = \frac{0.500 \text{ g}}{2.92 \times 10^{-3} \text{ mol}} = 171 \text{ g/mol}$$

The compound contains one nickel atom ($M = 58.69$ g/mol) and x CO molecules ($M = 28.01$ g/mol), so 171 g/mol $= 58.69$ g/mol $+ (x)(28.01$ g/mol)

$$x = \frac{171 \text{ g/mol} - 58.69 \text{ g/mol}}{28.01 \text{ g/mol}} = 4 \qquad \text{The formula is } Ni(CO)_4$$

2.109 At the top of Mount Everest, the atmospheric pressure is 250 Torr, much lower than the pressure at sea level, 760 Torr. A lower pressure corresponds to a smaller molecular density. Thus, when people say "the air is thin," they are referring to the fact that air is less dense at higher elevations.

2.111 Because Table 2-4 refers to dry air, the first step is to take into account the presence of H_2O at 50% relative humidity. The partial pressure of water must be subtracted from the total pressure to obtain the pressure due to dry air. Then $p_i = X_{dry\ air}\ p_{dry\ air}$.

From Table 2-5, p_{vap,H_2O} at 25 °C = 23.756 Torr. The pressure at sea level is 1 atm (i.e., 1.01325 bar).

$$p_{H_2O} = 23.756\ \text{Torr}\left(\frac{1\ \text{bar}}{750.06\ \text{Torr}}\right)\left(\frac{50\%}{100\%}\right) = 1.6 \times 10^{-2}\ \text{bar}$$

$$p_{dry\ air} = 765\ \text{Torr}\left(\frac{1\ \text{bar}}{750.06\ \text{Torr}}\right) - (1.6 \times 10^{-2}\ \text{bar}) = 1.004\ \text{bar}$$

$$p_{N_2} = (0.7808)(1.004\ \text{bar}) = 0.784\ \text{bar}$$

$$p_{O_2} = (0.2095)(1.004\ \text{bar}) = 0.210\ \text{bar}$$

$$p_{Ar} = (9.34 \times 10^{-3})(1.004\ \text{bar}) = 9.38 \times 10^{-3}\ \text{bar}$$

$$p_{CO_2} = (3.85 \times 10^{-4})(1.004\ \text{bar}) = 3.86 \times 10^{-4}\ \text{bar}$$

$$p_{Ne} = (1.82 \times 10^{-5})(1.004\ \text{bar}) = 1.83 \times 10^{-5}\ \text{bar}$$

$$p_{He} = (5.24 \times 10^{-6})(1.004\ \text{bar}) = 5.26 \times 10^{-6}\ \text{bar}$$

$$p_{CH_4} = (1.4 \times 10^{-6})(1.004\ \text{bar}) = 1.4 \times 10^{-6}\ \text{bar}$$

$$p_{Kr} = (1.14 \times 10^{-6})(1.004\ \text{bar}) = 1.14 \times 10^{-6}\ \text{bar}$$

2.113 This is an empirical formula problem, with p–V–T data added to allow calculation of the molar mass. First, determine the empirical formula from mass percentages. Consider 100 g of the compound, which contains the following amounts:

$$\text{C: } 71.22\ \text{g}\left(\frac{1\ \text{mol}}{12.011\ \text{g}}\right) = 5.930\ \text{mol C} \qquad \text{H: } 14.94\ \text{g}\left(\frac{1\ \text{mol}}{1.0079\ \text{g}}\right) = 14.82\ \text{mol H}$$

$$\text{N: } 13.84\ \text{g}\left(\frac{1\ \text{mol}}{14.007\ \text{g}}\right) = 0.9881\ \text{mol N}$$

Divide each by the smallest among them, 0.9881 mol N:

$$C: \left(\frac{5.930 \text{ mol C}}{0.9881 \text{ mol N}} \right) = 6.001 \text{ mol C/mol N, round to 6 C/N}$$

$$H: \left(\frac{14.82 \text{ mol H}}{0.9881 \text{ mol N}} \right) = 15.00 \text{ mol H/mol N}$$

Empirical formula: $C_6H_{15}N$, empirical $M = 101$ g/mol

Next, use the p–V–T data to determine the actual molar mass:

$$m = 250 \text{ mg} \left(\frac{10^{-3} \text{ g}}{1 \text{ mg}} \right) = 0.250 \text{ g} \qquad p = 435 \text{ Torr} \left(\frac{1 \text{ bar}}{750.06 \text{ Torr}} \right) = 0.580 \text{ bar}$$

$$V = 150 \text{ mL} \left(\frac{10^{-3} \text{ L}}{1 \text{ mL}} \right) = 0.150 \text{ L} \qquad T = 150 + 273 \text{ K} = 423 \text{ K}$$

$$M = \frac{mRT}{pV} = \frac{(0.250 \text{ g})(0.08314 \text{ L bar mol}^{-1} \text{ K}^{-1})(423 \text{ K})}{(0.580 \text{ bar})(0.150 \text{ L})} = 101 \text{ g/mol}$$

The molar mass is the same as the empirical molar mass, so the molecular formula is the same as the empirical formula.

2.115 The balloon will rise when its net density is less than that of the air that surrounds it. Thus,

$$m_{\text{air}} = m_{\text{He}} + 350 \text{ kg}$$
$$m_{\text{air}} - m_{\text{He}} = 350 \text{ kg}$$

However, the volume of air displaced is the same as the volume of He inside the balloon, so we have:

$$\rho_{\text{air}}V - \rho_{\text{He}}V = 350 \text{ kg}$$
$$V = \frac{350 \text{ kg}}{\rho_{\text{air}} - \rho_{\text{He}}}$$

The molar mass of dry air is calculated from the molecular weights of nitrogen and oxygen, the main constituents of dry air:

$$M_{\text{air}} = \left(\frac{78\%}{100\%} (28.02 \text{ g/mol}) \right) + \left(\frac{22\%}{100\%} (32.00 \text{ g/mol}) \right) = 28.9 \text{ g/mol}$$

The density of dry air is therefore:

$$\rho_{\text{air}} = \frac{pM}{RT} = \frac{(1 \text{ bar})(28.9 \text{ g/mol})}{(0.08314 \text{ L bar mol}^{-1} \text{ K}^{-1})(298)} = 1.17 \text{ g/L}$$

The density of helium gas under these conditions is

$$\rho_{He} = \frac{pM}{RT} = \frac{(1\,\text{bar})(4.003\,\text{g}\,/\,\text{mol})}{(0.08314\,\text{L bar mol}^{-1}\,\text{K}^{-1})(298)} = 0.162\,\text{g}\,/\,\text{L}$$

And so the volume is

$$V = \frac{350\,\text{kg}}{\rho_{air} - \rho_{He}} = \frac{350 \times 10^3\,\text{g}}{(1.17 - 0.162)\text{g}\,/\,\text{L}} = 3.47 \times 10^5\,\text{L}$$

2.117 As the mouse breathes, it inhales air containing oxygen. Some of this oxygen is used for metabolism, resulting in CO_2, which the mouse exhales. The amount of oxygen consumed is therefore equal to the amount of CO_2 exhaled. The solid KOH in the chamber absorbs all the CO_2 and any water the mouse exhales. Any reduction in pressure is therefore due to the removal of oxygen from the atmosphere.

The initial amount of gas in the chamber is

$$n_{initial} = \frac{p_{initial}V}{RT} = \frac{(765\,\text{Torr})\left(\dfrac{1\,\text{bar}}{750.06\,\text{Torr}}\right)(2.05\,\text{L})}{(0.08314\,\text{L bar K}^{-1}\text{mol}^{-1})(298\,\text{K})} = 0.08439\,\text{mol}$$

And after two hours, the amount of gas remaining is

$$n_{initial} = \frac{p_{initial}V}{RT} = \frac{(725\,\text{Torr})\left(\dfrac{1\,\text{bar}}{750.06\,\text{Torr}}\right)(2.05\,\text{L})}{(0.08314\,\text{L bar K}^{-1}\text{mol}^{-1})(298\,\text{K})} = 0.07998\,\text{mol}$$

The amount of oxygen consumed is therefore $0.08439 - 0.07998\,\text{mol} = 0.00441\,\text{mol}$.

$(0.00441\,\text{mol})(32.00\,\text{g/mol}) = 0.141\,\text{g}\ O_2$.

2.119 Dry air, according to Table 2-4, has a mole fraction of oxygen of 0.2095. The amount of oxygen in the pressurized tank is calculated as follows:

$$n_{O_2} = X_{O_2} n_{air} = X_{O_2}\left(\frac{p_{air}V_{air}}{RT}\right)$$

$$= 0.2095\left(\frac{(165\,\text{bar})(12.5\,\text{L})}{(0.08314\,\text{L bar K}^{-1}\text{mol}^{-1})(27 + 273)\,\text{K}}\right) = 82.69\,\text{mol}\ O_2$$

$(82.69\,\text{mol})(32.00\,\text{g/mol}) = 2646\,\text{g}\ O_2$.

At a rate of 14.0 g O_2 per minute, the tank will last $\dfrac{2646\,\text{g}\ O_2}{14.0\,\text{g/min}} = 189\,\text{min}$

Allowing 6 minutes for surfacing, and neglecting the time required to dive to depth, you could therefore stay at depth for 183 minutes.

2.121

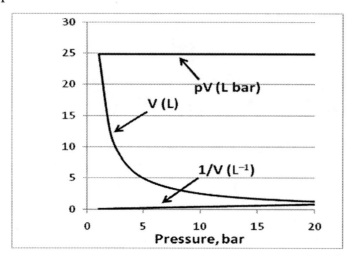

The plot shows that as p increases, V decreases. This is explained by the ideal gas equation, $V = \dfrac{nRT}{p}$.

The plot of pV versus p is linear and horizontal. According to the ideal gas equation, $pV = nRT$. As long as n and T are constant, the product pV will also be constant. This is also referred to as Boyle's Law.

The plot of $1/V$ versus p rises slowly with p. According to the ideal gas law, per mole, $\dfrac{1}{V} = \dfrac{p}{RT}$. $1/V$ is thus the concentration of the gas; that is, mol/L.

2.123 Solving this problem requires using both equation 2-3, which relates kinetic energy to molecular speed, and equation 2-4, which relates kinetic energy to temperature.

From equation 2-3, we can calculate the kinetic energy of a molecule moving at speed v:

$$E_{kinetic} = \frac{mv^2}{2} \text{ (per molecule)}$$

If a collection of atoms has average speed v, then $E_{kinetic}$ is the average kinetic energy of these atoms. For a mole of these atoms, the average kinetic energy is therefore:

$$E_{kinetic} = N_A \frac{mv^2}{2} \text{ (per mole)}$$

From equation 2-4, we found that the average molar kinetic energy is related to temperature:

$$\overline{E}_{kinetic} = \frac{3RT}{2}$$

Equating these two expressions for kinetic energy and solving for the temperature gives:

$$N_A \frac{mv^2}{2} = \frac{3RT}{2}$$
$$N_A mv^2 = 3RT$$
$$T = \frac{N_A mv^2}{3R} = \frac{Mv^2}{3R}$$

For cesium atoms, M = 0.1329 kg mol^{-1}. Thus,

$$T = \frac{0.1329 \text{ kg mol}^{-1}(0.005 \text{ m s}^{-1})^2}{3(8.314 \text{ J K}^{-1}\text{mol}^{-1})} = 1.3 \times 10^{-7} \text{ K}$$

Chapter 3 Energy and Its Conservation

Solutions to Problems in Chapter 3

3.1 Conservation of energy states that energy is neither created nor destroyed. It can only be transformed from one form into another:

(a) Total energy, which is conserved, is the sum of potential, kinetic, and thermal energies. On the tree, an apple possesses some gravitational potential energy. As an apple falls, that potential energy is converted into kinetic energy of motion.

(b) When an apple hits the ground, the impact transfers energy to molecules in the earth and in the apple. As a result, there is a slight increase in temperature; the kinetic energy of the apple has been converted into thermal energy.

3.3 Speed and kinetic energy are related through the equation, $E_{kinetic} = \frac{1}{2}mv^2$:

$$E_{kinetic} = \frac{(9.1094 \times 10^{-31} \text{ kg})(4.55 \times 10^5 \text{ m/s})^2}{2} \left(\frac{1 \text{ J}}{1 \text{ kg m}^2 \text{ s}^{-2}} \right) = 9.43 \times 10^{-20} \text{ J}$$

3.5 Electrical energy can be calculated using Equation 3-1: $E_{electrical} = k\frac{q_1 q_2}{r}$.

When distance is expressed in picometres (1 pm = 10^{-12} m) and charges are in electronic units, the constant in the equation is $k = 2.31 \times 10^{-16}$ J pm. Attractive energies are negative because when close together, the charges have lower potential energy than when far apart. The charges on the ions in this example are +1 and –1:

$$E_{electrical} = \frac{(2.31 \times 10^{-16} \text{ J pm})(+1)(-1)}{320 \text{ pm}} = -7.22 \times 10^{-19} \text{ J}$$

3.7 Energy comes in various forms: radiant, kinetic, potential, and thermal:

(a) Radiant energy is consumed and thermal energy is produced.
(b) Chemical potential energy is consumed and kinetic energy is produced.
(c) Chemical potential energy is consumed and thermal energy is produced.

3.9 Speed and kinetic energy are related through the equation, $v = \sqrt{\frac{2E_{kinetic}}{m}}$. The joule has units kg m^2 s^{-2}, so mass must be expressed in kg and v has units m/s:

$$m_{neutron} = 1.6749 \times 10^{-27} \text{ kg}$$

$$v = \sqrt{\frac{2(3.75 \times 10^{-23} \text{ J})}{1.6749 \times 10^{-27} \text{ kg}}} = 2.12 \times 10^2 \text{ m/s}$$

3.11 Calculate the temperature change resulting from an energy input using C values from Table 3-1 of your textbook and Equation 3-2:

$$q = nC\Delta T \qquad\qquad \Delta T = \frac{q}{nC}$$

(a) $n = \dfrac{m}{M} = \dfrac{10.0\ \text{g}}{26.98\ \text{g/mol}} = 0.3706\ \text{mol Al}$

$\Delta T = \dfrac{25.0\ \text{J}}{(0.3706\ \text{mol})(24.35\ \text{J mol}^{-1}\ °\text{C}^{-1})} = 2.77\ °\text{C}$

$T_f = T_i + \Delta T = 15.0 + 2.77 = 17.8\ °\text{C}$

(b) $n = \dfrac{m}{M} = \dfrac{25.0\ \text{g}}{26.98\ \text{g/mol}} = 0.9266\ \text{mol Al}$

$\Delta T = \dfrac{25.0\ \text{J}}{(0.9266\ \text{mol})(24.35\ \text{J mol}^{-1}\ °\text{C}^{-1})} = 1.108°\text{C}$

$T_f = T_i + \Delta T = 29.5 + 1.108 = 30.6\ °\text{C}$

(c) $n = \dfrac{m}{M} = \dfrac{25.0\ \text{g}}{107.9\ \text{g/mol}} = 0.2317\ \text{mol Ag}$

$\Delta T = \dfrac{25.0\ \text{J}}{(0.2317\ \text{mol})(25.351\ \text{J mol}^{-1}\ °\text{C}^{-1})} = 4.256\ °\text{C}$

$T_f = T_i + \Delta T = 29.5 + 4.256 = 33.8\ °\text{C}$

(d) $n = \dfrac{m}{M} = \dfrac{25.0\ \text{g}}{18.02\ \text{g/mol}} = 1.387\ \text{mol H}_2\text{O}$

$\Delta T = \dfrac{25.0\ \text{J}}{(1.387\ \text{mol})(75.291\ \text{J mol}^{-1}\ °\text{C}^{-1})} = 0.2394\ °\text{C}$

$T_f = T_i + \Delta T = 22.0 + 0.2394 = 22.2\ °\text{C}$

3.13 Calculate an energy change accompanying a temperature change using C values from Table 3-1 of your textbook and Equation 3-2:

$$q = nC\Delta T \qquad\qquad \Delta T = 95.0 - 23.0 = 72.0\ °\text{C}$$

$n_{\text{kettle}} = 1.35\ \text{kg}\left(\dfrac{1000\ \text{g}}{1\ \text{kg}}\right)\left(\dfrac{1\ \text{mol}}{55.85\ \text{g}}\right) = 24.2\ \text{mol Fe}$

$q_{\text{kettle}} = (24.2\ \text{mol})(25.10\ \text{J mol}^{-1}\ °\text{C}^{-1})(72.0\ °\text{C}) = 4.37 \times 10^4\ \text{J}$

$n_{\text{water}} = 2.75\ \text{kg}\left(\dfrac{1000\ \text{g}}{1\ \text{kg}}\right)\left(\dfrac{1\ \text{mol}}{18.02\ \text{g}}\right) = 153\ \text{mol water}$

$q_{\text{water}} = (153\ \text{mol})(75.291\ \text{J mol}^{-1}\ °\text{C}^{-1})(72.0\ °\text{C}) = 8.29 \times 10^5\ \text{J}$

3.15 When two objects at different temperatures are placed in contact, energy flows from the warmer to the cooler object until the two objects are at the same temperature. Let that temperature be T. The energy lost by the warmer object equals the energy gained by the cooler object:

$$\text{Water: } n = \frac{37.5 \text{ g}}{18.02 \text{ g/mol}} = 2.081 \text{ mol water}$$

$$q = (2.081 \text{ mol})(75.351 \text{ J mol}^{-1} \text{ °C}^{-1})(T - 20.5 \text{ °C}) = (156.8 \text{ J °C}^{-1})(T - 20.5 \text{ °C})$$

$$\text{Coin: } n = \frac{27.4 \text{ g}}{107.9 \text{ g/mol}} = 0.2539 \text{ mol Ag}$$

$$q = (0.2539 \text{ mol})(25.351 \text{ J mol}^{-1} \text{ °C}^{-1})(T - 100.0 \text{ °C}) = (6.437 \text{ J °C}^{-1})(T - 100.0 \text{ °C})$$

Substitute and solve for T:

$$(156.8 \text{ J °C}^{-1})(T - 20.5 \text{ °C}) = -(6.437 \text{ J °C}^{-1})(T - 100.0 \text{ °C})$$
$$156.8 \, T - 3214.5 \text{ °C} = -6.437 \, T + 643.7 \text{ °C}$$
$$163.2 \, T = 3858.2 \text{ °C} \qquad\qquad T = 23.6 \text{ °C}$$

3.17 Calculate the energy change for a temperature change using Equation 3-2 and C values from Table 3-1 of your textbook:

$$q = nC\Delta T \qquad\qquad \Delta T = 87.6 - 21.5 = 66.1 \text{ °C}$$

$$n_{water} = \frac{m}{M} = 475 \text{ mL}\left(\frac{1.00 \text{ g}}{1 \text{ mL}}\right)\left(\frac{1 \text{ mol}}{18.02 \text{ g}}\right) = 26.36 \text{ mol}$$

$$q_{water} = (26.36 \text{ mol})(75.291 \text{ J mol}^{-1} \text{ °C}^{-1})(66.1 \text{ °C}) = 1.31 \times 10^5 \text{ J} = 131 \text{ kJ}$$

3.19 First determine the energy difference (ΔE) between chicken and beef:

$$\text{Chicken: } 250 \text{ g}\left(\frac{6.0 \text{ kJ}}{1 \text{ g}}\right) = 1500 \text{ kJ} \qquad\qquad \text{Beef: } 250 \text{ g}\left(\frac{16 \text{ kJ}}{1 \text{ g}}\right) = 4000 \text{ kJ}$$

$$\Delta E = 4000 \text{ kJ} - 1500 \text{ kJ} = 2500 \text{ kJ}$$

A 55-kg person walking 6.0 km/h consumes 1090 kJ/h:

$$t = 2500 \text{ kJ}\left(\frac{1 \text{ h}}{1090 \text{ kJ}}\right) = 2.3 \text{ h}$$

Therefore, to consume the additional energy a person must walk

$$\text{distance} = 2.3 \text{ h}\left(\frac{6.0 \text{ km}}{1 \text{ h}}\right) = 14 \text{ km}$$

3.21 (a) In a combustion reaction, the products are CO_2 and H_2O:

$$C_7H_6O_2 + O_2 \rightarrow CO_2 + H_2O \text{ (unbalanced)}$$

Follow standard procedures to balance the equation. Give CO_2 a coefficient of 7 to balance C, H_2O a coefficient of 3 to balance H:

$$C_7H_6O_2 + O_2 \rightarrow 7\,CO_2 + 3\,H_2O$$

$$7\,C + 6\,H + 4\,O \rightarrow 7\,C + 6\,H + 17\,O$$

Balance O by giving O_2 a coefficient of 15/2, and then multiply by 2 to clear fractions:

$$2\,C_7H_6O_2 + 15\,O_2 \rightarrow 14\,CO_2 + 6\,H_2O$$

(b) Find energy per mole from the energy released by 1.350 g using the molar mass (122.1 g/mol):

$$\Delta E = \left(\frac{-35.61\ \text{kJ}}{1.350\ \text{g}}\right)\left(\frac{122.1\ \text{g}}{1\ \text{mol}}\right) = -3.221 \times 10^3\ \text{kJ/mol } C_7H_6O_2$$

(c) Fifteen moles of O_2 is consumed for each 2 moles of benzoic acid, so the energy released per mole of O_2 is

$$\Delta E = \left(\frac{-3.221 \times 10^3\ \text{kJ}}{1\ \text{mol acid}}\right)\left(\frac{2\ \text{mol acid}}{15\ \text{mol } O_2}\right) = -4.295 \times 10^2\ \text{kJ/mol } O_2$$

3.23 In Table 3-2, we find the following bond energies for H–X bonds where X is an element in Group 16:

Bond	H–O	H–S	H–Te
Energy (kJ/mol)	460	365	240

There is a trend: moving down the column, the bond energy decreases. From this trend, we predict that the H–Se bond has an energy between H–S and H–Te, around 300 kJ/mol (the experimental value is 315 kJ), and the H–Po bond will be weaker than H–Te, perhaps around 200 kJ/mol (no experimental value is available).

3.25 To estimate the energy change in a reaction, list the number of bonds of each type in reactants and products, and subtract the sum of average bond energies for products from the sum of average bond energies for reactants, using values from Table 3-2 of your textbook:

The reaction is $N_2 + 2\,O_2 \rightarrow N_2O_4$

Reactants			Products		
Bond	No.	Energy (kJ/mol)	Bond	No.	Energy (kJ/mol)
N≡N	1	945	N–N	1	160

O=O	2	495	N–O	2	200
			N=0	2	605

$$\Delta E_{\text{reaction}} \cong [1 \text{ mol}(945 \text{ kJ/mol}) + 2 \text{ mol}(495 \text{ kJ/mol})]$$
$$- [1 \text{ mol}(160 \text{ kJ/mol}) + 2 \text{ mol}(200 \text{ kJ/mol}) + 2 \text{ mol}(605 \text{ kJ/mol})]$$
$$\Delta E_{\text{reaction}} = 1935 \text{ kJ} - 1770 \text{ kJ} \cong 165 \text{ kJ}$$

3.27 Figure 3-13 in your textbook shows three paths for a reaction. One is the actual path. A second is decomposition of reactants into elements and then recombination of the elements into products. A third is breakage of all bonds in the reactants into atoms of the elements, and then reaction of the atoms to form all bonds in the products:

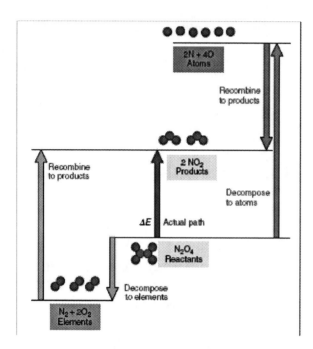

3.29 To estimate the energy change in a reaction, list the number of bonds of each type in the reagents, and subtract the sum of average bond energies for products from the sum of average bond energies for reactants, using values from Table 3-2 of your textbook:

$$2 \text{ H}_2 + \text{O}_2 \rightarrow 2 \text{ H}_2\text{O}$$

Reactants					**Products**
Bond	No.	Energy (kJ/mol)	Bond	No.	Energy (kJ/mol)
H–H	2	435	O–H	4	460
O=O	1	495			

$$\Delta E_{\text{reaction}} \cong [2 \text{ mol}(435 \text{ kJ/mol}) + 1 \text{ mol}(495 \text{ kJ/mol})] - [4 \text{ mol}(460 \text{ kJ/mol})] = -475 \text{ kJ}$$

$$3\ H_2 + N_2 \rightarrow 2\ NH_3$$

Reactants   **Products**

Bond	No.	Energy (kJ/mol)	Bond	No.	Energy (kJ/mol)
H–H	3	435	N–H	6	390
N≡N	1	945			

$$\Delta E_{reaction} \cong [3\ mol(435\ kJ/mol) + 1\ mol(945\ kJ/mol)] - [6\ mol(390\ kJ/mol)] = -90\ kJ$$

$$3\ H_2 + CO \rightarrow CH_4 + H_2O$$

Reactants   **Products**

Bond	No.	Energy (kJ/mol)	Bond	No.	Energy (kJ/mol)
H–H	3	435	C–H	4	415
C≡O	1	1070	O–H	2	460

$$\Delta E_{reaction} \cong [3\ mol(435\ kJ/mol) + 1\ mol(1070\ kJ/mol)] - [4\ mol(415\ kJ/mol)$$
$$+ 2\ mol(460\ kJ/mol)] = 2375\ kJ - 2580\ kJ = -205\ kJ$$

3.31 Calculate the heat accompanying a temperature change in a calorimeter using Equation 3-6 and the heat capacity C of the calorimeter: $q_{calorimeter} = C_{cal}\Delta T$

$$\Delta T = 23.8 - 21.6 = 2.2\ ^\circ C$$

The problem statement tells us to assume that $C_{cal} \cong C_{water}$:

$$C_{cal} \cong C_{water} = (155\ mL)\left(\frac{1.00\ g}{1\ mL}\right)\left(\frac{1\ mol}{18.02\ g}\right)(75.291\ J\ mol^{-1}\ ^\circ C^{-1}) = 647.6\ J\ ^\circ C^{-1}$$

$$q_{calorimeter} = (647.6\ J\ ^\circ C^{-1})(2.2\ ^\circ C) = 1.4 \times 10^3\ J$$

The heat gained by the calorimeter is heat lost by the solution process, giving

$$q_{solution} = -1.4 \times 10^3\ J$$

3.33 The heat capacity of a calorimeter can be found from energy and temperature data using Equation 3-6, $q_{calorimeter} = C_{cal}\Delta T$

Here, the heat released by the combustion process is absorbed by the calorimeter:

$$q_{calorimeter} = -q_{glucose} = -1.7500 \text{ g} \left(\frac{-15.57 \text{ kJ}}{1 \text{ g}} \right) = 27.25 \text{ kJ}$$

$$\Delta T = 23.34 - 21.45 = 1.89 \text{ °C}$$

$$C_{cal} = \frac{q_{calorimeter}}{\Delta T} = \frac{27.25 \text{ kJ}}{1.89 \text{ °C}} = 14.4 \text{ kJ/°C}$$

3.35 Calculate the energy released during combustion from the temperature increase and the total heat capacity of the calorimeter. Then convert to molar energy using molar mass:

$$q_{calorimeter} = C_{cal}\Delta T = (7.85 \text{ kJ/°C})(29.04 \text{ °C} - 24.65 \text{ °C}) = 34.46 \text{ kJ}$$

$$\Delta E_{reaction} = q_{reaction} = q_{calorimeter} = -34.46 \text{ kJ}$$

$$M = 8(12.01 \text{ g/mol}) + 10(1.0078 \text{ g/mol}) + 4(14.01 \text{ g/mol}) + 2(16.00 \text{ g/mol}) = 194.2 \text{ g/mol}$$

$$\Delta E_{molar} = \left(\frac{-34.46 \text{ kJ}}{1.35 \text{ g}} \right) \left(\frac{194.2 \text{ g}}{1 \text{ mol}} \right) = -4.96 \times 10^3 \text{ kJ/mol}$$

3.37 Expansion work can be calculated using Equation 3-8, $w_{sys} = -p_{ext}\Delta V_{sys}$. The external pressure opposing inflation of a balloon is atmospheric pressure, 1.00 atm:

$$V_f = 2.5 \text{ L}, \ V_i = 0.0 \text{ L}$$

$$\Delta V = 2.5 \text{ L} - 0.0 \text{ L} = 2.5 \text{ L} = 2.5 \times 10^{-3} \text{ m}^3$$

$$w_{sys} = -p_{ext}\Delta V_{sys} = -(1.00 \text{ bar}) \left(\frac{10^5 \text{ Pa}}{1 \text{ bar}} \right) (2.5 \times 10^{-3} \text{ m}^3) = -2.5 \times 10^2 \text{ J}$$

3.39 Standard enthalpy changes are calculated from Equation 3-12 using standard enthalpies of formation, which can be found in Appendix D of your textbook:

$$\Delta H°_{reaction} = \Sigma \text{coeff}_{products} \ \Delta H°_{f \, products} - \Sigma \text{coeff}_{reactants} \ \Delta H°_{f \, reactants}$$

(a) $\Delta H°_{reaction} = [2 \text{ mol}(-393.5 \text{ kJ/mol}) + 2 \text{ mol}(-285.83 \text{ kJ/mol})]$
$$-[1 \text{ mol}(52.4 \text{ kJ/mol}) + 3 \text{ mol}(0 \text{ kJ/mol})] = -1411.1 \text{ kJ}$$

(b) $\Delta H°_{reaction} = [1 \text{ mol}(0 \text{ kJ/mol}) + 3 \text{ mol}(0 \text{ kJ/mol})] - 2 \text{ mol}(-45.9 \text{ kJ/mol}) = 91.8 \text{ kJ}$

(c) $\Delta H°_{reaction} = [1 \text{ mol}(-2984.0 \text{ kJ/mol}) + 5 \text{ mol}(0 \text{ kJ/mol})]$
$$-[5 \text{ mol}(-277.4 \text{ kJ/mol}) + 4 \text{ mol}(0 \text{ kJ/mol})] = -1597.0 \text{ kJ}$$

(d) $\Delta H^{\circ}_{reaction} = [1 \text{ mol}(-910.7 \text{ kJ/mol}) + 4 \text{ mol}(-92.3 \text{ kJ/mol})]$

$\qquad - [1 \text{ mol}(-687.0 \text{ kJ/mol}) + 2 \text{ mol}(-285.83 \text{ kJ/mol})] = -21.2 \text{ kJ}$

3.41 Reaction energies are related to reaction enthalpies (Problem 3.39) through Equation 3-11:

$$\Delta H_{reaction} \cong \Delta E_{reaction} + RT\Delta(n)_{gases}$$

Since temperature does not change, we can rearrange the last term:

$$\Delta E_{reaction} \cong \Delta H_{reaction} - RT\Delta(n_{gases})$$

Begin by calculating the change in the number of moles of *gases*, and then use the rearranged equation to determine $\Delta E_{reaction}$:

(a) $\Delta n_{gases} = 2 - (3 + 1) = -2 \text{ mol}$

$$\Delta E_{reaction} \cong -1411.1 \text{ kJ} - (8.314 \text{ J mol}^{-1} \text{ K}^{-1})(298 \text{ K})(-2 \text{ mol})\left(\frac{10^{-3} \text{ kJ}}{1 \text{ J}}\right)$$

$$= -1411.1 + 4.96 = -1.406.1 \text{ kJ}$$

(b) $\Delta n_{gases} = 3 + 1 - 2 = 2 \text{ mol}$

$$\Delta E_{reaction} \cong 91.8 \text{ kJ} - (8.314 \text{ J mol}^{-1} \text{ K}^{-1})(298 \text{ K})(2 \text{ mol})\left(\frac{10^{-3} \text{ kJ}}{1 \text{ J}}\right) = 91.8 - 4.96 = 86.8 \text{ kJ}$$

(c) All compounds are solid, $\Delta n_{gases} = 0 \text{ mol}$　　　　$\Delta E_{reaction} \cong -1597.0 \text{ kJ}$

(d) $\Delta n_{gases} = 0 \text{ mol}$　　　　　　$\Delta E_{reaction} \cong -21.2 \text{ kJ}$

3.43 A formation reaction has elements in their standard states as reactants and 1 mole of a single substance as the product:

(a) $3 \text{ K } (s) + \text{P } (s) + 2 \text{ O}_2 (g) \rightarrow \text{K}_3\text{PO}_4 (s)$

(b) $2 \text{ C (graphite)} + 2 \text{ H}_2 (g) + \text{O}_2 (g) \rightarrow \text{CH}_3\text{CO}_2\text{H} (l)$

(c) $3 \text{ C (graphite)} + 9/2 \text{ H}_2 (g) + 1/2 \text{ N}_2 (g) \rightarrow (\text{CH}_3)_3\text{N} (g)$

(d) $2 \text{ Al } (s) + 3/2 \text{ O}_2 (g) \rightarrow \text{Al}_2\text{O}_3 (s)$

3.45 Standard enthalpy changes are calculated from Equation 3-12 using standard enthalpies of formation, which can be found in Appendix D of your textbook:

$$\Delta H^{\circ}_{reaction} = \sum \text{coeff}_{products} \Delta H^{\circ}_{f\,products} - \sum \text{coeff}_{reactants} \Delta H^{\circ}_{f\,reactants}$$

(a) $\Delta H^{\circ}_{reaction} = [4 \text{ mol}(91.3 \text{ kJ/mol}) + 6 \text{ mol}(-285.83 \text{ kJ/mol})]$

$\qquad -[4 \text{ mol}(-45.9 \text{ kJ/mol}) + 5 \text{ mol}(0 \text{ kJ/mol})] = -1166.2 \text{ kJ}$

(b) $\Delta H^{\circ}_{reaction} = [2 \text{ mol}(0 \text{ kJ/mol}) + 6 \text{ mol}(-285.83 \text{ kJ/mol})]$

$\qquad -[4 \text{ mol}(-45.9 \text{ kJ/mol}) + 3 \text{ mol}(0 \text{ kJ/mol})] = -1531.4 \text{ kJ}$

3.47 Figure 3-21 shows that the energy needs per capita in the computer age are 1×10^6 kJ/day. To determine the annual energy usage of Canada, multiply this need by the total population and the number of days in a year:

Annual energy usage = $(1 \times 10^6$ kJ/person-day)(365 days/yr)(33 212 696 persons)

$= 1 \times 10^{16}$ kJ

3.49 Use a ruler to measure the energy consumption for each source for the year 1975 on Figure 3-22. Here are the values:

Source	Amount (10^{18} kJ)	%
Wood	1.7	2.2
Hydro	3.4	4.3
Nuclear	2.1	2.7
Coal	13.8	17.5
Gas	20.7	26.2
Petroleum	37.2	47.1
Total	78.9	100

Use these values to construct a pie chart, which appears on the right.

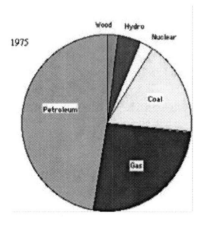

3.51 Because energy is a state function, the energy released (ΔE) when 1 g of gasoline burns is the same regardless of the conditions. The energy released can be set equal to the work done (w) plus the heat transferred (q). Work is done when an automobile accelerates, but no work is done when an automobile idles:

Heat released (idle) = Heat released (acceleration) + Work done (acceleration). Thus, more heat must be removed under idling conditions (this is part of the reason why automobiles tend to overheat in traffic jams).

3.53 The difference between $\Delta H_{reaction}$ and $\Delta E_{reaction}$ can be estimated using Equation 3-11:

$$\Delta H_{reaction} \cong \Delta E_{reaction} + RT\Delta(n)_{gases} \qquad \Delta H_{reaction} - \Delta E_{reaction} \cong RT\Delta(n)_{(gases)}$$

The reaction is $C \text{ (graphite)} + H_2O \text{ (}l\text{)} \rightarrow CO \text{ (}g\text{)} + H_2(g)$

$\Delta n_{gases} = 1 \text{ mol} + 1 \text{ mol} - 0 \text{ mol} = 2 \text{ mol gases}$

$\Delta H_{reaction} - \Delta E_{reaction} \cong (8.314 \text{ J mol}^{-1} \text{ K}^{-1})(298 \text{ K})(2 \text{ mol}) \cong 4.96 \times 10^3 \text{ J}$

3.55 To work a problem involving heat transfers, it is useful to set up a block diagram illustrating the process. In this problem, a copper block transfers energy to ice:

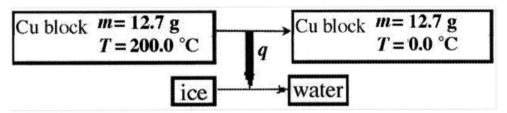

Thus, $q_{ice} = -q_{Cu}$

$$q_{Cu} = n_{Cu}C\Delta T \qquad q_{ice} = n_{ice}\Delta H_{fus}$$

Substituting gives

$$n_{ice}\Delta H_{fus} = -n_{Cu}C\Delta T$$

Here are the data needed for the calculation:

$$n_{Cu} = \frac{m}{M} = \frac{12.7 \text{ g}}{63.546 \text{ g/mol}} = 0.1999 \text{ mol}$$

$C = 24.435 \text{ J mol}^{-1} \text{ }^{\circ}\text{C}^{-1}$

$\Delta T = 0.0 \text{ }^{\circ}\text{C} - 200.0 \text{ }^{\circ}\text{C} = -200 ^{\circ}\text{C}$

$\Delta H_{fus} = 6.01 \text{ kJ/mol} = 6.01 \times 10^3 \text{ J/mol}$

Substitute and solve for the amount of ice that melts:

$$n_{ice} = \frac{-n_{Cu}C\Delta T}{\Delta H_{fus}} = -\frac{(0.1999 \text{ mol})(24.435 \text{ J mol}^{-1} \text{ }^{\circ}\text{C}^{-1})(-200 \text{ }^{\circ}\text{C})}{6.01 \times 10^3 \text{ J/mol}} = 0.1625 \text{ mol}$$

Finally, convert to mass:

$$m_{ice} = nM = (0.1625 \text{ mol})(18.02 \text{ g/mol}) = 2.93 \text{ g of ice melts}$$

3.57 (a) In a combustion reaction, a substance reacts with O_2 to form CO_2 and H_2O:

$$C_6H_{12}O_6 + O_2 \rightarrow CO_2 + H_2O$$

Balance the equation using standard procedures. Give CO_2 a coefficient of 6 and H_2O a coefficient of 6 to balance C and H:

$$C_6H_{12}O_6 + O_2 \rightarrow 6\ CO_2 + 6\ H_2O$$

$$6\ C + 12\ H + 8\ O \rightarrow 6\ C + 12\ H + 18\ O$$

Give O_2 a coefficient of 6 to balance O:

$$C_6H_{12}O_6\ (s) + 6\ O_2\ (g) \rightarrow 6\ CO_2\ (g) + 6\ H_2O\ (l)$$

(b) To determine the molar heat of combustion, multiply the heat released (negative, indicating an exothermic reaction) in burning 1 g by the molar mass:

$$\Delta H_{molar} = (-15.7\ \text{kJ/g})(180.15\ \text{g/mol}) = -2.83 \times 10^3\ \text{kJ/mol} = \Delta H^\circ_{reaction}$$

(c) Use the molar heat of combustion along with Equation 3-12 and data from Appendix D to determine the heat of formation of glucose:

$$\Delta H^\circ_{reaction} = \Sigma\, \text{coeff}_{products}\ \Delta H^\circ_{f\,products} - \Sigma\, \text{coeff}_{reactants}\ \Delta H^\circ_{f\,reactants}$$

$$-2.83 \times 10^3\ \text{kJ/mol} = [6\ \text{mol}(-393.5\ \text{kJ/mol}) + 6\ \text{mol}(-285.83\ \text{kJ/mol})]$$

$$-[6\ \text{mol}(0\ \text{kJ/mol}) + \Delta H^\circ_{f,\ glucose}]$$

$$\Delta H^\circ_{f,\ glucose} = -1.25 \times 10^3\ \text{kJ/mol}$$

3.59 To predict stability from bond energies, tabulate the number of bonds of each type and add up their contributions. Expand the structures in order to count bonds of each type:

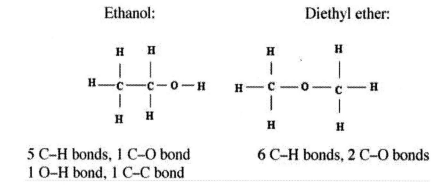

Ethanol:	Diethyl ether:
5 C–H bonds, 1 C–O bond 1 O–H bond, 1 C–C bond	6 C–H bonds, 2 C–O bonds

We can ignore the 5 C–H bonds and 1 C–O bond that the two compounds have in common, leaving O–H and C–C in ethanol vs. C–H and C–O in diethyl ether:

Ethanol: 460 kJ/mol (O–H) + 345 kJ/mol (C–C) = 805 kJ/mol

Diethyl ether: 415 kJ/mol (C–H) + 360 kJ/mol (C–O) = 775 kJ/mol

Based on average bond energies, ethanol is more stable by 30 kJ/mol.

3.61 When an airplane accelerates down a runway, its engines burn fuel, converting chemical energy into kinetic energy of motion. During takeoff, some kinetic energy is converted into gravitational potential energy. The net transformation is conversion of stored chemical energy of the airplane's fuel into kinetic energy and gravitational potential energy of the airplane (much of the chemical energy also is "wasted" as heat that is transferred to the air passing through the engines).

3.63 A molar heat capacity can be calculated from a temperature change using $q = nC\Delta T$:

$$\Delta T = 47.6\ °C - 28.0\ °C = 19.6\ °C \qquad n = \frac{52.5\ g}{118.71\ g/mol} = 0.4422\ mol$$

$$C = \frac{q}{n\Delta T} = \frac{(100.0\ J)}{(0.4422\ mol)(19.6\ °C)} = 11.5\ J\ mol^{-1}\ °C^{-1}$$

3.65 A calculation using average bond energies requires a list of all bonds in the reagents. To obtain an estimate for reaction energy per mole of material, use Equation 3-5:

$$\Delta E = \sum BE_{bonds\ broken} - \sum BE_{bonds\ formed}$$
$$\Delta E \cong \sum BE_{reactants} - \sum BE_{products}$$

Ethane reagents:	$H_3C–CH_3$	7/2 O_2	2 CO_2	3 H_2O
Bonds:	1 C–C, 6 C–H	7/2 (1 O=O)	2 (2 C=0)	3 (2 O–H)
Energies (kJ/mol):	345, 415	495	800	460

$$\Delta E \cong [1(345\ kJ/mol) + 6(415\ kJ/mol) + 7/2(495\ kJ/mol)]$$
$$-[4(800\ kJ/mol) + 6(460\ kJ/mol)] \cong -1392\ kJ/mol$$

$$\Delta E\ (per\ gram) = \frac{\Delta E}{M} = \frac{-1392\ kJ/mol}{30.08\ g/mol} = -46.3\ kJ/g$$

Ethylene reagents:	$H_2C=CH_2$	3 O_2	2 CO_2	2 H_2O
Bonds:	1 C=C, 4 C–H	3 (1 O=O)	2 (2 C=0)	2 (2 O–H)
Energies (kJ/mol):	615, 415	495	800	460

$$\Delta E \cong [1(615\ kJ/mol) + 4(415\ kJ/mol) + 3(495\ kJ/mol)] - [4(800\ kJ/mol) + 4(460\ kJ/mol)]$$

$$\Delta E \cong -1280\ kJ/mol$$

$$\Delta E\ (per\ gram) = \frac{\Delta E}{M} = \frac{-1280\ kJ/mol}{28.06\ g/mol} = -45.6\ kJ/g$$

Acetylene reagents:	$HC\equiv CH$	5/2 O_2	2 CO_2	H_2O
Bonds:	1 C≡C, 2 C-H	5/2 (1 O=O)	2 (2 C=0)	2 O–H
Energies (kJ/mol):	835, 415	495	800	460

$$\Delta E \cong [1(835 \text{ kJ/mol}) + 2(415 \text{ kJ/mol}) + 2.5(495 \text{ kJ/mol})]$$
$$-[4(800 \text{ kJ/mol}) + 2(460 \text{ kJ/mol})] = -1217 \text{ kJ/mol}$$

$$\Delta E \text{ (per gram)} = \frac{\Delta E}{M} = \frac{-1217 \text{ kJ/mol}}{26.04 \text{ g/mol}} = -46.7 \text{ kJ/g}$$

These estimates indicate that acetylene releases slightly more energy per unit mass than ethane or ethylene, but experimental values show that ethane releases the most. Remember that average bond energy calculations provide estimates, not exact values.

3.67 Kinetic energy is given by $E_{kinetic} = \frac{1}{2}mv^2$, expressed in SI units:

$$m = 2250 \text{ lb}\left(\frac{0.45359 \text{ kg}}{1 \text{ lb}}\right) = 1.021 \times 10^3 \text{ kg}$$

$$v = \left(\frac{57.5 \text{ km}}{\text{h}}\right)\left(\frac{10^3 \text{ m}}{1 \text{ km}}\right)\left(\frac{1 \text{ h}}{3600 \text{ s}}\right) = 15.97 \text{ m/s}$$

$$E_{kinetic} = \frac{(1.021 \times 10^3 \text{ kg})(15.97 \text{ m/s})^2}{2} = 1.30 \times 10^5 \text{ kgm}^2\text{s}^{-2} = 1.30 \times 10^5 \text{ J}$$

3.69 Standard enthalpy changes are calculated from Equation 3-12 using standard enthalpies of formation, which can be found in Appendix D of your textbook:

$$\Delta H^{\circ}_{reaction} = \sum \text{coeff}_{products} \Delta H^{\circ}_{f \, products} - \sum \text{coeff}_{reactants} \Delta H^{\circ}_{f \, reactants}$$

(a) $\Delta H^{\circ}_{reaction} = [2 \text{ mol}(-395.5 \text{ kJ/mol})]$
$$-[2 \text{ mol}(-296.8 \text{ kJ/mol}) + 1 \text{ mol}(0 \text{ kJ/mol})] = -197.4 \text{ kJ}$$

(b) $\Delta H^{\circ}_{reaction} = [1 \text{ mol}(11.1 \text{ kJ/mol})] - [2 \text{ mol}(33.2 \text{ kJ/mol})] = -55.3 \text{ kJ}$

(c) $\Delta H^{\circ}_{reaction} = [1 \text{ mol}(-1675.7 \text{ kJ/mol}) + 2 \text{ mol}(0 \text{ kJ/mol})]$
$$-[1 \text{ mol}(-824.2 \text{ kJ/mol}) + 2 \text{ mol}(0 \text{ kJ/mol})] = -851.5 \text{ kJ}$$

3.71 To estimate the energy change in a reaction, determine Lewis structures to obtain types of bonds, list the number of bonds of each type in reactants and products, and subtract the sum of average bond energies for products from the sum of average bond energies for reactants, using values from Table 3-2 of your textbook.

$$H–Cl + H_2C=CH_2 \rightarrow H_3C–CH_2Cl$$

Reactants			Products		
Bond	No.	Energy (kJ/mol)	Bond	No.	Energy (kJ/mol)

H–Cl	1	430	C–Cl	1	330
C–H	4	415	C–H	5	415
C=C	1	615	C–C	1	345

$$\Delta H^\circ_{\text{reaction}} \cong [1 \text{ mol}(430 \text{ kJ/mol}) + 4 \text{ mol}(415 \text{ kJ/mol}) + 1 \text{ mol}(615 \text{ kJ/mol})]$$

$$-[1 \text{ mol}(330 \text{ kJ/mol}) + 5 \text{ mol}(415 \text{ kJ/mol}) + 1 \text{ mol}(345 \text{ kJ/mol})] = -45 \text{ kJ}$$

$$Cl_2 + H_2C=CH_2 \rightarrow ClH_2C–CH_2Cl$$

Reactants   **Products**

Bond	No.	Energy (kJ/mol)	Bond	No.	Energy (kJ/mol)
Cl–Cl	1	240	C–Cl	2	330
C–H	4	415	C–H	4	415
C=C	1	615	C–C	1	345

$$\Delta H^\circ_{\text{reaction}} \cong [1 \text{ mol}(240 \text{ kJ/mol}) + 4 \text{ mol}(415 \text{ kJ/mol}) + 1 \text{ mol}(615 \text{ kJ/mol})]$$

$$-[2 \text{ mol}(330 \text{ kJ/mol}) + 4 \text{ mol}(415 \text{ kJ/mol}) + 1 \text{ mol}(345 \text{ kJ/mol})] = -150 \text{ kJ}$$

3.73 An enthalpy of formation can be calculated from the heat of a reaction provided all other enthalpies of formation are known. For this reaction, formation enthalpies of three of the four reagents appear in Appendix D of your textbook:

$$\Delta H^\circ_{\text{reaction}} = \Sigma \text{coeff}_{\text{products}} \, \Delta H^\circ_{\text{f products}} - \Sigma \text{coeff}_{\text{reactants}} \, \Delta H^\circ_{\text{f reactants}}$$

$$-3677 \text{ kJ} = [1 \text{ mol}(-2984.0 \text{ kJ/mol}) + 3 \text{ mol}(-296.8 \text{ kJ/mol})]$$

$$-[\Delta H^\circ_{\text{f, P}_4\text{S}_3} + 8 \text{ mol}(0 \text{ kJ/mol})]$$

$$\Delta H^\circ_{\text{f, P}_4\text{S}_3} = 3677 \text{ kJ} - 2984.0 \text{ kJ} - 890.4 \text{ kJ} = -197 \text{ kJ}$$

3.75 To work a problem involving heat transfers, it is useful to set up a block diagram illustrating the process. In this problem, the unknown metal transfers energy to water:

Thus, $q_{\text{metal}} = -q_{\text{water}} = -(nC\Delta T)_{\text{water}}$

$$n_{\text{water}} = \frac{m}{M} = \frac{80.0 \text{ g}}{18.02 \text{ g/mol}} = 4.44 \text{ mol water}$$

$C = 75.291 \text{ J mol}^{-1} \, ^\circ\text{C}^{-1}$ $\Delta T = 28.4 - 24.8 = 3.6 \, ^\circ\text{C}$

$q_{\text{metal}} = -q_{\text{water}} = -(4.44 \text{ mol})(75.291 \text{ J mol}^{-1} \, ^\circ\text{C}^{-1})(3.6 \, ^\circ\text{C}) = -1203 \text{ J}$

We can use $q = mC\Delta T$ and the mass of the metal to find its heat capacity:

$$\Delta T = 28.4 - 100.0 = -71.6 \,°C$$

$$C = \frac{q}{m\Delta T} = \frac{-1203 \text{ J}}{(44.0 \text{ g})(-71.6 \,°C)} = 0.382 \text{ J g}^{-1} \,°C^{-1}$$

This matches the heat capacity of Cu ($0.385 \text{ J g}^{-1} \,°C^{-1}$), so the metal is copper.

3.77 The enthalpy of a reaction can be calculated from Equation 3-12 and standard heats of formation (see Appendix D):

$$\Delta H^{°}_{\text{reaction}} = \sum \text{coeff}_{\text{products}} \, \Delta H^{°}_{\text{f products}} - \sum \text{coeff}_{\text{reactants}} \, \Delta H^{°}_{\text{f reactants}}$$

(a) The question asks for $\Delta H_{\text{reaction}}$ when 1 mole of hydrazine burns:

$$N_2H_4 \,(l) + 1/2 \, N_2O_4 \,(g) \rightarrow 3/2 \, N_2 \,(g) + 2 \, H_2O \,(g)$$

$$\Delta H_{\text{reaction}} = [1.5 \text{ mol}(0 \text{ kJ/mol}) + 2 \text{ mol}(-241.83 \text{ kJ/mol})]$$
$$- [1 \text{ mol}(50.6 \text{ kJ/mol}) + 0.5 \text{ mol}(11.1 \text{ kJ/mol})] = -539.8 \text{ kJ/mol}$$

(b) When hydrazine burns in oxygen, the reaction is:

$$N_2H_4 \,(l) + O_2 \,(g) \rightarrow N_2 \,(g) + 2 \, H_2O \,(g)$$

$$\Delta H_{\text{reaction}} = [1 \text{ mol}(0 \text{ kJ/mol}) + 2 \text{ mol}(-241.83 \text{ kJ/mol})]$$
$$- [1 \text{ mol}(50.6 \text{ kJ/mol}) + 1 \text{ mol}(0 \text{ kJ/mol})] = -534.3 \text{ kJ/mol}$$

The reaction with N_2O_4 is slightly more exothermic because that compound is slightly less stable than the elements from which it forms. (N_2O_4 is used in preference to O_2 because the reaction occurs on contact, whereas O_2 requires an ignition device.)

3.79 Consult Figure 3-22 to find information about energy sources in various years.

(a) Before 1850, there was only one major energy source, wood.

(b) Before 1980, natural gas production peaked in about 1972, at 23×10^{18} kJ/yr.

(c) To find the percentage due to fossil fuels, first measure the amounts from each source, and then take the appropriate fraction. Here are the amounts:

Source	Amount (10^{18} kJ)
Wood	1.8
Hydro	1.8
Nuclear	0.0
Coal	13.9

Gas	7.3
Petroleum	16.4
Total	41.2

$$\text{Fossil fuel \%} = \left(\frac{13.9 + 7.3 + 16.4}{41.2}\right)(100\%) = 91.2\%$$

3.81 The process shown is condensation of a gas to form a solid, the reverse of sublimation. Sublimation always is endothermic, as energy must be provided to overcome intermolecular forces of attraction; thus the depicted process is exothermic.

(a) ΔH_{sys} is negative (exothermic process)

(b) ΔE_{surr} is positive (surroundings must absorb the energy released in the process)

(c) $\Delta E_{universe} = 0$ (total energy always is conserved)

3.83 To work a problem involving heat transfers, it is useful to set up a block diagram illustrating the process. In this problem, a rhodium block transfers energy to ice:

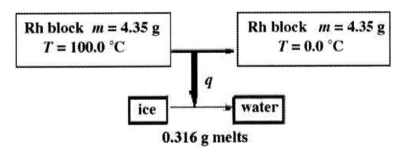

Thus, $q_{ice} = -q_{Rh}$ $\qquad$ $q_{Rh} = n_{Rh}C\Delta T$, and $q_{ice} = n_{ice}\Delta H_{fus}$

Substituting gives $\qquad n_{ice}\Delta H_{fus} = -n_{Rh}C\Delta T$

Solve for C, the heat capacity of rhodium:

$$C = \frac{-n_{ice}\Delta H_{fus}}{n_{Rh}\Delta T}$$

Here are the data needed for the calculation:

$$n_{Rh} = \frac{m}{M} = \frac{4.35 \text{ g}}{102.91 \text{ g/mol}} = 0.04227 \text{ mol} \qquad n_{ice} = \frac{m}{M} = \frac{0.319 \text{ g}}{18.01 \text{ g/mol}} = 0.01771 \text{ mol}$$

$$\Delta T = 0.0\ ^\circ C - 100.0\ ^\circ C = -100.0\ ^\circ C$$

$$\Delta H_{fus} = 6.01 \text{ kJ/mol} = 6.01 \times 10^3 \text{ J/mol}$$

Substitute and evaluate C:

$$C = \frac{-n_{ice}\Delta H_{fus}}{n_{Rh}\Delta T} = \frac{-(0.01771 \text{ mol})(6.01 \times 10^3 \text{ J/mol})}{(0.04227 \text{ mol})(-100.0\ ^\circ C)} = 25.2 \text{ J mol}^{-1}\ ^\circ C^{-1}$$

3.85 Data in Problem 3.57 indicate that glucose combustion releases energy amounting to 15.7 kJ/g. First determine how much glucose would be required at 100% efficiency, and then correct for inefficiency and sugar content:

$$\text{Mass required} = 220 \text{ kJ}\left(\frac{1 \text{ g glucose}}{15.7 \text{ kJ}}\right)\left(\frac{100\%}{30\%}\right)\left(\frac{100 \text{ g cereal}}{35 \text{ g glucose}}\right) = 133 \text{ g cereal}$$

3.87 To work a problem involving heat transfers, it is useful to set up a block diagram illustrating the process. In this problem, a silver spoon absorbs energy from coffee (water):

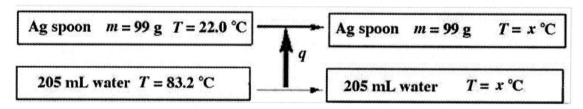

Thus, $q_{\text{water}} = -q_{\text{Ag}}$

$q_{\text{Ag}} = (nC\Delta T)_{\text{Ag}}$ and $q_{\text{water}} = (nC\Delta T)_{\text{water}}$

For Ag:

$$n = \frac{m}{M} = \frac{99 \text{ g}}{107.9 \text{ g/mol}} = 0.918 \text{ mol Ag}$$

$C_{\text{Ag}} = 25.351 \text{ J mol}^{-1} \text{ °C}^{-1}$ $\qquad \Delta T = x - 22.0 \text{ °C}$

For water:

$$n = \frac{m}{M} = 205 \text{ g}\left(\frac{1.00 \text{ g}}{1 \text{ mL}}\right)\left(\frac{1 \text{ mol}}{18.02 \text{ g}}\right) = 11.38 \text{ mol H}_2\text{O}$$

$C_{\text{water}} = 75.291 \text{ J mol}^{-1} \text{ °C}^{-1}$ $\qquad \Delta T = x - 83.2 \text{ °C}$

Substitute and solve for x:

$(11.38 \text{ mol})(75.291 \text{ J mol}^{-1} \text{ °C}^{-1})(x - 83.2 \text{ °C}) = -(0.918 \text{ mol})(25.351 \text{ J mol}^{-1} \text{ °C}^{-1})(x - 22.0 \text{ °C})$

$856.8 \, x - 71 \, 286.7 = -(23.27 \, x - 512)$

$856.8 \, x - 71 \, 286.7 = -23.27 \, x + 512$

$880.1 \, x = 71 \, 798.7$ $\qquad\qquad x = 81.6 \text{ °C}$

The final temperature of the coffee is 81.6° (not a very efficient way of cooling!). To determine whether an Al spoon would be more or less effective, compare nC for the two spoons:

$$\text{Al:} = 99 \text{ g}\left(\frac{1 \text{ mol}}{26.982 \text{ g}}\right)(24.35 \text{ J mol}^{-1} \text{ °C}^{-1}) = 89 \text{ J/°C}$$

$$Ag := 99\ g\left(\frac{1\ mol}{107.9\ g}\right)(25.351\ J\ mol^{-1}\ {}^{\circ}C^{-1}) = 23\ J/{}^{\circ}C$$

An Al spoon requires more energy per degree temperature increase, so an Al spoon would be more effective in cooling the coffee (but because the density of Al is smaller than that of Ag, an aluminum spoon *of the same mass* would be quite a bit larger in size).

3.89 The expression that always is correct for q is (e), $q = \Delta E - w$, which is derived from the first law of thermodynamics. For the others:

(a) $\Delta E \neq q$ when volume changes

(b) $\Delta H \neq q$ when pressure changes

(c) $q_v \neq q$ when volume changes

(d) $q_p \neq q$ when pressure changes

3.91 Determine the heat required to warm the soup from 25 °C to 85 °C (60 °C).

Knowing that C_m for water is 75.291 J mol^{-1} °C^{-1}

$$q_{soup} = 575\ ml\left(\frac{1\ g}{ml}\right)\left(\frac{1\ mol}{18.02\ g}\right)(75.291\ J\ mol^{-1}\ {}^{\circ}C^{-1})(60\ {}^{\circ}C) = 144\ kJ$$

To warm the soup you must add 144 J.

$$time = \frac{heat}{power} = \frac{144\ kJ}{750\ J/s} = 192\ s$$

Round to two significant figures, because ΔT is only known to two significant figures:

$t = 1.9 \times 10^2$ s (3.2 min)

3.93 To work a problem involving heat transfers, it is useful to set up a block diagram illustrating the process. In this problem, a copper block transfers energy to water:

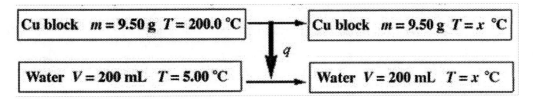

Thus, $q_{water} = -q_{Cu}$ $q_{Cu} = (nC\Delta T)_{Cu}$, and $q_{water} = (nC\Delta T)_{water}$

For Cu, $n = \dfrac{m}{M} = \dfrac{9.50\ g}{63.546\ g/mol} = 0.1495\ mol\ Cu$

$C = 24.435\ J\ mol^{-1}\ °C^{-1}$ $\Delta T = x - 200.0\ °C$

For water, $n = \dfrac{m}{M} = 200\ mL\left(\dfrac{1.00\ g}{1\ mL}\right)\left(\dfrac{1\ mol}{18.02\ g}\right) = 11.10\ mol\ water$

$C = 75.291\ J\ mol^{-1}\ °C^{-1}$ $\Delta T = (x - 5.00)\ °C$

Substitute and solve for x:

$(11.10\ mol)(75.291\ J\ mol^{-1}\ °C^{-1})(x - 5.00\ °C) = -(0.1495\ mol)(24.435\ J\ mol^{-1}\ °C^{-1})(x - 200.0\ °C)$

$835.7\,x - 4179 = -(3.653\,x - 730.6)$

$835.7\,x - 4179 = -3.653\,x + 730.6$

$839.4\,x = 4910$

$x = 5.85\ °C$

The final temperature of both the copper block and the water is 5.85 °C.

3.95 (a) Work done on the system compresses it, so the new figure should show a smaller volume (see figure below).

(b) An exothermic reaction increases the temperature of the system, leading to expansion and a larger volume:

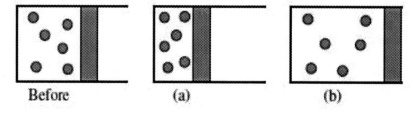

Before   (a)   (b)

3.97 To determine the missing information on this diagram, we must recognize that the starting materials (at the bottom of the diagram) are elements in their standard states.

Therefore, ΔH_1^0 and ΔH_2^0 are related to standard heats of formation.

For ΔH_1^0, 3 O_2 are simple spectators; the reaction is 2 N_2 + $O_2 \rightarrow$ 2 N_2O, for which

$$\Delta H_1^0 = 2(\Delta H_{f, N_2O}^°) = (2\ mol)(81.6\ kJ/mol) = 163.2\ kJ$$

We have a value for ΔH_2^0, but we need to identify the products. Consult Appendix D for $\Delta H_f^°$ values of nitrogen oxides and find a combination that adds to give 77.5 kJ. Values larger than 77.3 kJ/mol can be excluded, leaving these possibilities:

NO$_2$: 33.2 kJ/mol   N$_2$O$_4$: 11.1 kJ/mol   N$_2$O$_5$: 13.3 kJ/mol

By inspection, we see that 77.5 kJ = (2 mol)(33.2 kJ/mol) + (1 mol)(11.1 kJ/mol). Thus, the products are 2 NO$_2$ + N$_2$O$_4$, using up all 4 N atoms and all 8 O atoms.

From the diagram, we can see that $\Delta H_2^0 = \Delta H_3^0 + \Delta H_1^0$, from which

$$\Delta H_3^\circ = \Delta H_1^\circ - \Delta H_2^\circ = 163.2 - 77.5 = 85.7 \text{ kJ}.$$

3.99 (a) Knowing ΔH°, use Equation 3-11 to determin ΔE:

$$\Delta H_{\text{reaction}}^\circ \cong \Delta E_{\text{reaction}}^\circ + RT\Delta n_{\text{gases}} \qquad \Delta E_{\text{reaction}}^\circ \cong \Delta H_{\text{reaction}}^\circ - RT\Delta n_{\text{gases}}$$

In this reaction, $\Delta n_{\text{gases}} = 1$ mol

$$\Delta E_{\text{reaction}}^\circ \cong 39 \text{ kJ} - (8.314 \text{ J mol}^{-1} \text{ K}^{-1})(15 + 273 \text{ K})\left(\frac{10^{-3} \text{ kJ}}{1 \text{ J}}\right)(1 \text{ mol}) = 37 \text{ kJ}$$

(b) Use Equation 3-12 to evaluate the standard enthalpy of formation of the anion:

$$\Delta H_{\text{reaction}}^\circ = \Sigma \text{coeff}_{\text{products}} \Delta H_{\text{f products}}^\circ - \Sigma \text{coeff}_{\text{reactants}} \Delta H_{\text{f reactants}}^\circ$$

39 kJ = [(1 mol)(-1207.6 kJ/mol) + (1 mol)(-393.5 kJ/mol)

    + (1 mol)(-285.83 kJ/mol)] $-$ [(1 mol)(-543.0 kJ/mol) + (2 mol)($\Delta H_{\text{f, HCO}_3^-}^\circ$)]

39 kJ = -1886.93 kJ $-$ [-543.0 kJ + (2 mol)($\Delta H_{\text{f, HCO}_3^-}^\circ$)]

(2 mol)($\Delta H_{\text{f, HCO}_3^-}^\circ$) = -1383 kJ

$\Delta H_{\text{f, HCO}_3^-}^\circ = -692$ kJ/mol

3.101 (a) Determine the structural formula by converting the line structure, adding a C at each line end or apex, and then adding $-$H to give each C atom four bonds:

(b) To write the balanced combustion reaction, first determine the chemical formula of the acid by counting the atoms of each element in the structural formula:

$$C_{20}H_{32}O_2$$

Then balance the combustion reaction in the usual fashion:

$$C_{20}H_{32}O_2 \ (s) \ + \ 27 \ O_2 \ (g) \ \rightarrow \ 20 \ CO_2 \ (g) \ + \ 16 \ H_2O \ (l)$$

(c) Make use of Equation 3-5 and tabulated bond energies to estimate the energy released in this combustion reaction.

Bond inventory of arachidonic acid:

Type	C–H	C–C	C=C	C=O	C–O	O–H

#	31	15	4	1	1	1
Energy (kJ)	415	345	615	750	360	460
Total (kJ)	12 865	5175	2460	750	360	460

Total bond energy: 22 070 kJ

$$\Delta E_{reaction} = \Sigma BE_{bonds\ broken} - \Sigma BE_{bonds\ formed}$$

$$\Delta E_{reaction} = 22\ 070 + 27(495) - 20(2)(750) - 16(2)(460) = -9.29 \times 10^3 \text{ kJ/mol}$$

(d) First, calculate the heat required to warm the bear. Then determine the amount of arachidonic acid required to do the job:

$$q = mC\Delta T = 500 \text{ kg} \left(\frac{10^3 \text{ g}}{1 \text{ kg}}\right)\left(\frac{4.184 \text{ J}}{1 \text{ g °C}}\right)(25 \text{ °C} - 5 \text{ °C}) = 4.18 \times 10^7 \text{ J}$$

$$\text{Mass required} = 4.18 \times 10^7 \text{ J}\left(\frac{1 \text{ kJ}}{10^3 \text{ J}}\right)\left(\frac{1 \text{ mol}}{9.29 \times 10^3 \text{ kJ}}\right)\left(\frac{304.45 \text{ g}}{1 \text{ mol}}\right) = 1.4 \times 10^3 \text{ g}$$

3.103 The energy that is added to a gaseous monatomic substance goes entirely into translational energy of the atoms. The energy that is added to a gaseous diatomic substance is distributed among the translational, vibrational, and rotational energies of the molecules. Thus, a given amount of added energy increases the total translational energy (and the temperature) of a diatomic substance by a smaller amount, and so the heat capacities of diatomic gases are larger than those of monatomic substances.

3.105 The problem asks for the energy released from burning 1 gallon of ethanol. Begin by determining the balanced chemical equation using standard procedures. The unbalanced equation is

$$C_2H_5OH + O_2 \rightarrow CO_2 + H_2O$$

Give CO_2 a coefficient of 2 to balance C, and H_2O a coefficient of 3 to balance H:

$$C_2H_5OH + O_2 \rightarrow 2\ CO_2 + 3\ H_2O$$

This leaves 7 O atoms on the product side and 1 O atom in C_2H_5OH. Give O_2 a coefficient of 3 to balance the O atoms and obtain the balanced equation:

$$C_2H_5OH + 3\ O_2 \rightarrow 2\ CO_2 + 3\ H_2O$$

To determine the energy released by burning 1 gallon of ethanol, we first need the reaction energy, which we can determine using the average bond energies in Table 3-2 or by using the heats of formation in Appendix D. We show the bond energy result.

Bonds Broken	Energy (kJ/mol)	Bonds Formed	Energy (kJ/mol)
C–C	345	4 C=O	4(−800)
5 C–H	5(415)	6 H–O	6(−460)
C–O	360		
O–H	460		
3 O=O	3(495)		
Total	4725		−5960

$\Delta E_{reaction} = 4725$ kJ/mol $- 5960$ kJ/mol $= -1235$ kJ/mol

Now use stoichiometric calculations to determine the energy for 1 gallon of ethanol.

$$1 \text{ gallon} \left(\frac{3.785 \text{ L}}{1 \text{ gallon}} \right) \left(\frac{10^3 \text{ mL}}{1 \text{ L}} \right) \left(\frac{0.787 \text{ g}}{1 \text{ mL}} \right) \left(\frac{1 \text{ mol}}{46.08 \text{ g}} \right) \left(\frac{1.235 \times 10^3 \text{ kJ}}{1 \text{ mol}} \right) = 7.89 \times 10^4 \text{ kJ}$$

3.107 The reaction for combustion of methane is

$$CH_{4(g)} + 2\ O_{2(g)} \rightarrow CO_{2(g)} + 2\ H_2O_{(g)}$$

The enthalpy change for this reaction is

$$\Delta H_{comb} = \Delta H^{\circ}_f(CO_{2(g)}) + 2\ \Delta H^{\circ}_f(H_2O_{(g)}) - \Delta H^{\circ}_f(CH_{4(g)})$$

$$= -393.5 + 2(-241.8) - (-74.6) \text{ kJ mol}^{-1}$$
$$= -802.5 \text{ kJ mol}^{-1}$$

It is best to work in SI units:

$p = 250$ bar $= 250 \times 10^5$ Pa
$\Delta V = 4{,}000$ L $- 1 \times 10^6$ L $= -9.96 \times 10^5$ L $= -9.96 \times 10^2$ m^3

The work done to compress the gas is:

$w = -p\Delta V$

$$= -250 \times 10^5 \text{ Pa } (-9.96 \times 10^2 \text{ m}^3)$$
$$= 2.49 \times 10^{10} \text{ Pa m}^3$$
$$= 2.49 \times 10^{10} \text{ J}$$
$$= 2.49 \times 10^7 \text{ kJ}$$

(Note that 1 Pa m^3 = 1 N m^{-2} m^3 = 1 N m = 1 kg m s^{-2} m = 1 kg m^2 s^{-2} = 1 J)

Because 45% of the energy from combustion of the methane is lost as heat, the heat which must be generated by the reaction is 2.49×10^7 kJ / 0.45 = 5.53×10^7 kJ

This much energy would be derived by burning:

$$n_{CH_4} = \frac{5.53 \times 10^7 \, kJ}{802.5 \, kJ \, mol^{-1}} = 6.89 \times 10^4 \, mol \, CH_4$$

The volume of methane is found from the ideal gas equation:

$$V = \frac{nRT}{p} = \frac{6.89 \times 10^4 \, mol(0.08314 \, L \, bar \, K^{-1} mol^{-1})(25 + 273)K}{2.00 \, bar} = 8.53 \times 10^5 \, L$$

Chapter 4 Atoms and Light

Solutions to Problems in Chapter 4

4.1 Calculate the molar volume of each metal from its density and molar mass, and then convert to atomic volume using Avogadro's number. Then estimate atomic diameter from the atomic volume by assuming that each atom occupies a cubical volume:

$$(a)\, V_{atom} = \frac{[M(\text{g/mol})][10^{-3}\ \text{kg/g}]}{[\rho\,(\text{kg/m}^3)][N_A]}$$

(b) $d = (V)^{1/3}$

(c) Layer thickness = (# of atoms)(atomic diameter)

$$\text{Ag: } V_{atom} = \left(\frac{107.868\ \text{g}}{1\ \text{mol}}\right)\left(\frac{10^{-3}\ \text{kg}}{1\ \text{g}}\right)\left(\frac{1\ \text{m}^3}{1.050 \times 10^4\ \text{kg}}\right)\left(\frac{1\ \text{mol}}{6.022 \times 10^{23}\ \text{atom}}\right)$$

$$= 1.706 \times 10^{-29}\ \text{m}^3/\text{atom}$$

$$d \cong (V)^{1/3} = (1.706 \times 10^{-29}\ \text{m}^3)^{1/3} = 2.574 \times 10^{-10}\ \text{m}$$

$$\text{Thickness} = (6.5 \times 10^6\ \text{atoms})(2.574 \times 10^{-10}\ \text{m}) = 1.7 \times 10^{-3}\ \text{m}$$

$$\text{Pb: } V_{atom} = \left(\frac{207.2\ \text{g}}{1\ \text{mol}}\right)\left(\frac{10^{-3}\ \text{kg}}{1\ \text{g}}\right)\left(\frac{1\ \text{m}^3}{1.134 \times 10^4\ \text{kg}}\right)\left(\frac{1\ \text{mol}}{6.022 \times 10^{23}\ \text{atom}}\right)$$

$$= 3.034 \times 10^{-29}\ \text{m}^3/\text{atom}$$

$$d \cong (V)^{1/3} = (3.034 \times 10^{-29}\ \text{m}^3)^{1/3} = 3.119 \times 10^{-10}\ \text{m}$$

$$\text{Thickness} = (6.5 \times 10^6\ \text{atoms})(3.119 \times 10^{-10}\ \text{m}) = 2.0 \times 10^{-3}\ \text{m}$$

4.3 Gas pressure results from transfer of momentum between atoms and container walls, indicating that atoms have momentum and mass; mass spectrometers sort atomic and molecular ions according to their mass-to-charge ratios; all matter is made up of atoms and has mass.

4.5 At the atomic level, a layer of metal atoms appears as spheres nestled closely together:

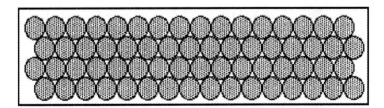

4.7 Use the equation, $v = \dfrac{c}{\lambda}$ to do these calculations:

(a) $v = \left(\dfrac{2.998 \times 10^8 \text{ m/s}}{4.33 \text{ nm}} \right) \left(\dfrac{1 \text{ nm}}{10^{-9} \text{ m}} \right) = 6.92 \times 10^{16} \text{ Hz}$

(b) $v = \dfrac{2.998 \times 10^8 \text{ m/s}}{2.35 \times 10^{-10} \text{ m}} = 1.28 \times 10^{18} \text{ Hz}$

(c) $v = \left(\dfrac{2.998 \times 10^8 \text{ m/s}}{735 \text{ mm}} \right) \left(\dfrac{10^3 \text{ mm}}{1 \text{ m}} \right) = 4.08 \times 10^8 \text{ Hz}$

(d) $v = \left(\dfrac{2.998 \times 10^8 \text{ m/s}}{4.57 \text{ μm}} \right) \left(\dfrac{10^6 \text{ μm}}{1 \text{ m}} \right) = 6.56 \times 10^{13} \text{ Hz}$

4.9 Use the equation, $\lambda = \dfrac{c}{v}$ to do these calculations:

(a) $\lambda = \left(\dfrac{2.998 \times 10^8 \text{ m/s}}{4.77 \text{ GHz}} \right) \left(\dfrac{1 \text{ GHz}}{10^9 \text{ Hz}} \right) = 6.29 \times 10^{-2} \text{ m}$

(b) $\lambda = \left(\dfrac{2.998 \times 10^8 \text{ m/s}}{28.9 \text{ kHz}} \right) \left(\dfrac{1 \text{ kHz}}{10^3 \text{ Hz}} \right) \left(\dfrac{10^2 \text{ cm}}{1 \text{ m}} \right) = 1.04 \times 10^6 \text{ cm}$

(c) $\lambda = \left(\dfrac{2.998 \times 10^8 \text{ m/s}}{60. \text{ Hz}} \right) \left(\dfrac{10^3 \text{ mm}}{1 \text{ m}} \right) = 5.0 \times 10^9 \text{ mm}$

(d) $\lambda = \left(\dfrac{2.998 \times 10^8 \text{ m/s}}{2.88 \text{ MHz}} \right) \left(\dfrac{10^6 \text{ μm}}{1 \text{ m}} \right) \left(\dfrac{1 \text{ MHz}}{10^6 \text{ Hz}} \right) = 1.04 \times 10^8 \text{ μm}$

4.11 The energy of one photon of light is described by the equation, $E = hv = \dfrac{hc}{\lambda}$. To obtain energy per mole of photons, multiply by Avogadro's number. Begin by converting the wavelength into metres and then determine the energy:

(a) $\lambda = 490.6 \text{ nm} \left(\dfrac{10^{-9} \text{ m}}{1 \text{ nm}} \right) = 4.906 \times 10^{-7} \text{ m}$

$E_{\text{photon}} = \dfrac{(6.626 \times 10^{-34} \text{ J s})(2.998 \times 10^{8} \text{ m/s})}{4.906 \times 10^{-7} \text{ m}} = 4.049 \times 10^{-19} \text{ J/photon}$

$E_{\text{molar}} = (4.049 \times 10^{-19} \text{ J/photon})(6.022 \times 10^{23} \text{ photons/mol}) \left(\dfrac{10^{-3} \text{ kJ}}{1 \text{ J}} \right) = 2.438 \times 10^{2} \text{ kJ/mol}$

(b) $\lambda = 25.5 \text{ nm} \left(\dfrac{10^{-9} \text{ m}}{1 \text{ nm}} \right) = 2.55 \times 10^{-8} \text{ m}$

$E_{\text{photon}} = \dfrac{(6.626 \times 10^{-34} \text{ J s})(2.998 \times 10^{8} \text{ m/s})}{2.55 \times 10^{-8} \text{ m}} = 7.79 \times 10^{-18} \text{ J/photon}$

$E_{\text{molar}} = (7.79 \times 10^{-18} \text{ J/photon})(6.022 \times 10^{23} \text{ photons/mol}) \left(\dfrac{10^{-3} \text{ kJ}}{1 \text{ J}} \right) = 4.69 \times 10^{3} \text{ kJ/mol}$

(c) $E_{\text{photon}} = (6.62608 \times 10^{-34} \text{ J s})(2.5437 \times 10^{10} \text{ s}^{-1}) = 1.6855 \times 10^{-23} \text{ J/photon}$

$E_{\text{molar}} = (1.6855 \times 10^{-23} \text{ J/photon})(6.022 \times 10^{23} \text{ photons/mol}) \left(\dfrac{10^{-3} \text{ kJ}}{1 \text{ J}} \right) = 1.0150 \times 10^{-2} \text{ kJ/mol}$

4.13 First, calculate the energy of one photon; then use $E_{\text{total}} = nE_{\text{photon}}$ to determine n (the number of photons):

$\lambda = 337.1 \text{ nm} \left(\dfrac{10^{-9} \text{ m}}{1 \text{ nm}} \right) = 3.371 \times 10^{-7} \text{ m}$

$E_{\text{photon}} = \dfrac{(6.626 \times 10^{-34} \text{ J s})(2.998 \times 10^{8} \text{ m/s})}{3.371 \times 10^{-7} \text{ m}} = 5.893 \times 10^{-19} \text{ J/photon}$

$n = \left(\dfrac{10.0 \text{ mJ}}{5.893 \times 10^{-19} \text{ J/photon}} \right) \left(\dfrac{10^{-3} \text{ J}}{1 \text{ mJ}} \right) = 1.70 \times 10^{16} \text{ photons}$

4.15 (a) Convert from molar energy to energy per photon by dividing by Avogadro's number, and then use $E = \dfrac{hc}{\lambda}$ to find $\lambda = \dfrac{hc}{E}$:

$E_{\text{photon}} = \left(\dfrac{745 \text{ kJ/mol}}{6.022 \times 10^{23} \text{ mol}^{-1}} \right) \left(\dfrac{10^{3} \text{ J}}{1 \text{ kJ}} \right) = 1.237 \times 10^{-18} \text{ J}$

$\lambda = \dfrac{(6.626 \times 10^{-34} \text{ J s})(2.998 \times 10^{8} \text{ m/s})}{1.237 \times 10^{-18} \text{ J}} = 1.61 \times 10^{-7} \text{ m}$

$v = \dfrac{c}{\lambda} = \dfrac{2.998 \times 10^{8} \text{ m/s}}{1.61 \times 10^{-7} \text{ m}} = 1.86 \times 10^{15} \text{ Hz}$

(b) $E_{photon} = 3.55 \times 10^{-19}$ J

$$\lambda = \frac{(6.626 \times 10^{-34} \text{ J s})(2.998 \times 10^8 \text{ m/s})}{3.55 \times 10^{-19} \text{ J}} = 5.60 \times 10^{-7} \text{ m}$$

$$v = \frac{c}{\lambda} = \frac{2.998 \times 10^8 \text{ m/s}}{5.60 \times 10^{-7} \text{ m}} = 5.35 \times 10^{14} \text{ Hz}$$

4.17 (a) $\lambda = \dfrac{c}{v} = \dfrac{2.998 \times 10^8 \text{ m/s}}{1.30 \times 10^{15} \text{ s}^{-1}} = 2.31 \times 10^{-7} \text{ m}$

(b) First determine the energy of the photon, then use $E_{binding} = E_{photon} - E_{kinetic}$

$$E_{photon} = hv = (6.626 \times 10^{-34} \text{ J s})(1.30 \times 10^{15} \text{ s}^{-1}) = 8.61 \times 10^{-19} \text{ J}$$

$$E_{binding} = (8.61 \times 10^{-19} \text{ J}) - (5.2 \times 10^{-19} \text{ J}) = 3.4 \times 10^{-19} \text{ J}$$

(c) The longest wavelength that will eject electrons corresponds to a photon with energy equal to the binding energy: $\lambda = \dfrac{hc}{E_{binding}}$

$$\lambda = \frac{(6.626 \times 10^{-34} \text{ J s})(2.998 \times 10^8 \text{ m/s})}{3.4 \times 10^{-19} \text{ J}} = 5.8 \times 10^{-7} \text{ m}$$

4.19 Refer to Figure 4-9 for an example of an energy-level diagram. The binding energy for cesium (Problem 4.17) is 3.4×10^{-19} J, and the binding energy for chromium (Problem 4.18) is 7.21×10^{-19} J. Your energy-level diagram should show that chromium has a substantially greater binding energy than cesium. Note, if both metals are exposed to photons of the same wavelength, the photoelectrons of Cs will have substantially greater kinetic energy than those emitted by Cr:

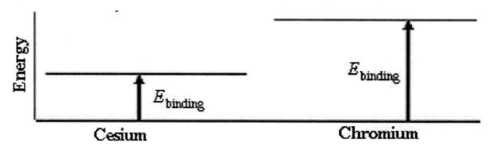

4.21 Use Figure 4-4 to answer these questions:

(a) Radio waves have wavelengths from 1 to 1000 m.

(b) Light with a wavelength of 5.8×10^{-7} m is yellow.

(c) A frequency of 4.5×10^8 Hz lies in the radar region.

4.23 Figure 4-14 does not show an emission wavelength directly to the ground state from the state that emits 404 nm light, but the combination of emissions at 436 nm and

254 nm connects this state to the ground state. Use $E = \dfrac{hc}{\lambda}$ for each photon:

$$\Delta E \text{ (kJ/mol)} = \frac{hcN_A}{\lambda_1} + \frac{hcN_A}{\lambda_2} = hcN_A\left(\frac{1}{\lambda_1} + \frac{1}{\lambda_2}\right)$$

$$\Delta E =$$

$$(6.626 \times 10^{-34} \text{ J s})(2.998 \times 10^8 \text{ m/s})(6.022 \times 10^{23} \text{ mol}^{-1})\left(\frac{1}{436 \text{ nm}} + \frac{1}{254 \text{ nm}}\right)\left(\frac{10^9 \text{ nm}}{1 \text{ m}}\right)$$

$$\Delta E = 7.45 \times 10^5 \text{ J/mol}$$

Convert to kJ/mol:

$$\Delta E = (7.45 \times 10^5 \text{ J/mol})\left(\frac{10^{-3} \text{ kJ}}{1 \text{ J}}\right) = 745 \text{ kJ/mol}$$

4.25 Energies and wavelengths for transitions in hydrogen atoms can be calculated from the equation for hydrogen atom energy levels:

$$E_n = \frac{-2.18 \times 10^{-18} \text{ J}}{n^2}$$

$$\Delta E_{8-1} = E_8 - E_1 = \left(\frac{-2.18 \times 10^{-18} \text{ J}}{8^2}\right) - \left(\frac{-2.18 \times 10^{-18} \text{ J}}{1^2}\right) = 2.146 \times 10^{-18} \text{ J}$$

$$\lambda = \frac{hc}{E} = \frac{(6.626 \times 10^{-34} \text{ J s})(2.998 \times 10^8 \text{ m/s})}{2.146 \times 10^{-18} \text{ J}} = 9.257 \times 10^{-8} \text{ m}$$

Convert to nm:

$$\lambda = 9.257 \times 10^{-8}\left(\frac{10^9 \text{ nm}}{1 \text{ m}}\right) = 92.57 \text{ nm}$$

$$\Delta E_{9-1} = E_9 - E_1 = \left(\frac{-2.18 \times 10^{-18} \text{ J}}{9^2}\right) - \left(\frac{-2.18 \times 10^{-18} \text{ J}}{1^2}\right) = 2.153 \times 10^{-18} \text{ J}$$

$$\lambda = \frac{hc}{E} = \frac{(6.626 \times 10^{-34} \text{ J s})(2.998 \times 10^8 \text{ m/s})}{2.153 \times 10^{-18} \text{ J}} = 9.226 \times 10^{-8} \text{ m}$$

Convert to nm:

$$\lambda = (9.226 \times 10^{-8})\left(\frac{10^9 \text{ nm}}{1 \text{ m}}\right) = 92.26 \text{ nm}$$

These photons lie in the ultraviolet region of the spectrum (see Figure 4-4).

4.27 As Figure 4-13 shows, absorption transitions must originate in the $n = 1$ level and can end on any higher level. From higher levels, in contrast, transitions can occur to any lower level, not just the $n = 1$ level. Thus there are many more possibilities for emission than for absorption from the ground state.

4.29 Use the mass of the electron and Avogadro's number to calculate the molar mass:

$$m_{molar} = (9.1093897 \times 10^{-31} \text{ kg})\left(\frac{10^3 \text{ g}}{1 \text{ kg}}\right)(6.022 \times 10^{23} \text{ mol}^{-1})$$

$$= 5.486 \times 10^{-4} \text{ g/mol}$$

4.31 To determine the wavelength of an electron from the kinetic energy, first determine the speed of the electron and then convert to wavelength. The speed of an electron is determined using $v = \sqrt{\dfrac{2E_{kinetic}}{m}}$. Remember, for the units to work out, mass must be in kg and the energy in J such that $\dfrac{\text{J}}{\text{kg}} = \dfrac{\text{kg m}^2 \text{ s}^{-2}}{\text{kg}} = \text{m}^2/\text{s}^2$. For wavelength calculations, use $\lambda = \dfrac{h}{mv}$:

(a) $v = \sqrt{\dfrac{2(1.15 \times 10^{-19} \text{ J})}{9.109 \times 10^{-31} \text{ kg}}} = 5.02 \times 10^5 \text{ m/s}$

$\lambda = \dfrac{6.626 \times 10^{-34} \text{ J s}}{(9.109 \times 10^{-31} \text{ kg})(5.02 \times 10^5 \text{ m/s})} = 1.45 \times 10^{-9} \text{ m or 1.45 nm}$

(b) $E_{electron} = \left(\dfrac{3.55 \text{ kJ/mol}}{6.022 \times 10^{23} \text{ mol}^{-1}}\right)\left(\dfrac{10^3 \text{ J}}{1 \text{ J}}\right) = 5.90 \times 10^{-21} \text{ J}$

$v = \sqrt{\dfrac{2(5.90 \times 10^{-21} \text{ J})}{9.109 \times 10^{-31} \text{ kg}}} = 1.14 \times 10^5 \text{ m/s}$

$\lambda = \dfrac{6.626 \times 10^{-34} \text{ J s}}{(9.109 \times 10^{-31} \text{ kg})(1.14 \times 10^5 \text{ m/s})} = 6.38 \times 10^{-9} \text{ m or 6.38 nm}$

(c) $E_{electron} = \dfrac{7.45 \times 10^{-3} \text{ J/mol}}{6.022 \times 10^{23} \text{ mol}^{-1}} = 1.24 \times 10^{-26} \text{ J}$

$v = \sqrt{\dfrac{2(1.24 \times 10^{-26} \text{ J})}{9.109 \times 10^{-31} \text{ kg}}} = 1.65 \times 10^2 \text{ m/s}$

$\lambda = \dfrac{6.626 \times 10^{-34} \text{ J s}}{(9.109 \times 10^{-31} \text{ kg})(1.65 \times 10^2 \text{ m/s})} = 4.41 \times 10^{-6} \text{ m or 4.41 } \mu\text{m}$

:

4.33 First determine the speed of the electron from its wavelength, and then use the speed to determine the kinetic energy. Kinetic energy and wavelength of an electron are related through

$$\lambda = \frac{h}{mv} \text{ and } E_{kinetic} = \frac{1}{2}mv^2$$

(a) $\lambda = 3.75 \text{ nm}\left(\dfrac{10^{-9}\text{m}}{1 \text{ nm}}\right) = 3.75 \times 10^{-9} \text{ m}$

$$v = \frac{6.626 \times 10^{-34} \text{ J s}}{(9.109 \times 10^{-31} \text{ kg})(3.75 \times 10^{-9} \text{ m})} = 1.94 \times 10^5 \text{ m/s}$$

$$E_{kinetic} = \frac{(9.109 \times 10^{-31} \text{ kg})(1.94 \times 10^5 \text{ m/s})^2}{2} = 1.71 \times 10^{-20} \text{ J}$$

(b) $v = \dfrac{6.626 \times 10^{-34} \text{ J s}}{(9.109 \times 10^{-31} \text{ kg})(4.66 \text{ m})} = 1.56 \times 10^{-4} \text{ m/s}$

$$E_{kinetic} = \frac{(9.109 \times 10^{-31} \text{ kg})(1.56 \times 10^{-4} \text{ m/s})^2}{2} = 1.11 \times 10^{-38} \text{ J}$$

(c) $\lambda = 8.85 \text{ mm}\left(\dfrac{10^{-3} \text{ m}}{1 \text{ mm}}\right) = 8.85 \times 10^{-3} \text{ m}$

$$v = \frac{6.626 \times 10^{-34} \text{ J s}}{(9.109 \times 10^{-31} \text{ kg})(8.85 \times 10^{-3} \text{ m})} = 8.22 \times 10^{-2} \text{ m/s}$$

$$E_{kinetic} = \frac{(9.109 \times 10^{-31} \text{ kg})(8.22 \times 10^{-4} \text{ m/s})^2}{2} = 3.08 \times 10^{-33} \text{ J}$$

4.35 The designation $6p$ specifies that $n = 6$ and $l = 1$; the other two quantum numbers, m_l and m_s, can take any of their possible values. Remember that m_l must be an integer with value between $+l$ and $-l$, whereas m_s is either $+\dfrac{1}{2}$ or $-\dfrac{1}{2}$. There are six valid sets:

n	l	m_l	m_s
6	1	+1	$+\dfrac{1}{2}$
6	1	+1	$-\dfrac{1}{2}$
6	1	0	$+\dfrac{1}{2}$
6	1	0	$-\dfrac{1}{2}$
6	1	−1	$+\dfrac{1}{2}$
6	1	−1	$-\dfrac{1}{2}$

4.37 Remember that l is restricted to positive integers less than n, m_l is restricted to integers between $-l$ and $+l$, and $m_s = +\dfrac{1}{2}$ or $-\dfrac{1}{2}$. For $n = 3$, l can be 0 with $m_l = 0$, l can be 1 with $m_l = +1$, 0, or -1; and l can be 2 with $m_l = +2$, $+1$, 0, -1, or -2. For each possibility, $m_s = +\dfrac{1}{2}$ or $-\dfrac{1}{2}$.

4.39 Remember that n must be a positive integer, l is restricted to zero and positive integers less than n, m_l is restricted to integers between $-l$ and $+l$, and $m_s = +\dfrac{1}{2}$ or $-\dfrac{1}{2}$.

(a) Non-existent: m_s must be $+\dfrac{1}{2}$ or $-\dfrac{1}{2}$; (b) actual; (c) non-existent: l must be less than n; and (d) actual.

4.41 The designation $3d$ specifies that $n = 3$ and $l = 2$; the largest possible value for m_l is $+l$, in this case $+2$; and "spin up" means $m_s = +\dfrac{1}{2}$. Thus, the values are 3, 2, 2, $+\dfrac{1}{2}$.

4.43 The designation $n = 2$, $l = 1$ refers to a $2p$ orbital. This orbital has no radial nodes, so its electron density rises and then falls. Figure 4-22 shows a contour diagram. For this orbital, an electron density drawing looks very similar to a contour drawing.

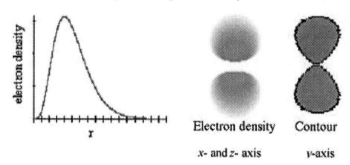

4.45 The p_y orbital runs lengthwise along the y-axis. It has the same view along the x-axis and z-axis. Along the y-axis, you will only see one lobe of the orbital:

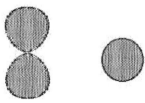

4.47 Orbital (a) has the shape characteristic of a p orbital, whereas orbitals (b) and (c) have shapes characteristic of d orbitals. Because orbital (a) appears smaller than the $3s$ orbital, it has $n < 3$ and is a $2p$ orbital; because orbitals (b) and (c) have similar sizes as the $3s$ orbital, they have $n = 3$ and are $3d$ orbitals: (a) $2p$ orbital, $n = 2$, $l = 1$; (b) and (c) $3d$ orbital, $n = 3$, $l = 2$.

4.49 Electron density plots accurately show how the wave function varies along one axis, but they fail to show whether or not an orbital has spherical symmetry, and they do not directly show electron densities. Thus, they fail to give a good picture of how an orbital appears in three dimensions.

4.51 A contour drawing of an *s* orbital is a spherical surface. The orbitals should grow larger as **n** increases:

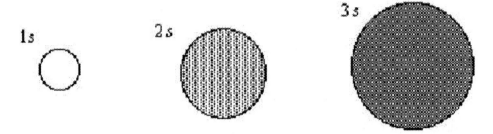

4.53 Each picture should show a molecule absorbing a photon and fragmenting appropriately:

N$_2$:

O$_2$:

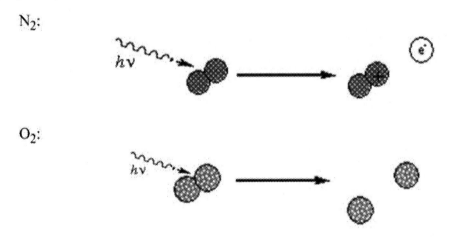

4.55 Calculate the fragmentation energy using $E = \dfrac{hcN_A}{\lambda}$:

$$\lambda = 340 \text{ nm}\left(\frac{10^{-9} \text{ m}}{1 \text{ nm}}\right) = 3.40 \times 10^{-7} \text{m}$$

$$hcN_A = (6.626 \times 10^{-34} \text{ J s})\left(\frac{10^{-3} \text{ kJ}}{1 \text{ J}}\right)(2.998 \times 10^8 \text{ m/s})(6.022 \times 10^{23} \text{ mol}^{-1})$$

$$hcN_A = 1.196 \times 10^{-4} \text{ kJ m mol}^{-1}$$

The minimum energy to fragment an ozone molecule is

$$\Delta E = \frac{1.196 \times 10^{-4} \text{ kJ m mol}^{-1}}{3.40 \times 10^{-7} \text{ m}} = 352 \text{ kJ/mol}$$

The kinetic energy of the fragments is the difference between the energy of the photon causing the fragmentation and the minimum energy required for fragmentation:

$$\lambda = 250 \text{ nm}\left(\frac{10^{-9}\text{ m}}{1 \text{ mn}}\right) = 2.50 \times 10^{-7} \text{ m}$$

$$E_{\text{kinetic}} = E_{\text{photon}} - \Delta E$$

$$= \left(\frac{(6.626 \times 10^{-34} \text{ J s})(2.998 \times 10^8 \text{ m/s})}{2.50 \times 10^{-7} \text{ m}}\right) - \left(\frac{352 \times 10^3 \text{ J/mol}}{6.022 \times 10^{23} \text{ mol}^{-1}}\right)$$

$$E_{\text{kinetic}} = (7.95 \times 10^{-19} \text{ J}) - (5.85 \times 10^{-19} \text{ J}) = 2.10 \times 10^{-19} \text{ J}$$

4.57 Use information in the Focus 4-2: Chemistry and the Environment to determine which region of the atmosphere absorbs the given wavelengths:

(a) Below 200 nm, O_2, N_2, and O_3 all absorb effectively (thermosphere).

(b) From 240-310 nm, only O_3 absorbs effectively (stratosphere).

4.59 (a) According to Figure 4-25, a pressure of 10^{-3} bar occurs at ~50 km.

(b) The balloon is on the border of the stratosphere and mesosphere.

(c) The chemistry of this region is the formation of ozone.

4.61 To calculate photon properties, use $\lambda v = c$ and the equations for photons in Table 4-1:

(a) $\lambda = \dfrac{c}{v} = \dfrac{2.998 \times 10^8 \text{ m/s}}{4.5 \times 10^{13} \text{ s}^{-1}} = 6.7 \times 10^{-6} \text{ m}$

(b) $E = hv = (6.626 \times 10^{-34} \text{ J s})(4.5 \times 10^{13} \text{ s}^{-1}) = 3.0 \times 10^{-20} \text{ J}$

(c) $t = \dfrac{\text{distance}}{\text{speed}} = \left(\dfrac{4.60 \times 10^5 \text{ m}}{2.998 \times 10^8 \text{ m/s}}\right) = 1.53 \text{ s}$

4.63 Calculate the wavelength using $\lambda = \dfrac{hcN_A}{E}$:

$$hcN_A = (6.626 \times 10^{-34} \text{ J s})(2.998 \times 10^8 \text{ m/s})\left(\frac{10^{-3} \text{ kJ}}{1 \text{ J}}\right)(6.022 \times 10^{23} \text{ mol}^{-1})$$

$$hcN_A = 1.196 \times 10^{-4} \text{ kJ m mol}^{-1}$$

$$\lambda = \frac{1.196 \times 10^{-4} \text{ kJ m mol}^{-1}}{243 \text{ kJ/mol}} = 4.92 \times 10^{-7} \text{ m or 492 nm}$$

This wavelength is in the visible region of the spectrum, which reaches the surface of the Earth. Thus, Cl_2 in the troposphere will be dissociated by sunlight.

4.65 A 4*p* electron must have *n* = 4, *l* = 1, but m_l has three possible values and m_s has two possible values. There are six possible sets of values:

n	*l*	*m*	m_s
4	1	1	$+\dfrac{1}{2}$
4	1	1	$-\dfrac{1}{2}$
4	1	0	$+\dfrac{1}{2}$
4	1	0	$-\dfrac{1}{2}$
4	1	−1	$+\dfrac{1}{2}$
4	1	−1	$-\dfrac{1}{2}$

4.67 When the frequency is doubled and the amplitude is kept the same, the peaks are narrower and occur twice as often:

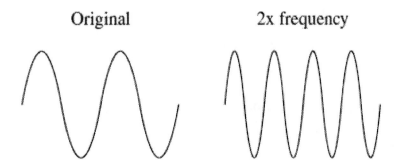

Original   2x frequency

4.69 The longest wavelength that can eject an electron from the metal surface is the wavelength that just overcomes the binding energy. Calculate the wavelength using

$$\lambda = \frac{hcN_A}{E}$$

$$hcN_A = 1.196 \times 10^{-4} \text{ kJ m mol}^{-1}$$

$$\lambda = \frac{1.196 \times 10^{-4} \text{ kJ m mol}^{-1}}{216.4 \text{ kJ/mol}} = 5.527 \times 10^{-7} \text{ m or } 552.7 \text{ nm}$$

4.71 The photoelectric effect is described by an energy balance equation:

$$E_{kinetic} = E_{photon} - E_{binding}$$

(a) $E_{binding} = E_{photon} - E_{electron} = (6.00 \times 10^{-19} \text{ J}) - (2.70 \times 10^{-19} \text{ J}) = 3.30 \times 10^{-19} \text{ J}$

(b) $\lambda = \dfrac{hc}{E} = \dfrac{(6.626 \times 10^{-34} \text{ J s})(2.998 \times 10^{8} \text{ m/s})}{6.00 \times 10^{-19} \text{ J}} = 3.31 \times 10^{-7} \text{ m}$

(c) For electrons, $\lambda = \dfrac{h}{mv}$:

$$v = \sqrt{\dfrac{2E_{kinetic}}{m}} = \sqrt{\dfrac{2(2.70 \times 10^{-19} \text{ J})}{9.109 \times 10^{-31} \text{ kg}}} = 7.70 \times 10^{5} \text{ m/s}$$

$$\lambda = \dfrac{(6.626 \times 10^{-34} \text{ kg m}^2 \text{ s}^{-1})}{(9.109 \times 10^{-31} \text{ kg})(7.70 \times 10^{5} \text{ m/s})} = 9.45 \times 10^{-10} \text{ m}$$

4.73 For electromagnetic radiation, $E = h\nu$ and $\lambda = \dfrac{c}{\nu}$:

$$\lambda = \left(\dfrac{2.998 \times 10^{8} \text{ m/s}}{27.3 \text{ MHz}} \right)\left(\dfrac{1 \text{ MHz}}{10^{6} \text{ Hz}} \right)\left(\dfrac{1 \text{ Hz}}{1 \text{ s}^{-1}} \right) = 11.0 \text{ m}$$

$$E = (6.626 \times 10^{-34} \text{ J s})(27.3 \text{ MHz})\left(\dfrac{10^{6} \text{ Hz}}{1 \text{ MHz}} \right)\left(\dfrac{1 \text{ s}^{-1}}{1 \text{ Hz}} \right) = 1.81 \times 10^{-26} \text{ J}$$

4.75 Make use of the figure legends to determine the answers to the questions:

(a) At an altitude of 60 km, $p \sim 2 \times 10^{-4}$ bar

(b) Examples are N_2, O_2, and O_3

(c) At a pressure of 8 Torr (or 0.0107 bar), the altitude is ~35 km

(d) Stratosphere

4.77 Calculate the frequencies using $\nu = \dfrac{c}{\lambda}$ and the energies using $E = \dfrac{hcN_A}{\lambda}$:

$hcN_A = 1.196 \times 10^{-4} \text{ kJ m mol}^{-1}$. The calculations for 487 nm are shown:

$$\lambda = 487 \text{nm}\left(\dfrac{1 \text{ m}}{10^{9} \text{ nm}} \right) = 4.87 \times 10^{-7} \text{ m}$$

$$\nu = \dfrac{2.998 \times 10^{8} \text{ m/s}}{4.87 \times 10^{-7} \text{ m}} = 6.16 \times 10^{14} \text{ s}^{-1}$$

$$E = \dfrac{hcN_A}{\lambda} = \dfrac{1.196 \times 10^{-4} \text{ kJ m mol}^{-1}}{4.87 \times 10^{-7} \text{ m}} = 246 \text{ kJ/mol}$$

λ (nm)	487	514	543	553	578
v (10^{14} s^{-1})	6.16	5.83	5.52	5.42	5.19
E (kJ/mol)	246	233	220	216	207

4.79 Energies and frequencies for transitions in hydrogen atoms can be calculated from the equation for hydrogen atom energy levels:

$$E_n = \frac{-2.18 \times 10^{-18} \text{ J}}{n^2}$$

$$\Delta E_{5-1} = E_5 - E_1 = \left(\frac{-2.18 \times 10^{-18} \text{ J}}{5^2}\right) - \left(\frac{-2.18 \times 10^{-18} \text{ J}}{1^2}\right) = 2.09 \times 10^{-18} \text{ J}$$

$$\lambda = \frac{hc}{E} = \frac{(6.626 \times 10^{-34} \text{ J s})(2.998 \times 10^8 \text{ m/s})}{2.09 \times 10^{-18} \text{ J}} = 9.50 \times 10^{-8} \text{ m}$$

$$\lambda = 9.50 \times 10^{-8} \text{ m}\left(\frac{10^9 \text{ nm}}{1 \text{ m}}\right) = 95.0 \text{ nm}$$

These photons fall in the ultraviolet region.

4.81 According to Figure 4-4, visible light has wavelengths in the 4×10^{-7} m to 7.2×10^{-7} m range. The highest frequency of visible light is

$$v = \frac{2.998 \times 10^8 \text{ m/s}}{4 \times 10^{-7} \text{m}} \cong 7.5 \times 10^{14} \text{ s}^{-1}.$$ This is smaller than the threshold frequency, so Mg cannot be used in photoelectric devices.

4.83 Volume is related to radius through the equation, $V = \frac{4}{3}\pi r^3$:

$$V_{atom} = \frac{4\,\pi(10^{-10} \text{ m})^3}{3} = 4 \times 10^{-30} \text{ m}^3 \qquad V_{nucleus} = \frac{4\,\pi(10^{-15} \text{ m})^3}{3} = 4 \times 10^{-45} \text{ m}^3$$

The fraction of the volume that the nucleus occupies is $\dfrac{4 \times 10^{-45} \text{ m}^3}{4 \times 10^{-30} \text{ m}^3} = \dfrac{1}{1 \times 10^{15}}$.

4.85 First calculate the energy of one photon of wavelength 510 nm. Then divide the detection limit by this energy to find the number of photons:

$$\lambda = 510 \text{ nm} \left(\frac{10^{-9} \text{ m}}{1 \text{ nm}} \right) = 5.10 \times 10^{-7} \text{ m}$$

$$E_{photon} = \frac{hc}{\lambda} = \frac{(6.626 \times 10^{-34} \text{ J s})(2.998 \times 10^8 \text{ m/s})}{5.10 \times 10^{-7} \text{ m}} = 3.90 \times 10^{-19} \text{ J /photon}$$

$$\# = \frac{2.35 \times 10^{-18} \text{ J}}{3.90 \times 10^{-19} \text{ J/photon}} = 6 \text{ photons}$$

4.87 The wavelengths associated with particles are calculated from $\lambda = \dfrac{h}{mv}$:

$$5.000\% \text{ of the speed of light is } v = \left(\frac{5.000\%}{100\%} \right)(2.998 \times 10^8 \text{ m/s}) = 1.499 \times 10^7 \text{ m/s}$$

$$\text{electron: } \lambda = \frac{h}{mv} = \frac{6.626 \times 10^{-34} \text{ J s}}{(9.109 \times 10^{-31} \text{ kg})(1.499 \times 10^7 \text{ m/s})} = 4.853 \times 10^{-11} \text{ m}$$

$$\text{proton: } \lambda = \frac{h}{mv} = \frac{6.626 \times 10^{-34} \text{ J s}}{(1.673 \times 10^{-27} \text{ kg})(1.499 \times 10^7 \text{ m/s})} = 2.642 \times 10^{-14} \text{ m}$$

4.89 The energy-level diagram shows the energy differences among the various levels. By examining it, we can see that the smallest energy difference is associated with Photon$_1$ and the largest energy difference with Photon$_2$. The energy of a photon is inversely proportional to its wavelength, so Photon$_1$ has the longest wavelength and Photon$_2$ the shortest wavelength. Thus, the assignments are as follows:

$$\text{Photon}_1 = 565 \text{ nm } (d \text{ to } c) \qquad \text{Photon}_2 = 121 \text{ nm } (c \text{ to } b)$$
$$\text{Photon}_3 = 152 \text{ nm } (b \text{ to } a)$$

$$\text{Level } b: \lambda = 152 \text{ nm} \left(\frac{10^{-9} \text{ m}}{1 \text{ nm}} \right) = 1.52 \times 10^{-7} \text{ m}$$

$$E = \frac{hc}{\lambda} = \frac{(6.626 \times 10^{-34} \text{ J s})(2.998 \times 10^8 \text{ m/s})}{1.52 \times 10^{-7} \text{ m}} = 1.31 \times 10^{-18} \text{ J}$$

$$\text{Level } c: \lambda = 121 \text{ nm} \left(\frac{10^{-9} \text{ m}}{1 \text{ nm}} \right) = 1.21 \times 10^{-7} \text{ m}$$

$$E = (1.31 \times 10^{-18} \text{ J}) + \frac{(6.626 \times 10^{-34} \text{ J s})(2.998 \times 10^8 \text{ m/s})}{1.21 \times 10^{-7} \text{ m}} = 2.95 \times 10^{-18} \text{ J}$$

$$\text{Level } d: \lambda = 565 \text{ nm} \left(\frac{10^{-9} \text{ m}}{1 \text{ nm}} \right) = 5.65 \times 10^{-7} \text{ m}$$

$$E = (2.95 \times 10^{-18} \text{ J}) + \frac{(6.626 \times 10^{-34} \text{ J s})(2.998 \times 10^8 \text{ m/s})}{5.65 \times 10^{-7} \text{ m}} = 3.30 \times 10^{-18} \text{ J}$$

4.91 (a) Calculate the energies of the photons using $E = \dfrac{hc}{\lambda}$

$$E_{488} = \frac{(6.626 \times 10^{-34} \text{ J s})(2.998 \times 10^{8} \text{ m/s})}{4.88 \times 10^{-7} \text{ m}} = 4.07 \times 10^{-19} \text{ J}$$

$$E_{514} = \frac{(6.626 \times 10^{-34} \text{ J s})(2.998 \times 10^{8} \text{ m/s})}{5.14 \times 10^{-7} \text{ m}} = 3.86 \times 10^{-19} \text{ J}$$

(b) Each transition leaves the ion in the same state, so the state from which 488 nm emission occurs must be slightly higher in energy than that from which 514 nm emission occurs:

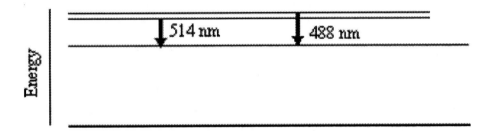

(c) Calculate frequency and wavelength using $v = \dfrac{E}{h}$ and $\lambda = \dfrac{c}{v}$:

$$v = \frac{E}{h} = \frac{2.76 \times 10^{-18} \text{ J}}{6.626 \times 10^{-34} \text{ J s}} = 4.17 \times 10^{15} \text{ s}^{-1}$$

$$\lambda = \frac{c}{v} = \frac{2.998 \times 10^{8} \text{ m/s}}{4.17 \times 10^{15} \text{ s}^{-1}} = 7.19 \times 10^{-8} \text{ m or 71.9 nm}$$

4.93 (a) Calculate photon energies using $E = \dfrac{hcN_A}{\lambda}$, $hcN_A = 1.196 \times 10^{-4} \text{ kJ m mol}^{-1}$

$$E_{589.6} = \frac{1.196 \times 10^{-4} \text{ kJ m mol}^{-1}}{5.896 \times 10^{-7} \text{ m}} = 202.8 \text{ kJ/mol}$$

$$E_{590.0} = \frac{1.196 \times 10^{-4} \text{ kJ m mol}^{-1}}{5.900 \times 10^{-7} \text{ m}} = 202.7 \text{ kJ/mol}$$

(b) The two levels are very close together at about 203 kJ/mol above the ground state:

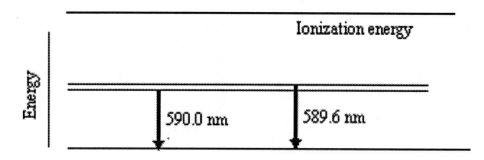

(c) The energy required to ionize the atom from this state is the difference in energies: $E = 486 - 202.7 = 283$ kJ/mol. Calculate the wavelength using

$$\lambda = \frac{hcN_A}{E}$$

$$\lambda = \frac{1.196 \times 10^{-4} \text{ kJ m mol}^{-1}}{283 \text{ kJ/mol}} = 4.23 \times 10^{-7} \text{ m or 423 nm}$$

4.95 (a) Binding energy is the energy required to eject an electron from a metal. For Metal A, electrons are not ejected until 4.0×10^{14} s^{-1} is reached, and therefore the binding energy is the energy corresponding to this frequency:

$$E = h\nu = (6.626 \times 10^{-34} \text{ J s})(4.0 \times 10^{14} \text{ s}^{-1}) = 2.7 \times 10^{-19} \text{ J}$$

Similarly for Metal B at 6.5×10^{14} s^{-1}

$$E = h\nu = (6.626 \times 10^{-34} \text{ J s})(6.5 \times 10^{14} \text{ s}^{-1}) = 4.3 \times 10^{-19} \text{ J}$$

Therefore, Metal B has the higher binding energy because higher frequency photons are required to generate photoelectrons.

(b) Kinetic energy = photon energy − binding energy

$$\lambda = 125 \text{ nm}\left(\frac{10^{-9} \text{ m}}{1 \text{ nm}}\right) = 1.25 \times 10^{-7} \text{ m}$$

$$E_{photon} = \frac{hc}{\lambda} = \frac{(6.626 \times 10^{-34} \text{ J s})(2.998 \times 10^8 \text{ m/s})}{1.25 \times 10^{-7} \text{ m}} = 1.59 \times 10^{-18} \text{ J}$$

Hence, using the photon energy above and the binding energy calculated in part (a), the kinetic energies are

Metal A: $(1.59 \times 10^{-18} \text{ J}) - (0.27 \times 10^{-18} \text{ J}) = 1.32 \times 10^{-18}$ J

Metal B: $(1.59 \times 10^{-18} \text{ J}) - (0.43 \times 10^{-18} \text{ J}) = 1.16 \times 10^{-18}$ J

(c) The wavelength range over which electrons can be ejected from one metal but not the other corresponds to the frequency range of 4.0×10^{14} s^{-1} to 6.5×10^{14} s^{-1}:

At 4.0×10^{14} s^{-1}, $\lambda = \dfrac{c}{v} = \dfrac{2.998 \times 10^8 \text{ m/s}}{4.0 \times 10^{14} \text{ s}^{-1}} = 7.5 \times 10^{-7}$ m or 750 nm

At 6.5×10^{14} s^{-1}, $\lambda = \dfrac{c}{v} = \dfrac{2.998 \times 10^8 \text{ m/s}}{6.5 \times 10^{14} \text{ s}^{-1}} = 4.6 \times 10^{-7}$ m or 460 nm

The wavelength range is 460 nm to 750 nm.

4.97 This problem combines hydrogen atom emission properties with photoelectron emission. Use the equation for hydrogen atom energy levels for the first two calculations, and the equation for the photoelectric effect for the last two calculations.

(a) $E_4 = \dfrac{-2.18 \times 10^{-18} \text{ J}}{4^2} = -1.36 \times 10^{-19}$ J

(b)

$E_{\text{photon}} = \Delta E_{\text{atom}} = E_4 - E_2 = (-1.36 \times 10^{-19} \text{ J}) - \left(\dfrac{-2.18 \times 10^{-18} \text{ J}}{2^2} \right) = 4.09 \times 10^{-19}$ J

(c) $\lambda = \dfrac{hc}{E_{\text{photon}}} = \dfrac{\left(6.626 \times 10^{-34} \text{ J s} \right)\left(2.998 \times 10^8 \text{ m/s} \right)}{4.09 \times 10^{-19} \text{ J}} = 4.86 \times 10^{-7}$ m or 486 nm

(d) $v = \sqrt{\dfrac{2E_{\text{kinetic}}}{m}} = \sqrt{\dfrac{2\left(0.86 \times 10^{-19} \text{ J} \right)}{9.109 \times 10^{-31} \text{ kg}}} = 4.3 \times 10^5$ m/s

$\lambda = \dfrac{h}{mv} = \dfrac{\left(6.626 \times 10^{-34} \text{ J s} \right)}{\left(9.109 \times 10^{-31} \text{ kg} \right)\left(4.3 \times 10^5 \text{ m/s} \right)} = 1.7 \times 10^{-9}$ m

4.99 In order to construct an energy-level diagram, first determine the energies of the various photons:

$E_{\text{photon}} = \dfrac{hc}{\lambda} \left(\dfrac{\left(6.626 \times 10^{-34} \text{ J s} \right)\left(2.998 \times 10^8 \text{ m/s} \right)}{\lambda \,(\text{in nm})} \right) \left(\dfrac{10^9 \text{ nm}}{1 \text{ m}} \right) = \dfrac{1.986 \times 10^{-16} \text{ J nm}}{\lambda \,(\text{in nm})}$

λ (nm)	422.7	272.2	239.9	671.8	504.2
E (10^{-19} J)	4.699	7.296	8.278	2.956	3.940

Construct an energy-level diagram showing the relationships among the various levels. Absorptions are upward transitions, emissions are downward transitions:

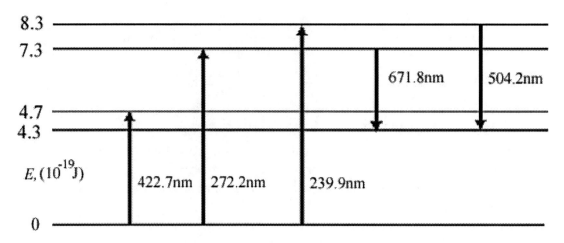

(a) The state reached by 504.2 nm emission is 4.341×10^{-19} J above the lowest state:

$$\lambda = \frac{1.986 \times 10^{-16} \text{ J nm}}{\Delta E} = \frac{1.986 \times 10^{-16} \text{ J nm}}{4.341 \times 10^{-19} \text{ J}} = 457.5 \text{ nm}$$

(b) $\Delta E = (7.296 - 4.698) \times 10^{-19}$ J $= 2.598 \times 10^{-19}$ J

$$\lambda = \frac{1.986 \times 10^{-16} \text{ J nm}}{\Delta E} = \frac{1.986 \times 10^{-16} \text{ J nm}}{2.598 \times 10^{-19} \text{ J}} = 764.4 \text{ nm}$$

(c) ΔE (272.2 nm – 671.8 nm) $= (7.296 - 2.955) \times 10^{-19}$ J $= 4.341 \times 10^{-19}$ J

ΔE (239.9 nm – 504.2 nm) $= (8.281 - 3.940) \times 10^{-19}$ J $= 4.341 \times 10^{-19}$ J

The 272.2 nm – 671.8 nm sequence and the 239.9 nm – 504.2 nm sequence both lead to the state at 4.341×10^{-19} J above the ground state.

4.101 Equation 4-5 has the form $E_n = -\dfrac{x}{n^2}$ where E_n is the energy of the electron in quantum level n, and x is the constant we want to calculate. We could therefore plot E_n versus $1/n^2$ and the slope of this line will be –x.

For each emission line, we need therefore to calculate its energy from the wavelength data $\left(E = \dfrac{hc}{\lambda} \right)$ and plot E_n versus $1/n^2$. We know the values of n because the lowest energy transition must arise from an electron falling from the third to the second level, hence n = 3 for $\lambda = 656.2$ nm, and the other emission lines arise from transitions from successively higher levels (4, 5 and 6).

λ, nm	E, J	n	$1/n^2$
656.2	3.03×10^{-19}	3	0.111
486.1	4.09×10^{-19}	4	0.0625
434.0	4.58×10^{-19}	5	0.0400
410.2	4.85×10^{-19}	6	0.0277

The plot looks like:

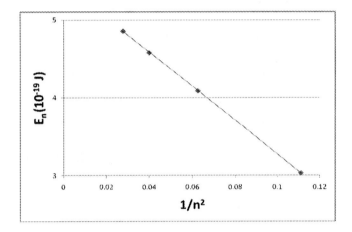

The slope of the line can be calculated using the two endpoints:

$$\text{slope} = \frac{(4.85 \times 10^{-19} - 3.03 \times 10^{-19})\,\text{J}}{0.0277 - 0.111} = -2.19 \times 10^{-18}\,\text{J}$$

and so the constant is calculated as 2.19×10^{-18} J, very close to the accepted value of 2.18×10^{-18} J.

Chapter 5 Atomic Energies and Periodicity

Solutions to Problems in Chapter 5

5.1 Orbital stability in multielectron atoms depends on the value of n (stability decreases with increasing value), Z (stability increases with increasing value), l (stability decreases with increasing value), and amount of screening (stability increases as other electrons are removed):

(a) He $1s$ is more stable than He $2s$ because of its lower value of n.

(b) Kr $5s$ is more stable than Kr $5p$ because of its lower value of l.

(c) He$^+$ $2s$ is more stable than He $2s$ because it is less screened by $1s$ electrons.

5.3 A hydrogen atom contains just one electron, so there is no screening effect. In the absence of screening, orbital energy depends only on n and Z, so all $n = 3$ orbitals have identical energy. In a helium atom, on the other hand, an electron in an $n = 3$ orbital is screened from the nucleus by the second electron, and the amount of screening decreases as l increases.

5.5 (a) The ionization energy of the He $2p$ orbital (0.585×10^{-18} J) is not much larger than that of the H $2p$ orbital (0.545×10^{-18} J). The similar values indicate nearly equal effective nuclear charges due to nearly complete screening; in the absence of screening, the He $2p$ orbital would have four times the ionization energy as the H $2p$ orbital.

(b) The ionization energy of the He$^+$ $2p$ orbital (2.18×10^{-18} J) is four times larger than that of the H $2p$ orbital (0.545×10^{-18} J). A four-fold increase when Z doubles indicates a Z^2 dependence.

5.7

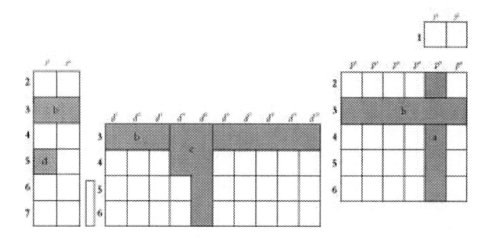

5.9 The element below lead in the periodic table would have atomic number 114 (count from Ds, element 110: two elements needed to complete the *d* block, two more to reach Column 14 in the *p* block). The block that is filling would match that of lead, but with one higher *n* value: 7*p*.

5.11 Element 117 would fall directly below astatine in Column 17, in Row 7.

5.13 The column location of an element is the indicator of how many valence electrons it has. Remember that *s* electrons as well as those in the filling block count as valence electrons: O, fourth column of *p* block, 4 *p* + 2 *s* = 6 valence electrons; V, third column of *d* block, 3 *d* + 2 *s* = 5 valence electrons; Rb, 1 valence electron; Sn, second column of *p* block, 2 *s* + 2 *p* = 4 valence electrons; and Cd, end of *d* block, 2 valence electrons.

5.15 Use the aufbau principle to determine the orbital occupancies and quantum numbers for the valence electrons in a configuration. When there are equivalent configurations, unpair as many electrons as possible in accordance with Hund's rule.

(a) Be, 4 electrons, [He] $2s^2$:

n	l	m_l	m_s
2	0	0	$+\frac{1}{2}$
2	0	0	$-\frac{1}{2}$

(b) O, 8 electrons, [He] $2s^2\,2p^4$:

n	l	m_l	m_s
2	0	0	$+\frac{1}{2}$
2	0	0	$-\frac{1}{2}$
2	1	1	$+\frac{1}{2}$
2	1	0	$+\frac{1}{2}$
2	1	−1	$+\frac{1}{2}$
2	1	1	$-\frac{1}{2}$

(c) Ne, 10 electrons, [He] $2s^2 2p^6$:

n	l	m_l	m_s
2	0	0	$+\dfrac{1}{2}$
2	0	0	$-\dfrac{1}{2}$
2	1	1	$+\dfrac{1}{2}$
2	1	1	$-\dfrac{1}{2}$
2	1	0	$+\dfrac{1}{2}$
2	1	0	$-\dfrac{1}{2}$
2	1	−1	$+\dfrac{1}{2}$
2	1	−1	$-\dfrac{1}{2}$

(d) P, 15 electrons, [Ne] $3s^2 3p^3$:

n	l	m_l	m_s
3	0	0	$+\dfrac{1}{2}$
3	0	0	$-\dfrac{1}{2}$
3	1	1	$+\dfrac{1}{2}$
3	1	0	$+\dfrac{1}{2}$
3	1	−1	$+\dfrac{1}{2}$

5.17 Atoms with unpaired electrons in their configurations are paramagnetic. Of the atoms in Problem 5.15, O and P have unpaired electrons. Here are the orbital occupancy diagrams for the least stable occupied orbitals:

5.19 (a) This is Pauli-forbidden: *s* orbitals can hold no more than two electrons.

(b) This is Pauli-forbidden: *s* orbitals can hold no more than two electrons.

(c) This is an excited-state configuration: the 1*s* and 2*s* orbitals are not full.

(d) This is an excited-state configuration: the 2*s* orbital is not full.

(e) This is the ground-state configuration.

(f) This configuration uses a non-existent orbital: there is no 1*p* orbital.

(g) This configuration uses a non-existent orbital: there is no 2*d* orbital.

5.21 The element with more unpaired electrons will have the higher spin. Mo has ground-state configuration [Kr] $5s^1 4d^5$; that of Tc is [Kr] $5s^2 4d^5$:

5.23 Write configurations using the standard filling procedure, but watch for exceptions:

(a) C: $Z = 6$, [He] $2s^2 2p^2$

(b) Cr: $Z = 24$, exception, [Ar] $4s^1 3d^5$

(c) Sb: $Z = 51$, [Kr] $5s^2 4d^{10} 5p^3$

(d) Br: $Z = 35$, [Ar] $4s^2 3d^{10} 4p^5$

5.25 The ground state for N is $1s^2 2s^2 2p^3$. Excited states with no electron having $n > 2$ can be formed by moving electrons out of the 1*s* and/or 2*s* orbital and placing them in the 2*p* orbital. There are seven ways to do this:

$1s^1 2s^2 2p^4$; $2s^2 2p^5$; $1s^1 2s^1 2p^5$; $1s^2 2s^1 2p^4$; $1s^2 2p^5$; $2s^1 2p^6$; and $1s^1 2p^6$

5.27 Ionization energy decreases with increasing $\boldsymbol{n}$ and increases with increasing Z. Cs, which has the largest $\boldsymbol{n}$ value, has the smallest ionization energy; K, which has the next larger $\boldsymbol{n}$ value, has the next smallest ionization energy; because of its larger Z value, Ar has a larger ionization energy than Cl: $\text{Ar} > \text{Cl} > \text{K} > \text{Cs}$.

5.29 The value of IE_2 is almost 10 times that of IE_1, indicating that the second electron removed is a core electron rather than a valence electron. The elements in Column 1 contain only one valence electron, so this is a Column 1 element. An electron affinity around −50 kJ/mol matches this assignment (see Figure 5-17). (The element is Cs.)

5.31 The valence configurations are N, $2s^2 2p^3$; Mg, $3s^2$; and Zn, $4s^2 3d^{10}$:

The added electron adds to an already-occupied orbital. Electron–electron repulsion makes this an unfavourable process.

The added electron adds to the next higher orbital, which is much higher in energy.

The added electron adds to the next higher orbital, which is much higher in energy.

5.33 Br^- has 36 electrons. The following are isoelectronic ions with less than four units of net charge: As^{3-}, Se^{2-}, Rb^+, Sr^{2+}, and Y^{3+}. For isoelectronic ions, size decreases with increasing Z: $Y^{3+} < Sr^{2+} < Rb^+ < Br^- < Se^{2-} < As^{3-}$.

5.35 Stable anions form from elements in Columns 16 (dianions) and 17 (monoanions). Stable cations form from various metals: Ca, Cu, Cs, and Cr, all metals, may be found in ionic compounds as cations.

Cl may be found in ionic compounds as a –1 anion.

C is not found as an atomic ion.

5.37 To form $[K^+I^-]$ from neutral gaseous atoms, we must remove an electron from potassium, add an electron to iodine, and then form the coulombic interaction. The sequence of reactions for doing this calculation is (all gas phase):

$$K \rightarrow K^+ + e^-, \Delta E = IE_1 = 418.8 \text{ kJ/mol}$$

$$I + e^- \rightarrow I^-, \Delta E = EA = -295.3 \text{ kJ/mol}$$

$$K^+ + I^- \rightarrow KI, \Delta E = IE_1 = E_{electrical} \text{ (use Equation 3-1):}$$

$$E_{electrical} = \frac{k(q^+)(q^-)}{r} = \frac{(1.389 \times 10^5 \text{ kJ pm mol}^{-1})(+1)(-1)}{(133 \text{ pm} + 220 \text{ pm})} = -393 \text{ kJ/mol}$$

The total energy is the sum of the above energies:

$$\Delta E_{total} = (418.8 \text{ kJ/mol}) + (-295.3 \text{ kJ/mol}) + (-393 \text{ kJ/mol}) = -270. \text{ kJ/mol}$$

5.39 (a) The larger the ionic charges, the larger the lattice energy, so $Ba^{3+}O^{3-}$ would have the greatest lattice energy. (b) The first ionization energy always is smallest, and the first electron affinity always is most negative, so Ba^+O^- would have the least energy to form the ions. (c) The compound that actually exists is $Ba^{2+}O^{2-}$. The gain in lattice energy for this compound more than offsets the additional energy required to form the ions, but the energy required to remove the third electron from Ba is prohibitive.

5.41 First identify the chemical equation for the formation of LiF (*s*):

$$Li \,(s) + 1/2 \, F_2 \,(g) \rightarrow LiF \,(s)$$

Now determine the steps needed for this reaction. Li (Column 1) forms a +1 cation and F (Column 17) forms a –1 anion:

(1) $Li \,(s) \rightarrow Li \,(g)$ $\Delta E = \Delta H_{vap} - 2.5 = 159.3 - 2.5 = 156.8$ kJ/mol

(2) $Li \,(g) \rightarrow Li^+ \,(g) + e^-$ $\Delta E = IE = 520.2$ kJ/mol

(3) $1/2 \, F_2 \,(g) \rightarrow F \,(g)$ $\Delta E = 1/2$ bond energy $= \dfrac{1}{2} (155$ kJ/mol$) = 77.5$ kJ/mol

(4) $F \,(g) + e^- \rightarrow F^- \,(g)$ $\Delta E = EA = -328.0$ kJ/mol

(5) $F^- \,(g) + Li^+ \,(g) \rightarrow LiF \,(s)$ $\Delta E = -$lattice energy $= -1036$ kJ/mol

Summing up all of these energies gives the overall energy change for the formation of lithium fluoride:

$$(156.8 + 520.2 + 77.5 - 328.0 - 1036) \text{ kJ/mol} = -610 \text{ kJ/mol}$$

5.43 A Born–Haber diagram should show the energy of vapourization of the metal, ionization energy of the metal, energy to break the molecular bonds, electron affinity of anions, and the lattice energy that brings the ions of the salt together:

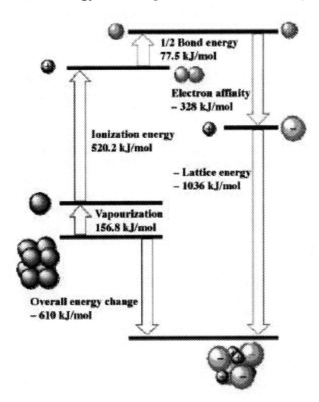

5.45 Use periodic variations in lattice energy to predict the expected value. The lattice energy values decrease by nearly the same amount from K^+ to Rb^+ as they decrease from Rb^+ to Cs^+. We predict that the decrease from Rb_2O to Cs_2O will be about the same as from K_2O to Rb_2O, 75 kJ/mol. The predicted value is $2163 - 75 = 2100$ kJ/mol, rounded to two significant figures because we do not expect this prediction to be exact.

5.47 Non-metals are found in the upper right portion of the periodic table, metalloids along a diagonal running through the *p* block, and all other elements are metals. In Column 16, Te is classified as a metalloid. The elements above it, O, S, and Se, are non-metals, and the element below it, Po, is a metal.

5.49 Non-metals are found in the upper right portion of the periodic table, metalloids along a diagonal running through the *p* block, and all other elements are metals. Thus, C and Cl are non-metals, and Ca, Cu, Cs, and Cr are metals.

5.51 The metalloids occupy a diagonal strip across the *p* block of the periodic table. The metals immediately to the left of this strip may be metalloid-like: Al, Ga, Sn, and Bi.

5.53 (a) When orbitals are nearly equal in energy, exceptions to the normal filling order exist. Normally, the $4s$ orbital fills immediately after $3p$, starting with element 19. Exceptions that indicate nearly equal energies of $3d$ and $4s$ are Cr, $4s^1 3d^5$; and Cu, $4s^1 3d^{10}$.

(b) The first two elements in Column 6, Cr and Mo, have $s^1 d^5$ configurations, whereas the second two elements in this column, W and Sg, have $s^2 d^4$ configurations.

(c) The $6d$ and $5f$ orbitals fill in the second *f* block, the actinides. In this block, four elements have both these orbitals partly filled, indicating nearly equal energy: Pa, U, Np, and Cm.

5.55 To determine the correct configuration of a cation, determine the configuration of the neutral atom, and then remove valence electrons, removing *s* electrons before *d* electrons:

Mn (25 electrons): $1s^2 2s^2 2p^6 3s^2 3p^6 4s^2 3d^5$

Mn^{2+} (remove two $4s$ electrons from Mn): $1s^2 2s^2 2p^6 3s^2 3p^6 3d^5$

The least stable occupied orbitals are the five $3d$ orbitals, among which electrons are distributed to produce maximum spin:

n	l	m_l	m_s
3	2	2	$+\dfrac{1}{2}$
3	2	1	$+\dfrac{1}{2}$

3	2	0	$+\dfrac{1}{2}$
3	2	−1	$+\dfrac{1}{2}$
3	2	−2	$+\dfrac{1}{2}$

5.57 Total electron spin depends on the number of unpaired electrons. In partially filled shells, each unpaired electron contributes $\dfrac{1}{2}$ to the total spin:

P: [Ne] $3s^2\,3p^3$; each $3p$ orbital is half filled, so net spin $= 3(\dfrac{1}{2}) = \dfrac{3}{2}$

Br⁻: [Kr]; all orbitals are filled, so net spin $= 0$

Cu⁺: [Ar] $3d^{10}$; all orbitals are filled, so net spin $= 0$

5.59 Ionization energy decreases with increasing value of n and, for the same value of n, increases with increasing Z (increasing electrical attraction). Na is the only species in this set with an $n = 3$ valence electron, so it has the smallest IE. Na⁺ has the largest Z, so it has the largest IE. O has its last two electrons paired, so electron–electron repulsion reduces its IE below that of N despite its higher Z value: Na $<$ O $<$ N $<$ Ne $<$ Na⁺.

5.61 Size increases with increasing n value and, for the same value of n, decreases with increasing Z (increasing electrical attraction). Cl, Cl⁻, and K⁺ all have $n = 3$ for the valence electrons, whereas Br⁻ has $n = 4$, so Br⁻ is the largest of this set. Among the others, the extra electron in Cl⁻ makes it larger than Cl, and K⁺ has higher Z than Cl: Br⁻ $>$ Cl⁻ $>$ Cl $>$ K⁺.

5.63 Use the periodic table to determine the correct configurations:

$Z = 9$ is F: [He] $2s^2\,2p^5$; $Z = 20$ is Ca: [Ar] $4s^2$; and $Z = 33$ is As: [Ar] $4s^2\,3d^{10}\,4p^3$

5.65 Use the values in Table 5-4 to calculate the averages. The average difference between alkali fluorides and alkali chlorides is 125 kJ/mol, and the average difference between alkali bromides and alkali iodides is 42.6 kJ/mol. Both sets of values show the same periodic trend, decreasing with the size of the alkali cation. The decreases occur because a larger cation cannot get as close to the anion as can a smaller cation.

5.67 Remember that in the transition metal cations, the ns orbital is less stable than $(n–1)d$:

Cu^+ is $4s^0 3d^{10}$

$3d$ ⇅ ⇅ ⇅ ⇅ ⇅ $4s$——

Mn^{2+} is $4s^0 3d^5$

$3d$ ↑ ↑ ↑ ↑ ↑ $4s$——

Au^{3+} is $6s^0 5d^8$

$5d$ ⇅ ⇅ ⇅ ↑ ↑ $6s$——

5.69 (a) In a one-electron atom or ion, orbital energy depends only on **n** and *Z*, both of which are the same for the hydrogen atom 2*s* and 2*p* orbitals.

(b) In multielectron atoms, orbitals with the same **n** value but different *l* values are screened to different extents. The orbital with the lower *l* value is less screened, and hence more stable.

5.71 This problem provides an exercise in graph reading:

(a) The greatest drop within a row is from element 80 (Hg) to 81 (Tl).

(b) Cs, element 55, has the lowest value shown in the figure.

(c) The value changes least from *Z* = 56 to 71, *Z* = 39 to 46, and *Z* = 20 to 29.

(d) Elements with values between 925 and 1050 kJ/mol are 15 (P), 16 (S), 33 (As), 34 (Se), 53 (I), 80 (Hg), 85 (At), and 86 (Rn).

5.73 Unpaired electrons occur only among the valence orbitals. Construct the configuration of neutral S, and then remove one electron to form S^+ and add one electron to form S^-:

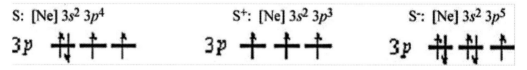

S: [Ne] $3s^2 3p^4$ S^+: [Ne] $3s^2 3p^3$ S^-: [Ne] $3s^2 3p^5$

$3p$ ⇅ ↑ ↑ $3p$ ↑ ↑ ↑ $3p$ ⇅ ⇅ ↑

S^+, with three unpaired electrons, has more unpaired than S or S^-.

5.75 The ground state places electrons in the most stable orbitals in a way that is consistent with the Pauli principle and maximizes electron spin. Thus, (c) is the ground-state configuration. Views (a) and (b) are excited states, but (d) is non-existent because it has two electrons with identical descriptions (3*s*, spin $+\frac{1}{2}$).

5.77 Use periodic trends and the periodic table to identify the elements involved:

(a) Mg has the second smallest radius among alkaline earths, and S has an anion that is isoelectronic with Ar, the Row 3 noble gas: MgS.

(b) K, beginning of Row 4, has a cation isoelectronic with Ar, at the end of Row 3; and the Row 2 element with the highest electron affinity is F: KF.

(c) Be is the alkaline earth element with the highest second ionization energy, and it combines in 1:2 ratio with elements from Row 17: $BeCl_2$.

5.79 Identify the orbitals using shapes and relative sizes. Given that c has $n = 3$: a and b are both spherical, with b smaller than a; a is the same size as c, so a is $3s$ and b is $2s$; c has the unique shape of the $3d_{z^2}$ orbital:

(a) b is most stable (smallest n value), and a is more stable than c.

(b) $n = 3, l = 0, m_l = 0, m_s = +\dfrac{1}{2}$; $n = 3, l = 0, m_l = 0, m_s = -\dfrac{1}{2}$.

(c) Any element from magnesium to calcium has $3s^2$ but $3d^0$: $Z = 12$ to $Z = 20$.

(d) The $3d$ transition metal cations have partially occupied $3d$ orbitals, for example, Fe^{3+}.

(e) There are eight other $n = 3$ orbitals: $3s$, three $3p$, and four additional $3d$.

(f) Whenever an electron is removed from a doubly-occupied orbital, electron–electron repulsion decreases for the remaining electron, so the orbital becomes smaller.

5.81 In an excited-state configuration, one electron is shifted to a less stable orbital. Start by constructing the ground-state configuration, and then move an electron from the most stable occupied to the least stable unoccupied orbital:

Be: $[He]\ 2s^2$; the next orbital is $2p$: $[He]\ 2s^1\ 2p^1$

O^{2-}: $[He]\ 2s^2\ 2p^6$; the next orbital is $3s$: $[He]\ 2s^2\ 2p^5\ 3s^1$

Br^-: $[Ar]\ 4s^2\ 3d^{10}\ 4p^6$; the next orbital is $5s$: $[Ar]\ 4s^2\ 3d^{10}\ 4p^5\ 5s^1$

Ca^{2+}: $[Ar]$; the next orbital is $4s$: $[Ne]\ 3s^2\ 3p^5\ 4s^1$

Sb^{3+}: $[Kr]\ 4d^{10}\ 5s^2$; the next orbital is $5p$: $[Kr]\ 4d^{10}\ 5s^1\ 5p^1$

5.83 Examine the configurations of the atoms and ions to determine the reason for the ionization energy differences:

Li: $1s^2\ 2s^1$ Be: $1s^2\ 2s^2$

Li^+: $1s^2$ Be^+: $1s^2\ 2s^1$

The first ionization energies involve the $2s$ orbital for both atoms, so Be, with a larger Z, has a larger IE. The second electron removed from Li is a core electron, so the IE is much greater than the second IE for Be.

5.85 Consult Figures 5-6, and 5-1 for examples of this kind of drawing.

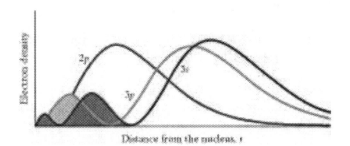

The shaded portions of the 3*s* and 3*p* indicate the regions that are ineffectively screened by the 2*p* orbital. 3*s* and 3*p* electrons have a greater probability of being in the larger areas mostly outside the 2*p* orbital. Thus, the 2*p* orbital effectively screens electrons in both the 3*s* and 3*p* orbitals.

5.87 Francium is an alkali metal (Column 1) whose properties should closely resemble those of cesium: low *IE*, highly reactive, forms a cation with +1 charge, soft metal, melts at a low temperature.

5.89 Removing electrons from any atom or ion reduces screening and electron–electron repulsion and therefore stabilizes the orbitals. The Li^{2+} cation has a more stable 2*s* orbital.

5.91 Consult any of the graphs showing how electron density varies with distance from the nucleus. Notice that the density value decreases gradually. At what distance from the nucleus has the electron density decreased sufficiently that we consider it to be zero? That question has no unambiguous answer.

5.93 If screening were less important than the change in principal quantum number, the filling order would be 1*s*, 2*s*, 2*p*, 3*s*, 3*p*, 3*d*, 4*s*, 4*p*, 4*d*, 4*f*, 5*s*, etc. Each successive row of the periodic table would be longer: 2, 8, 18, 32, 50. Here is the resulting table:

N	s^1	s^2	p^1	p^2	p^3	p^4	p^5	p^6	d^1	d^2	d^3	d^4	d^5	d^6	d^7	d^8	d^9	d^{10}	*f* block	*g* block
1	1	2																		
2	3	4	5	6	7	8	9	10												
3	11	12	13	14	15	16	17	18	19	2	21	22	23	24	25	26	27	28		
4	29	30	31	32	33	34	35	36	37	3	39	40	41	42	43	44	45	46	47–60	
5	61	62	63	64	65	66	67	68	69	7	71	72	73	74	75	76	77	78	72–92	93–120

The first 18 elements have the same configurations as the actual elements do, but element 19 has d^1 rather than s^1 configuration, so it is not an alkali metal. Instead, elements 29 and 61 are alkali metals, along with elements 3 and 11. Elements 17 and 35 would still be "halogens," because each could add one electron to complete its *p* orbitals. Elements with d^9 configurations also might behave like halogens. Elements 2 and 10 would still be noble gases, but elements 18 and 36 might be chemically reactive because of the accessibility of their *d* orbitals. The chemistry of the most abundant elements would be pretty much the same as in our actual world, because the first 18 elements would have the same configurations. The transition metals would be quite different, however. Would this periodic table be easier or harder to discover? Either answer can be supported. On the one hand, the table looks more regular, with each successive row being longer. On the other hand, the regular patterns shown by the halogens would not be as obvious. Lothar Meyer might have had an easier time explaining his plots of atomic volumes, but Mendeleev probably would have had more difficulty identifying chemical regularities.

5.95 Begin with $CaCl_2$.

First identify the chemical equation:

$$Ca(s) + Cl_2(g) \rightarrow CaCl_2(s)$$

Now determine the steps needed for this reaction.

(1) $Ca(s) \rightarrow Ca(g)$ $\Delta E = \Delta H_{vap} - 2.5 = 178 - 2.5 = 175.5$ kJ

(2) $Ca(g) \rightarrow Ca^+(g) + e^-$ $\Delta E = IE_1 = 589.8$ kJ/mol

(3) $Ca^{2+}(g) \rightarrow Ca^+(g) + e\text{-}$ $\Delta E = IE_2 = 1145$ kJ/mol

(4) $Cl_2(g) \rightarrow 2\,Cl(g)$ $\Delta E = $ bond energy $= 240$ kJ/mol

(5) $2\left[Cl(g)^+ + e^- \rightarrow Cl^-(g)\right]$ $\Delta E = 2EA = 2(\text{-}348.8 \text{ kJ/mol}) = \text{-}697.6$ kJ/mol

(6) $2\,Cl^-(g) + Ca^{2+}(g) \rightarrow CaCl_2(s)$ $\Delta E = -$lattice energy $= -2223$ kJ/mol

Summing up all of these energies gives the overall energy change for the formation of $CaCl_2$:

$(175.5 + 589.8 + 1145 + 240 - 697.6 - 2223)$ kJ/mol $= -770.$ kJ/mol

Now consider CaCl. The chemical equation is

$$Ca(s) + 1/2\,Cl_2(g) \rightarrow CaCl(s)$$

Next, determine the steps needed for this reaction:

(1) $Ca(s) \rightarrow Ca(g)$ $\Delta E = \Delta H_{vap} - 2.5 = 178 - 2.5 = 175.5$ kJ/mol

(2) $Ca(g) \rightarrow Ca^+(g) + e^-$ $\Delta E = IE_1 = 589.8$ kJ/mol

(3) $1/2\left[Cl_2(g) \rightarrow 2Cl(g)\right]$ $\Delta E = 1/2$ bond energy $= 120$ kJ/mol

(4) $Cl(g) + e^- \rightarrow Cl(g)$ $\Delta E = EA = -348.8$ kJ/mol

(5) $Cl^-(g) + Ca^+(g) \rightarrow CaCl(s)$ $\Delta E = $ -lattice energy $= -720$ kJ/mol

Summing up all of these energies gives the overall energy change for the formation of

CaCl:

$$(175.5 + 589.8 + 120 - 348.8 - 720) \text{ kJ/mol} = -184 \text{ kJ/mol}$$

The compound most likely to form is the one that releases more energy (the more stable compound). Because the energy change for $CaCl_2$ is more negative than that for CaCl, $CaCl_2$ is the more stable compound.

5.97 The periodic table would contain 50% more columns in Morspin compared with our universe because of the extra possible spin value. All other laws would remain the same, so orbital shapes and periodic trends would be the same as in our universe.
(a) Elements 18 and 30 would be in the same column of the periodic table. Element 18, with a smaller valence *n* value, would have the larger ionization energy.
(b) Elements 15 and 17^{2+} would have the same number of electrons, so the one with smaller *Z*, element 15, would have the larger radius.
(c) Elements 47 and 48 both would fall in the *p* block. Element 47 would have p^2 configuration whereas element 48 would have p^3 configuration. An electron added to element 48 would be added to an already-occupied orbital, which would generate significant electron–electron repulsion, so element 47 would have a larger (more negative) electron affinity than element 48.

5.99 The periodic table in Morspin would contain 50% more columns compared with our universe because of the extra possible spin value. All other laws would remain the same, so orbital shapes and periodic trends would be the same as in our universe.
(a) The noble gases are at the ends of the rows: the first three would have 3, 15, and 27 electrons.

(b) Element 4 would have one valence 2*s* electron: $n = 2, l = 0, m_l = 0, m_s = +\dfrac{1}{2}$;

element 7 would have three valence 2*s* electrons and one valence 2*p* electron:

n	*l*	m_l	m_s
2	0	0	$+\dfrac{1}{2}$
2	0	0	$-\dfrac{1}{2}$
2	0	0	0
2	1	1	$+\dfrac{1}{2}$

Element 32 would have three valence 4*s* and two valence 3*d* electrons:

n	*l*	m_l	m_s
4	0	0	$+\dfrac{1}{2}$
4	0	0	$-\dfrac{1}{2}$
4	0	0	0

| 3 | 2 | 2 | $+\dfrac{1}{2}$ |
| 3 | 2 | 1 | $+\dfrac{1}{2}$ |

(c) There would be 12 $n = 3$ electrons in Morspin and only 8 $n = 3$ electrons in our universe. Screening occurs in Morspin just as in our universe, and the more inner electrons there are, the more effectively screened the 4*s* electron is. Therefore, the 4*s* electron in Morspin would be more effectively screened.

5.101 AUFBAU predicts that this element will have the electronic configuration [Rn] $5f^{14}6d^{10}7s^{2}7p^{2}$. All electrons except the two in the 7*p* orbital are paired (i.e. are in orbitals with another electron of opposing spin). However, the two in the 7*p* orbitals will be unpaired according to Hund's rule, and so their spins will be additive, making this atom paramagnetic.

The doubly charged ion will have the electronic configuration [Rn] $5f^{14}6d^{10}7s^{2}$, which has all spins paired, and so will be diamagnetic.

Chapter 6 Fundamentals of Chemical Bonding

Solutions to Problems in Chapter 6

6.1 Determine a configuration from the position of an element in the periodic table. The electrons with the highest principal quantum number will be involved in bond formation:

(a) O: $1s^2\, 2s^2\, 2p^4$; its six $n = 2$ electrons will be involved in bond formation.

(b) P: $1s^2\, 2s^2\, 2p^6\, 3s^2\, 3p^3$; its five $n = 3$ electrons will be involved in bond formation.

(c) B: $1s^2\, 2s^2\, 2p^1$; its three $n = 2$ electrons will be involved in bond formation.

(d) Br: $1s^2\, 2s^2\, 2p^6\, 3s^2\, 3p^6\, 3d^{10}\, 4s^2\, 4p^5$; its seven $n = 4$ electrons will be involved in bond formation.

6.3 One valence $5p$ orbital of the iodine atom is directed along the axis between the two atoms and can overlap with the $1s$ orbital of the hydrogen atom to form a bond:

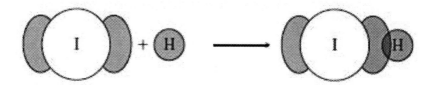

6.5 The Group number tells us how many valence electrons any element has:

(a) Al, Group 13, 13 − 10 = 3 valence electrons

(b) As, Group 15, 15 − 10 = 5 valence electrons

(c) F, Group 17, 17 − 10 = 7 valence electrons

(d) Sn, Group 14, 14 − 10 = 4 valence electrons

6.7 Bonds form by sharing electrons in valence orbitals. The configuration of Li is [He] $2s^1$. The electron in the $2s$ orbital of a lithium atom and the electron in the $1s$ orbital of a hydrogen atom are shared between the two atoms, forming a bond between the nuclei:

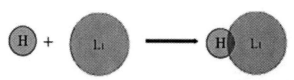

6.9 Electronegativities describe the tendency of each element to attract bonding electrons from another. In any pair, the element with higher electronegativity attracts bonding electrons more strongly. Use Figure 6-7 to obtain the electronegativity of each element:

(a) N (3.0) will attract electrons more than C (2.5)

(b) S (2.5) will attract electrons more than H (2.1)

(c) I (2.5) will attract electrons more than Zn (1.6)

(d) S (2.5) will attract electrons more than As (2.0)

6.11 Electronegativity differences determine the direction of bond polarities. The more electronegative atom has the negative charge:

(a) $^{\delta+}Si-O^{\delta-}$; (b) $^{\delta-}N-C^{\delta+}$; (c) $^{\delta+}Cl-F^{\delta-}$, and (d) $^{\delta-}Br-C^{\delta+}$

6.13 Bond polarity increases with the difference in electronegativity of the bond-forming elements. In these compounds, H is the less electronegative element, so bond polarity increases with the electronegativity of the other element: the electronegativity order is P < S < N < O, so the order of bond polarity is PH_3 < H_2S < NH_3 < H_2O.

6.15 Determine numbers of valence electrons from the position of each element in the periodic table, adding one for each negative charge and subtracting one for each positive charge.

(a) H_3PO_4: each H has one valence electron, P (Group 15) has five, and each O (Group 16) has six, for a total of 32 valence electrons.

(b) $(C_6H_5)_3C^+$: each H has one valence electron and each C (Group 14) has four; subtract one for the positive charge, giving a total of 90 valence electrons.

(c) $(NH_2)_2CO$: each H has one valence electron, each N (Group 15) has five, C (Group 14) has four, and O has six, for a total of 24 valence electrons.

(d) SO_4^{2-}: S and O (Group 16) have six valence electrons; add two for the two negative charges, giving a total of 32 valence electrons

6.17 Build a molecular framework from information contained in the chemical formula. Each chemical bond requires two valence electrons:

(a) $(CH_3)_3CBr$: There are 13 bonds, which require 26 valence electrons:

(b) $(CH_3CH_2CH_2)_2NH$: There are 21 bonds, which require 42 valence electrons:

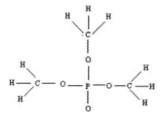

(c) HClO₃: remember that H bonds to O in oxoacids. There are four bonds, which require eight valence electrons:

(d) OP(OCH₃)₃: the parentheses identify three OCH₃ groups bonded to P. There are 16 bonds, which require 32 valence electrons:

6.19 Use the procedure in your textbook for determining the Lewis structures.

1. Each species has eight valence electrons.

2. Each species has its H atoms bonded to N, and each N–H bond requires two electrons.

3. There are no outer atoms other than H.

4. Place remaining electrons on the N atom:

(a)   (b)   (c)

Optimize electron configurations of the inner atoms: N has an octet in each structure, so all three are correct.

6.21 Use the procedure in your textbook for determining the Lewis structures.

1. Count the valence electrons.

   (a) PBr₃: P (Group 15) has five electrons, Br (Group 17) has seven, for a total of 26

   (b) SiF₄: Si (Group 14) has four electrons, F (Group 17) has seven, for a total of 32

(c) BF$_4^-$: B (Group 13) has three electrons, F (Group 17) has seven, plus one for the anion's charge, for a total of 32

2. Assemble the bonding framework. In each of these molecules, the unique atom is inner. Each bond uses two valence electrons:

(a) PBr$_3$: three P–Br bonds use six electrons

(b) SiF$_4$: four Si–F bonds use eight electrons

(c) BF4$^-$: four B–F bonds use eight electrons

3. Add three non-bonding electron pairs to all non-hydrogen outer atoms.

(a) PBr$_3$: three pairs on each Br atom use 18 electrons; there are two left

(b) SiF$_4$: three pairs on each F atom use 24 electrons; there are none left

(c) BF$_4^-$: three pairs on each F atom use 24 electrons; there are none left

4. Add any remaining electrons to the inner atom:

5. Optimize electron configurations of the inner atoms: P and Si have FC = 0, so structures (a) and (b) are correct. B has FC = –1, but there is no way to reduce this charge, so structure (c) is correct.

6.23 Use the procedure in your textbook for determining the Lewis structures.

1. Count the valence electrons.

(a) H$_3$CNH$_2$: C (Group 14) has four electrons, N (Group 15) has five, and each H has one, for a total of 14

(b) CF$_2$Cl$_2$; C (Group 14) has four electrons, F and Cl (Group 17) have seven, for a total of 32

(c) OF$_2$: O (Group 16) has six electrons, F (Group 17) has seven, for a total of 20

2. Assemble the bonding framework. Each bond uses two valence electrons.

(a) The formula of H$_3$CNH$_2$ indicates its framework:

Six bonds use 12 valence electrons, leaving two to be placed

(b) CF_2Cl_2: C is inner, with four bonds to the outer halogens that use eight electrons

(c) OF_2: two O–F bonds use four electrons

3. Add three non-bonding electron pairs to all non-hydrogen outer atoms:

(a) H_3CNH_2: there are no outer atoms other than H; there are two electrons left

(b) CF_2Cl_2: three pairs on each halogen atom use 24 electrons; there are none left

(c) OF_2: three pairs on each F use 12 electrons; there are four left

4. Add any remaining electrons to an inner atom:

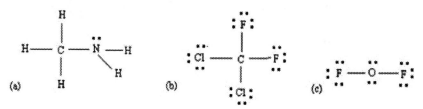

5. Optimize electron configurations of the inner atoms: All inner atoms are from the second row and have octets, so these are the correct structures.

6.25 Use the procedure in your textbook for determining the Lewis structures.

1. Count valence electrons:

(a) $(CH_3)_2CO$ has $3(4) + 6(1) + 6 = 24$ valence electrons

(b) CH_3CN has $2(4) + 3(1) + 5 = 16$ valence electrons

(c) CH_2CHCH_3 has $3(4) + 6(1) = 18$ valence electrons

(d) CH_3CHNH has $2(4) + 5(1) + 5 = 18$ valence electrons

2. Use the chemical formula to determine the framework:

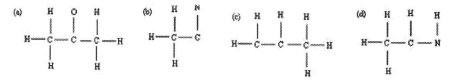

3. Add three electron pairs to each outer atom except H. Only (a) and (b) have such atoms:

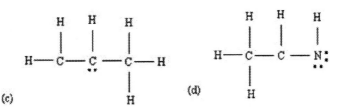

At the end of this step, all the valence electrons have been placed for (a) and (b), but there are two electrons left to place on (c) and (d).

4. Place remaining electrons on inner atoms, starting with the most electronegative:

5. Optimize electron configurations of the inner atoms: each structure has an inner atom with less than an octet of electrons. Shift lone pairs to make multiple bonds until all inner atoms have octets:

6. There are no equivalent structures.

6.27 Use the procedure in your textbook for determining Lewis structures.

1. Count valence electrons:

    (a) IF_5: 6(7) = 42 valence electrons

    (b) SO_3: 4(6) = 24 valence electrons

    (c) $OPCl_3$: 6 + 5 + 3(7) = 32 valence electrons

    (d) XeF_2: 8 + 2(7) = 22 valence electrons

2. Use the chemical formula to determine the framework. Each bond uses two valence electrons.

    (a) IF_5: I is inner and forms five bonds to F atoms, using 10 valence electrons

    (b) SO_3: S is inner and forms three bonds to O atoms, using six valence electrons

    (c) $OPCl_3$: P is inner and forms four bonds to the outer atoms, using eight valence electrons

(d) XeF_2: Xe is inner and forms two bonds to F atoms, using four valence electrons

3. Add three electron pairs to each outer atom except H.

   (a) Three pairs on each of five F atoms uses 30 more electrons, leaving $42 - 10 - 30 = 2$ electrons

   (b) Three pairs on each of three O atoms uses 18 more electrons, leaving $24 - 6 - 18 = 0$ electrons

   (c) Three pairs on each of four outer atoms uses 24 more electrons, leaving $32 - 8 - 24 = 0$ electrons

   (d) Three pairs on each of two F atoms uses 12 more electrons, leaving $22 - 4 - 12 = 6$ electrons

4. Add remaining electrons to inner atoms. Each structure has only one inner atom, so all the remaining electrons must go on that atom regardless of octet considerations:

5. Optimize electron configurations of the inner atoms: Each inner atom is beyond Row 2, so optimize based on formal charge. In structures (a) and (d), the formal charge of the inner atom is zero, so these are the correct structures. The S atom has a formal charge of $6 - 3 = 3$, so make three double bonds, one from each O atom. The P atom has a formal charge of $5 - 4 = 1$, so make a double bond to O, which has a formal charge of -1:

6. There are no equivalent structures.

6.29 Use the procedure in your textbook for determining the Lewis structures.

   1. Count valence electrons, adding one for each negative charge:

      (a) NO_3^-: $5 + 3(6) + 1 = 24$ valence electrons

(b) HSO_4^-: $5(6) + 1 + 1 = 32$ valence electrons

(c) CO_3^{2-}: $4 + 3(6) + 2 = 24$ valence electrons

(d) ClO_2^-: $7 + 2(6) + 1 = 20$ valence electrons

2. Use the chemical formula to determine the framework. Each bond uses two valence electrons:

(a) NO_3^-: N is inner and forms bonds to three O atoms, using six valence electrons

(b) HSO_4^-: S is inner and forms bonds to four O atoms, one of which bonds to H, using 10 valence electrons

(c) CO_3^{2-}: C is inner and forms bonds to three O atoms, using six valence electrons

(d) ClO_2^-: Cl is inner and forms bonds to two Cl atoms, using four valence electrons

3. Add three electron pairs to each outer atom except H.

(a) Three pairs on each of three O atoms uses 18 electrons, leaving $24 - 6 - 18 = 0$ electrons.

(b) Three pairs on each of three O atoms (the fourth is inner, bonded to H and S) uses 18 electrons, leaving $32 - 10 - 18 = 4$ electrons.

(c) Three pairs on each of three O atoms uses 18 electrons, leaving $24 - 6 - 18 = 0$ electrons.

(d) Three pairs on each of 2 two atoms uses 12 electrons, leaving $20 - 4 - 12 = 4$ electrons.

4. Add remaining electrons to inner atoms to complete octets.

(a) No electrons to add

(b) Add the four electrons to the inner O, the more electronegative atom

(c) No electrons to add

(d) Add the four electrons to Cl:

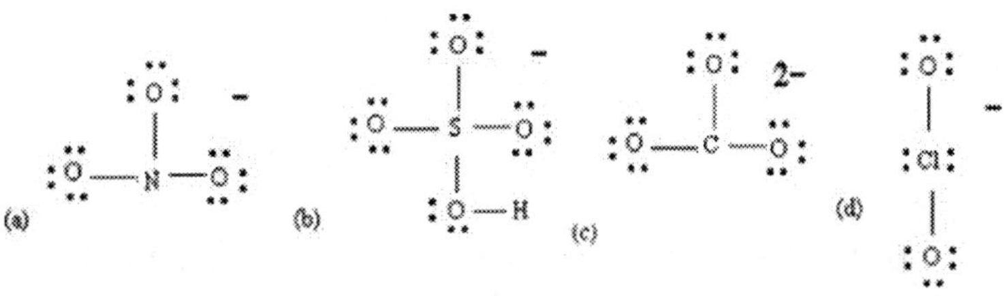

5. Optimize electron configurations of the inner atoms. The inner atoms in (a) and

(c) are second row, so complete their octets. The inner atoms in (b) and (d) are third row, so reduce their formal charges to zero by making two double bonds to each:

6. The outer oxygen atoms all are equivalent, so all but (d) have three equivalent structures.

6.31 Molecular shapes are determined by the steric numbers (SN) of inner atoms, which can be found from the Lewis structures. The Lewis structures of these molecules, determined in Problems 6.21 and 6.23, show that each inner atom has SN = 4, so each has tetrahedral molecular group geometry:

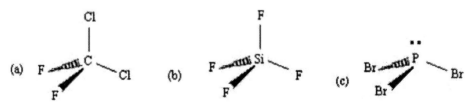

(a) CF_2Cl_2 has tetrahedral shape, with F atoms at two apices and Cl atoms at the other two; (b) SiF_4 has tetrahedral shape; (c) PBr_3, with a lone pair of electrons on the inner atom, has trigonal pyramidal shape

6.33 Construct the bonding framework from the formula, knowing that the more electronegative atoms (Cl) will be in outer positions. In 1,2-dichloroethane, each carbon atom is bonded to the other, and there is one Cl atom bonded to each carbon atom. There are $2(7) + 2(4) + 4(1) = 26$ valence electrons. The bonding framework contains seven bonds, and there are three lone pairs on each Cl atom, accounting for all the valence electrons. Thus, the bonding framework is the correct Lewis structure of 1,2-dichloroethane. Each carbon atom has SN = 4 and tetrahedral geometry. Your ball-and-stick model should reflect this:

6.35 To draw structural isomers of an alkane, start with the "straight chain" compound, and then rearrange the bonding arrangement to make other isomers. It is convenient to work with only the carbon skeleton. You can add H atoms to complete the Lewis structures:

6.37 Determine the Lewis structure following the standard procedures.

1. There are $2(4) + 7(1) + 5 = 20$ valence electrons.

2. The chemical formula indicates that the bonding framework includes two –CH_3 groups attached to the nitrogen atoms. The bonding framework contains 9 bonds = 18 electrons, leaving two valence electrons:

3. All the outer atoms are hydrogen, so no electrons can be placed on outer atoms.

4. Place the remaining pair to the nitrogen atom. The resulting structure has an octet around N, so it is the correct Lewis structure.

Each inner has SN = 4, so the electron group geometry about each inner atom is tetrahedral. The shape about the N atom is trigonal pyramidal, like ammonia:

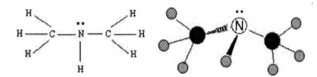

6.39 Determine the Lewis structure following the standard procedures:

1) There are $4(4) + 4 + 12(1) = 32$ valence electrons.

2) The chemical formula indicates that the bonding framework includes four $-CH_3$ groups attached to the silicon atom. The bonding framework has 16 bonds = 32 electrons. All electrons are placed, so this structure is the correct Lewis structure. Each inner atom has SN = 4, and the geometry about each inner atom is tetrahedral:

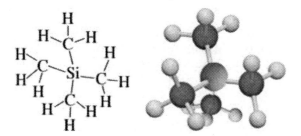

6.41 Determine molecular shapes from electron pair geometry, taking account of lone pairs:

(a) Two lone pairs and three ligands gives SN = 5 and trigonal bipyramidal electron group geometry. Two equatorial positions are occupied by the lone pairs, so this is T-shaped.

(b) A steric number of 5 means trigonal bipyramidal electron group geometry. One equatorial position is occupied by the lone pair, so this is seesaw shaped.

(c) A steric number of 3 means trigonal planar electron group geometry, and the molecular shape is the same when there are no lone pairs.

(d) A steric number of 6 means octahedral electron group geometry. One lone pair gives the shape of a square pyramid.

6.43 Determine the Lewis structures following the standard procedures.

GeF_4:

1. There are $4 + 4(7) = 32$ valence electrons.

2. Four electron pairs are used in forming the bonding framework, leaving $32 - 4(2) = 24$ electrons.

3. Place six electrons around each outer F atom, leaving $24 - 4(6) = 0$ electrons.

4. There are no remaining electrons.

5) The resulting structure has $FC_{Ge} = 4 - 4 = 0$, so the structure is correct.

SeF_4:

1. There are 6 + 4(7) = 34 valence electrons.

2. Four electron pairs are used in forming the bonding framework, leaving 34 – 4(2) = 26 electrons.

3. Place three pairs of electrons around each outer F atom, leaving 26 – 4(6) = 2 remaining electrons.

4. Place the remaining two electrons on the inner Se atom.

5. The resulting structure has FC_{Se} = 6 – 4 –2 = 0, so the structure is correct.

XeF_4:

1. There are 8 + 4(7) = 36 valence electrons.

2. Four electron pairs are used in forming the bonding framework, leaving 36 – 4(2) = 28 electrons.

3. Place three pairs of electrons around each outer F atom, leaving 28 – 4(6) = 4 remaining electrons.

4. Place the remaining four electrons on the inner Xe atom.

5. The resulting structure has FC_{Xe} = 8 – 4 – 4 = 0, so the structure is correct.

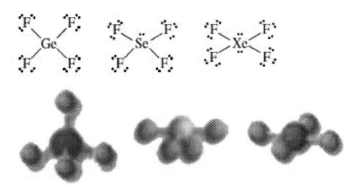

In GeF_4, SN_{Ge} = 4, so this molecule is tetrahedral. In SeF_4, SN_{Se} = 5 with a lone pair, so this molecule has seesaw geometry. In XeF_4, SN_{xe} = 6 with two lone pairs, so this molecule has square planar geometry.

6.45 Determine the Lewis structures following the standard procedures:

SO_2:

1. There are 3(6) = 18 valence electrons.

2. Two electron pairs are used in forming the bonding framework, leaving 18 – 2(2) = 14 electrons.

3. Place six electrons around each outer O atom, leaving 14 – 2(6) = 2 electrons.

4. Place the remaining pair on the S atom.

5. The resulting structure has $FC_S = 6 - 4 = 2$. Make two double bonds to complete the Lewis structure.

SbF_5:

1. There are $5 + 5(7) = 40$ valence electrons.

2. Five electron pairs are used in forming the bonding framework, leaving $40 - 5(2) = 30$ electrons.

3. Place six electrons around each outer F atom, leaving $30 - 5(6) = 0$ electrons.

4. There are no electrons left to place on the inner atom.

5. The resulting structure has $FC_{Sb} = 5 - 5 = 0$, so the structure is correct.

ClF_4^+:

1. There are $7 + 4(7) - 1 = 34$ valence electrons.

2. Four electron pairs are used in forming the bonding framework, leaving $34 - 4(2) = 26$ electrons.

3. Place six electrons around each outer F atom, leaving $26 - 4(6) = 2$ electrons.

4. Place the remaining electron pair on the inner Cl atom.

5. The resulting structure has $FC_{Cl} = 7 - 4 - 2 = +1$, the same as the overall charge, so the structure is correct.

ICl_4^-:

1. There are $5(7) + 1 = 36$ valence electrons.

2. Four electron pairs are used in forming the bonding framework, leaving $36 - 4(2) = 28$ electrons.

3. Place six electrons around each outer F atom, leaving $28 - 4(6) = 4$ electrons.

4. Place the remaining electron pairs on the inner I atom.

5. The resulting structure has $FC_I = 7 - 4 - 4 = -1$, the same as the overall charge, so the structure is correct.

(a) SN = 3, giving trigonal planar electron group geometry. One lone pair gives a bent shape, with ideal angle of 120°. (b) SN = 5, giving trigonal bipyramidal shape and ideal angles of 90° and 120°. (c) SN = 5, giving trigonal bipyramidal electron group geometry. One lone pair gives a seesaw shape, with ideal angles of 90° and 120°. (d) SN = 6, giving octahedral electron group geometry. Two lone pairs result in a square planar shape, with ideal bond angles of 90°.

6.47 Determine Lewis structures using the standard procedures. (Here we do not number the steps.) Only asymmetric molecules have dipole moments. Use the Lewis structures to determine steric numbers and ascertain whether or not the molecules are asymmetric.

SiF_4: There are $4 + 4(7) = 32$ valence electrons. Four pairs are used in the bonding framework, and three pairs are placed on each outer F atom. This leaves $FC_{Si} = 4 - 4 = 0$, so this is the correct Lewis structure.

H_2S: There are $6 + 2 = 8$ valence electrons. Two pairs are used in the bonding framework, and the remaining four electrons are placed on the S atom.

XeF_2: There are $8 + 2(7) = 22$ electrons. Two pairs are used in the bonding framework, three pairs are placed on each outer F atom, and the remaining six electrons are placed on the inner Xe atom. The resulting structure has $FC_{Xe} = 8 - 8 = 0$, so this is the correct Lewis structure.

$GaCl_3$: There are $3 + 3(7) = 24$ valence electrons. Three pairs are used in the bonding framework. Add three pairs to each outer atom (using the remaining electrons). Gallium is in the fourth row, Group 13. Its formal charge is $3 - 3 = 0$, so this is the correct structure.

NF_3: There are $5 + 3(7) = 26$ valence electrons. Three pairs are used in the bonding framework. Add three pairs to each outer atom. The remaining pair goes on the inner N atom, giving N an octet, so this is the correct structure:

SiF₄ is tetrahedral and symmetric; it has no dipole moment; H_2S, like H_2O, is bent and has a dipole moment; XeF_2 has SN = 5, with its three lone pairs in equatorial positions, so the molecule is linear without a dipole moment; $GaCl_3$ has SN = 3, so the molecule is trigonal planar and has no dipole moment; NF_3, like NH_3, is pyramidal and has a dipole moment.

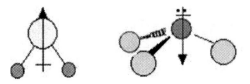

6.49 The Lewis structure of CO_2 shows no lone pairs on the C atom, giving SN = 2 and linear shape. The two C=O bonds point opposite each other, so bond polarities cancel. The Lewis structure of SO_2 (Problem 6.45) shows a lone pair on the S atom, resulting in a bent molecule whose polar bonds do not cancel each other.

6.51 Determine Lewis structures in the usual fashion:

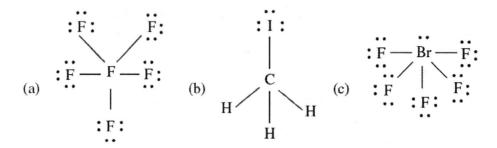

(a) PF_5 has no lone pairs on its inner atom, and all its ligands are identical, so it is trigonal bipyramidal with ideal bond angles of 90° and 120°.

(b) CH_3I has SN = 4 and tetrahedral electron group, with no lone pairs, but the very large I atom repels the H atoms, making the I–C–H angles larger than 109.5° and the H–C–H angles smaller than 109.5°.

(c) BrF_5 has octahedral electron group geometry and a lone pair on Br, making the F–Br–F angles to the F atom at the apex smaller than 90°:

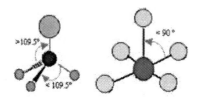

6.53 The determinants of bond length, in order of importance, are principal quantum number of the valence orbitals, bond order, and bond polarity. Among these bonds, H–N is shortest because it involves an $n = 1$ orbital and Cl–N is longest because it involves an $n = 3$ orbital. Of the three bonds involving $n = 2$ orbitals, the N–N single bond is longest and the polar C≡N triple bond is shortest: H–N < C≡N < N≡N < N–N < Cl–N.

6.55 Bond strengths, like bond lengths, depend on principal quantum number, bond order, and bond polarity, but the correlation is not as strong as for bond lengths, so tabulated values must be consulted. Here are the values from Table 3-2 of your textbook, listed in increasing order:

Bond	C–C	<	H–N	<	C=C	<	C=O	<	N≡N
Energy (kJ/mol)	345		390		615		750		945

Strong bonding and bond polarity of the $n = 1$ orbital makes H–N > C–C; multiple bonding makes C=C > H–N; bond polarity makes C=O > C=C; and multiple bonding makes N≡N > C=O.

6.57 Determine Lewis structures following the standard procedures. Draw ball-and-stick models that illustrate the molecular geometries determined by the steric numbers of the inner atoms.

(a) Cl_2O has $2(7) + 6 = 20$ valence electrons. Its Lewis structure has two bonding pairs, three lone pairs on each chlorine atom, and two lone pairs on the inner oxygen atom. $SN_O = 4$, leading to tetrahedral electron pair geometry and bent molecular shape:

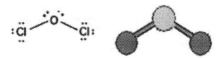

(b) C_6H_6 has $6(4) + 6(1) = 30$ valence electrons. After constructing the bonding framework, six electrons remain to be placed. Place these on alternate carbon atoms; then complete octets around the other carbon atoms by shifting lone pairs to make double bonds. There are two equivalent resonance structures:

Provisional structure   resonance structures

$SN_C = 3$, leading to trigonal planar geometry about each carbon atom and a flat, hexagonal molecular shape.

(c) C_2H_4O has $2(4) + 4(1) + 6 = 18$ electrons. When the bonding framework is complete, four electrons remain. These are placed on the oxygen atom. The three bond angles of the triangular ring must sum to 180°, even though SN = 4 for each atom.

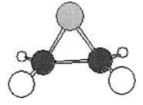

6.59 Determine the Lewis structures using standard procedures. Each of the molecules in this problem has four halogen atoms bonded to an inner atom. Both CI_4 and $SiCl_4$ have 32 electrons and SeF_4 has 34 valence electrons. The 32-electron species have the same Lewis structure and the same geometry.

CI_4 and $SiCl_4$:

1. There are $6 + 4(7) = 32$ valence electrons.

2. Four electron pairs are used in forming the bonding framework, leaving $32 - 4(2) = 24$ electrons.

3. Place three pairs of electrons around each outer atom, leaving $24 - 4(6) = 0$ electrons.

4. There are no remaining electrons.

5. The resulting structure has $FC_{Si} = 4 - 4 = 0$ and C has an octet, so the structures are correct.

SeF_4:

1. There are $6 + 4(7) = 34$ valence electrons.

2. Four electron pairs are used in forming the bonding framework, leaving $34 - 4(2) = 26$ electrons.

3. Place six electrons around each outer F atom, leaving $26 - 4(6) = 2$ electrons.

4. Place the remaining two electrons on the inner Se atom.

5. The resulting structure has $FC_{Se} = 6 - 4 - 2 = 0$, so the structure is correct.

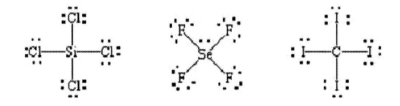

The 32-electron species have SN = 4 and are tetrahedral; SeF_4 is a seesaw.

6.61 To find exceptions to normal trends, consult the appropriate listings in tabulated values. Table 6-1 lists bond lengths by n value. For $n = 2$, it lists four $X–X$ bonds (C–C, N–N, O–O, and F–F) and three $X–Y$ bonds (C–N, C–O, and C–F). Consult Table 3-2 to find the exceptions:

(a) C–N is shorter but weaker than C–C; (b) C–N is longer but stronger than N–N.

6.63 Determine the number of different isomers by constructing the different possible geometric possibilities. Two X atoms can be placed at opposite ends of one Cartesian axis; all structures with this arrangement are equivalent, because the molecule can be rotated about an axis to superimpose these positions. The three X atoms can all be placed at right angles to one another, giving a different isomer. Thus, there are two isomers:

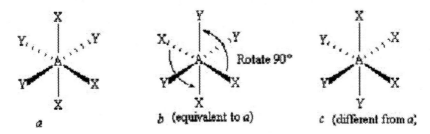

6.65 The empirical chemical formula of the silicon–oxygen network of zircon, an orthosilicate, is SiO_4^{4-}. Each silicon atom has SN = 4 and tetrahedral geometry, and each O atom is outer, so the molecular geometry of the SiO_4^{4-} anions is tetrahedral. Orthosilicates are ionic, containing discrete (not connected) SiO_4^{4-} anions and metal cations.

6.67 Determine Lewis structures following the standard procedures.

(a) Bromate (BrO_3^-) contains $7 + 3(6) + 1 = 26$ valence electrons. Three pairs are required for the bonding framework, nine pairs go on the oxygen atoms, and the remaining pair is placed on the bromine atom. The resulting structure has $FC_{Br} = +2$. Shift two electron pairs to make two double bonds and reduce FC_{Br} to 0:

(b) Nitrite (NO_2^-) contains $5 + 2(6) + 1 = 18$ valence electrons. Two pairs are required for the bonding framework, six pairs go on the O atoms, and the remaining pair is placed on N. Shift one electron pair to make a double bond and complete the octet on nitrogen. There are two equivalent structures. $FC_N = 5–5 = 0$:

(c) Phosphate (PO_4^{3-}) contains $5 + 4(6) + 3 = 32$ valence electrons. Four pairs are required for the bonding framework and 12 pairs go on the oxygen atoms, giving a provisional structure with $FC_P = 5 – 4 = +1$. Shift one electron pair to make a double bond and reduce the formal charge to zero. There are four resonance structures.

There are two additional equivalent structures

(d) Hydrogen carbonate (HCO_3^-) contains $1 + 4 + 3(6) + 1 = 24$ valence electrons. Four pairs are required for the bonding framework, three pairs go on each outer oxygen atom, and the remaining two pairs go on the inner oxygen atom. Shift one electron pair from an outer O atom to form a double bond to C and complete its octet. There are two resonance structures:

6.69 Consult Table 6-2 for the correspondence between molecular shapes and lone pairs.

(a) This is the seesaw shape, derived from the trigonal bipyramid when there is one lone pair. The ideal bond angles are 90° and 120°, and an example is SF_4. (b) This is the square planar shape, derived from the octahedron when there are two lone pairs, which take opposing axial positions. The ideal and actual bond angles are 90°, and an example is XeF_4. (c) This is the tetrahedron, the shape associated with SN = 4 and no lone pairs. The ideal bond angles are 109.5°, and an example is CH_4.

6.71 The determinants of bond length, in order of relative importance, are principal quantum number of the valence orbitals, bond order, and bond polarity. The bonds found in these molecules are H–O, H–C, C=O, and C≡N. Of these, H–O and H–C are shorter than the others because they form from an $n = 1$ orbital; H–O is shorter than H–C because it is a more polar bond. C=O has bond order 2 and is longer than C≡N because bond order is a stronger determinant of length than bond polarity.

6.73 Follow the standard procedure to obtain the Lewis structure of a molecule, and then use the Lewis structure to determine the geometry around inner atoms.

(a) This molecule has 4 N (5 e⁻) + 2 C (4 e⁻) + 2 O (6 e⁻) + 4 H (1 e⁻) = 44 valence electrons, 22 of which are used for the 11 bonds in the framework. Add six electrons around each outer O atom, and then place the remaining five pairs on inner N atoms, which are more electronegative than C atoms, giving a provisional structure in which the three second-row atoms marked with *3* have only three pairs of electrons. Complete their octets by shifting electron pairs to form double bonds in the final Lewis structure.

Provisional structure   Final structure

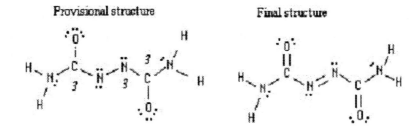

(b) The end N atoms have SN = 4 and tetrahedral electron pair geometry. Because of their lone pairs, these N atoms have trigonal pyramidal geometries. The C atoms and the centre N atoms have SN = 3 and trigonal planar electron pair geometry. The geometry about the C atoms is the same as their electron pair geometry, whereas the lone pairs on the centre N atoms result in bent shapes about these atoms.

6.75 VSEPR theory predicts tetrahedral bond angles (109.5°) for both these molecules. A bond angle of 104.5° is a bit smaller than this, because of the larger repulsion of lone pairs. An angle of 92.2° indicates that the bonding can be well described using *p* orbitals rather than spacing the electron pairs as far apart as possible. Space-filling models of the two molecules show that the smaller oxygen atom cannot accommodate two H atoms at right angles, but the larger S atom can.

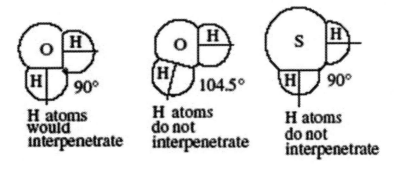

6.77 The electronegativity of hydrogen (2.1) is the smallest among the elements involved in these chemical bonds, so the ranking of bond polarity is from smallest to largest electronegativity of the bonding partner. Remember that electronegativity decreases down a column and increases across a row: Si–H < C–H < N–H < O–H < F–H.

6.79 All positions around a tetrahedral centre are equivalent, so there is only one isomer of CH_2Br_2. The two structures show different views of the same compound.

6.81 The six-membered ring of benzyne requires bond angles of 120°. The two carbon atoms involved in the triple bond have SN = 2, for which the optimum VSEPR angle is 180°. Thus, these atoms react readily, either by adding another bonded atom to generate SN = 3 or by breaking the ring so the bond angle can become 180°.

6.83 Determine the Lewis structures using standard procedures. Both molecules have 2(5) + 6 = 16 valence electrons. Two pairs are required for the bonding framework, and each outer atom receives three pairs:

$$:\ddot{N}\!\!-\!\!N\!\!-\!\!\ddot{O}: \qquad\qquad :\ddot{N}\!\!-\!\!O\!\!-\!\!\ddot{N}:$$

Because these molecules contain only Row 2 atoms, we need to complete the octets. This can be accomplished by shifting two electron pairs to make double bonds:

N-N-O   N-O-N

$$\ddot{N}{=}\kern-2pt{=}\ddot{N}{=}\kern-2pt{=}\ddot{O}{\cdot\cdot} \qquad \ddot{N}{=}\kern-2pt{=}O{=}\kern-2pt{=}\ddot{N}$$

Molecules have dipole moments only if they are non-symmetrical. A linear N–O–N structure would not have a dipole moment, because the N–O dipole would exactly cancel the O–N dipole. Thus, N_2O must have the structure N–N–O if it is to have a dipole moment.

6.85 A molecule has a dipole moment if it is non-symmetrical. In isomer *a*, the two polar C–Cl bonds point in opposite directions and cancel, thus *a* is non-polar. Both *b* and *c* are polar, because their dipoles do not cancel; *b* has the larger dipole moment, because its dipoles point more nearly in the same direction.

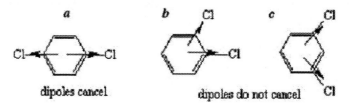

6.87 The Lewis structures of molecules with formula XF_3 show octets around the inner atom and $FC_X = 0$, making them stable. Compounds with formula XF_5 also have $FC_X = 0$ but have five electron pairs associated with the inner atom. This is possible for phosphorus, a third-row element that has *d* orbitals available for bonding. It is not possible for nitrogen, a second-row element that lacks valence *d* orbitals:

6.89 The shape of a molecule indicates the steric number of its inner atom and the number of atoms bonded to the inner atom. Consult Table 6-2 of your textbook for details: Tellurium, in Column 16 of the periodic table, has six valence electrons, and each fluorine contributes seven valence electrons. Species with an even number of fluorine atoms are neutral, but those with an odd number must have one unit of charge (+1 or –1) in order to contain an even number of electrons.

(a) Bent indicates TeF_2.

(b) T shape indicates SN = 5 on Te, three F atoms, and two lone pairs on Te, for a total of $4 + 3(8) = 28$ valence electrons. TeF_3 would have $6 + 3(7) = 27$ valence electrons, so the formula of this species must be TeF_3^-.

(c) Square pyramid indicates SN = 6 on Te, five F atoms, and one lone pair on Te, for a total of 2 + 5(8) = 42 valence electrons. TeF_5 would have 6 + 5(7) = 41 valence electrons, so the formula of this species must be TeF_5^-.

(d) Trigonal bipyramid indicates SN = 5 on Te and five F atoms, for a total of 5(8) = 40 valence electrons. TeF_5 would have 6 + 5(7) = 41 valence electrons, so the formula of this species must be TeF_5^+.

(e) Octahedron indicates SN = 6 and six F atoms, formula TeF_6.

(f) Seesaw indicates SN = 5 on Te, four F atoms, and one lone pair on Te, formula TeF_4.

6.91 (a) Square planar XY_2Z_2 molecules may have like atoms adjacent to or opposite each other:

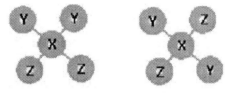

(b) When like atoms are opposite each other, their bond polarities oppose each other and cancel, so this isomer has no dipole moment. When like atoms occupy adjacent positions, the bond polarities do not entirely cancel, generating a net dipole moment:

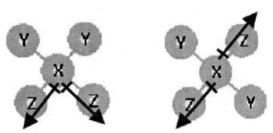

6.93 A compound XY_7 is possible if X has d orbitals available for bonding, but it is sterically crowded because of the large number of Y atoms around the central X atom. For this compound to exist, Y must be as small as possible and X must be as large as possible. Thus, Y is the halogen with the smallest size, fluorine; and X is iodine, the halogen with the largest size (apart from astatine, which has no stable isotopes). The compound is IF_7.

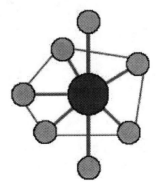

6.95 Build Lewis structures for transition metal complexes following the same rules as for other compounds, using the additional guidelines described in the problem. Here are the structures:

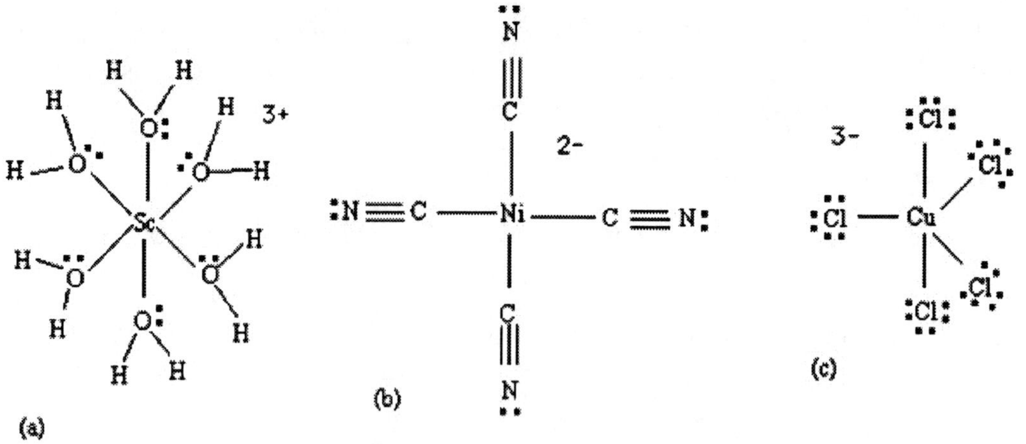

(a) The transition metal ion has SN = 6 and octahedral shape. Each O atom has SN = 4 and a lone pair, giving it trigonal pyramidal shape.

(b) The transition metal ion has SN = 4, but we are told that it is not tetrahedral. Instead, it is square planar. Each C atom has SN = 2 and linear shape.

(c) The transition metal ion has SN = 5 and trigonal bipyramidal shape.

6.97 To expand a line structure, place a C atom at every vertex and line end, and then add H atoms until each C atom has four bonds. Add lone pairs to complete octets on outer atoms (except H) and Row 2 atoms:

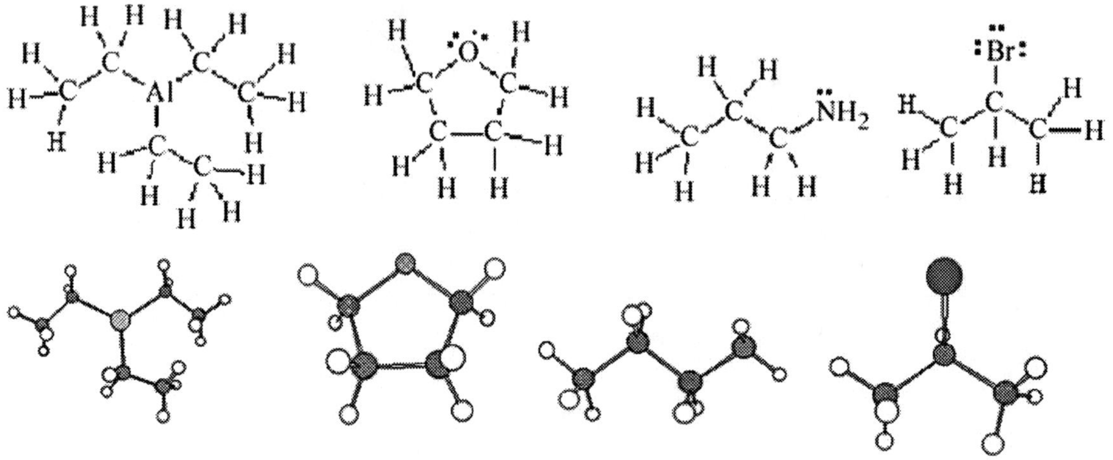

In this set of compounds, aluminum has SN = 3 and trigonal planar shape. All the other inner atoms have SN = 4 and tetrahedral electron group geometries. All the carbon atoms have tetrahedral shapes, the nitrogen atom is triangular pyramidal, and the oxygen atom is bent.

6.99 All three minerals have four oxygen atoms bonded to each silicon atom. Silica is a continuous covalent network of Si–O bonds where every oxygen atom is an inner atom and is bonded to two silicon atoms, with an empirical formula of SiO_2. Orthosilicates, however, contain discrete SiO_4^{4-} anions where every oxygen atom is an outer atom.

Metasilicate minerals contain anions that are covalent networks of Si–O bonds like silica. In metasilicates, however, some oxygen atoms are inner (bonded to two silicon atoms), whereas others are outer (bonded to only one silicon atom).

6.101 The dipole moment of F_2O arises from the contributions of both O-F dipoles added vectorially and will point from the O atom to a point between the two F atoms:

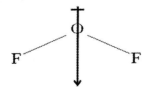

As the molecule rotates faster and the bond angle increases, we expect the dipole moment to decrease because the molecule is becoming more linear. If we denote the dipole moment of each O-F bond as x, and the bond angle as θ, we can calculate the total dipole moment in terms of x and θ:

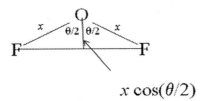

$$x \cos(\theta/2)$$

The contribution to the total dipole moment from each O-F bond is $x \cos(\theta/2)$, so the total dipole moment is $2x \cos(\theta/2)$. With a bond angle of $103.3°$, this is $2x \cos(103.3°/2) = 1.241 x$. With the bond angle increased to $107.0°$, this is $2x \cos(107.0°/2) = 1.190 x$. The dipole moment therefore decreases by the ratio $1.190/1.241 = 0.959$, and the new dipole moment must be $0.959(0.297 \text{ D}) = 0.285$ D.

Chapter 7 Theories of Chemical Bonding

Solutions to Problems in Chapter 7

7.1 A diatomic halogen molecule forms its bond by overlap of the valence p orbitals that point along the bond axis. In Br_2, the valence orbitals have $n = 4$, so the bond forms by overlap of two $4p$ orbitals.

7.3 Hydrogen always uses its $1s$ orbital to form bonds. A Group 1 element has only one valence electron, in an ns orbital. Lithium is a second-row element, with $n = 2$ for its valence electron. Thus, the bond in LiH forms by overlap of the hydrogen $1s$ orbital and the lithium $2s$ orbital:

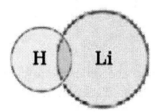

7.5 Bond angles near 90° signal interactions of valence p orbitals from the inner atom. An outer halogen atom always uses one of its valence p orbitals to form a bond. Antimony has $n = 5$ valence orbitals, and fluorine has $n = 2$ valence orbitals, so each of the three bonds in SbF_3 can be described as resulting from overlap between a $5p$ orbital from Sb and a $2p$ orbital from F. There are three identical bonds that point at near-right angles to one another:

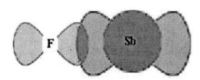

7.7 The steric number of an inner atom uniquely determines its hybridization:

(a) SN = 2 + 2 = 4, requiring sp^3 hybridization; (b) SN = 3 + 1 = 4, requiring sp^3 hybridization; (c) SN = 3 + 0 = 3, requiring sp^2 hybridization; and (d) SN = 5 + 1 = 6, requiring $sp^3 d^2$ hybridization.

7.9 The name of a hybrid orbital set uses the letters of the atomic orbitals, with superscripts indicating the number of each type, used to make the set: (a) sp^3; (b) sp; and (c) $sp^3 d^2$.

7.11 The steric number of an inner atom uniquely determines its hybridization. Use the Lewis structure of the molecule to identify steric numbers:

(a) SN = 4, sp^3 hybrids; (b) SN = 5, $sp^3 d$ hybrids; (c) SN = 3, sp^2 hybrids; and (d) SN = 2, sp hybrids.

7.13 The steric number of an inner atom uniquely determines its hybridization. Use the Lewis structure of the molecule to find steric numbers: (a) SN = 4, sp^3 hybrids; (b) SN = 5, $sp^3 d$ hybrids; (c) SN = 6, $sp^3 d^2$ hybrids; and (d) SN = 3, sp^2 hybrids.

7.15 A description of bonding begins with the Lewis structure of the molecule, from which steric numbers of inner atoms identify hybridization. Outer atoms use atomic orbitals for bond formation. Here is the Lewis structure of chloroform:

$$
\begin{array}{c}
:\ddot{Cl}: \\
| \\
H-C-\ddot{Cl}: \\
| \\
:\ddot{Cl}:
\end{array}
$$

The carbon atom has SN = 4 and uses sp^3 hybrids to form three C–Cl bonds and one C–H bond. Each chlorine atom uses a $3p$ orbital for bond formation and has lone pairs in the $3s$ and two other $3p$ orbitals. The H atom uses its $1s$ orbital to form the C–H bond. The geometry about the carbon atom is tetrahedral. The bonds can be depicted as follows (the white region around the Cl atom represents its core electrons):

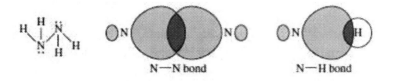

7.17 A description of bonding begins with the Lewis structure of the molecule, from which steric numbers of inner atoms translate into hybridization. Outer atoms use atomic orbitals for bond formation. The N atoms in hydrazine have SN = 4, use sp^3 hybrids, and have tetrahedral geometry: four sp^3–$1s$ N–H σ bonds, one sp^3–sp^3 N–N σ bond, and two lone pairs in sp^3 hybrids. Here is the Lewis structure (showing the approximate geometry) and sketches of the bonds:

7.19 To determine the bonding pattern from a line structure, first convert the line
structure into a molecular structure by adding C and H atoms. Molecular structures
contain enough information to deduce the steric numbers of C, N, and O inner
atoms without determining the complete Lewis structure. Acetone has three inner
C atoms. Two have only single bonds (one C–C and three C–H); these have SN =
4. The atom bonded to O has one π bond and SN = 3. In the entire molecule, there
are six sp^3–$1s$ C–H σ bonds, two sp^3–sp^2 C–C σ bonds, one sp^2–$2p$ C–O σ bond,
one π bond between C and O from $2p$-$2p$ overlap, and two lone pairs in $2s$ and $2p$
orbitals on O. Here are sketches of the bonding orbitals:

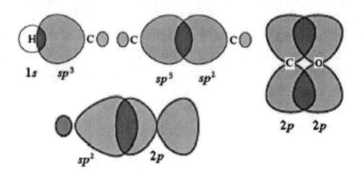

7.21 The line structure of neocembrene in Focus 7–1: Chemistry and Life shows that it
contains only inner carbon atoms and outer hydrogen atoms. Some of the inner
atoms have all single bonds, SN = 4, sp^3 hybridization, and tetrahedral geometry;
others have one double bond, SN = 3, sp^2 hybridization, and trigonal planar
geometry. The line structure presented here shows the SN values for the 20 carbon
atoms:

7.23 We determine the steric number of each carbon atom from the line structures of the
compounds. These are shown on the line structures:

1,4-Pentadiene   1-Pentyne   Cyclopentene

C atoms designated "4" have only single bonds, SN = 4, sp^3 hybridization, and
tetrahedral geometry. Those with double bonds, designated "3," have SN = 3, sp^2
hybridization, and trigonal planar geometry. Two C atoms, designated "2," have

two double bonds, *sp* hybridization, and linear geometry. The σ bonds are described by overlap of hybrid orbitals, with each C–H bond described as a hybrid overlapping with a hydrogen 1*s* orbital. The π bonds are described by side-by-side overlap of 2*p* orbitals. In 1-pentyne, the two C atoms designated "2" have three bonds: a σ bond formed from *sp* hybrids and two π bonds from unhybridized 2*p* orbitals, at right angles to each other, as in acetylene. Here are sketches of the different types of bonds:

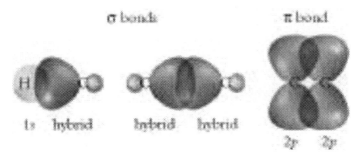

7.25 Examine the shape and orientation of a pair of orbitals to determine what kind of bond, if any, it will form: (a) $2p_z$ and $2p_z$ point toward each other along the bond axis and form a σ bond; (b) $2p_y$ and $2p_x$ lie in different planes and do not form a bond; (c) sp^3 and $2p_z$ point toward each other along the bond axis and form a σ bond; and (d) $2p_y$ and $2p_y$ are in the same plane and overlap side-by-side to form a π bond.

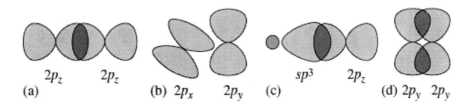

7.27 The orbital energy-level diagram for first-row diatomic species has just two levels, each of which can hold two electrons. The relative bond strengths can be deduced from the bond orders, which are calculated using Equation 7-1 of your textbook:

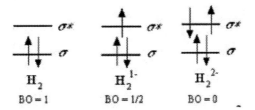

From weakest to strongest bond, the order is $H_2^{2-} < H_2^- < H_2$.

7.29 The stability of a diatomic molecule is determined by the number of bonding and antibonding electrons. To compare species, determine how many valence electrons each species possesses, and then place the electrons in the available orbitals, following the Pauli and aufbau principles. Here are the configurations for 9 and 10 valence electrons:

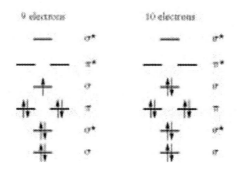

(a) CO has 10 valence electrons. To make CO^+, an electron is removed from a bonding orbital, which destabilizes the ion, so CO has a stronger bond.

(b) N_2 has 10 valence electrons. To make N_2^+, an electron is removed from a bonding orbital, which destabilizes the ion, so N_2 has a stronger bond.

(c) CN^- has 10 valence electrons. To make CN, an electron is removed from a bonding orbital, which destabilizes the species, so CN^- has a stronger bond.

7.31 When two orbitals from different atoms interact, they form two MOs, one showing additive overlap and the other showing subtractive overlap. Additive overlap generates a bonding MO, whereas subtractive overlap generates an antibonding MO. Axial overlap gives σ orbitals, and off-axis overlap gives π orbitals:

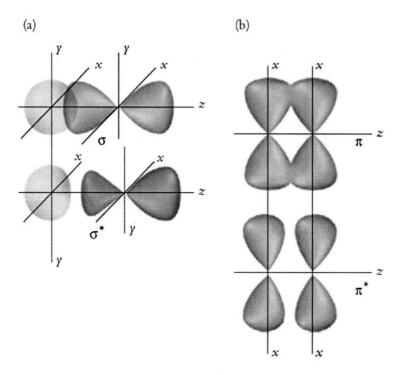

7.33 Determine how many valence electrons the ion possesses, then place the electrons in the available orbitals, following the Pauli and aufbau principles. Stability is determined by the number of bonding and antibonding electrons. ClO^- has 14

valence electrons, six from O, seven from Cl, plus one due to its charge:

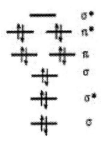

There are eight bonding electrons and six antibonding electrons, for a net of two bonding electrons and a single bond for this ionic species.

7.35 Determine the Lewis structure using the usual procedures. The molecule has $2(6) + 4 = 16$ valence electrons:

$$\ddot{S} = C = \ddot{S}$$

The bonding pattern for CS_2 is like that for CO_2, except that the S atoms have $n = 3$ valence orbitals rather than $n = 2$ orbitals. The inner atom can be described using *sp* hybrids, which overlap with *3p* orbitals from the S atoms to form two σ bonds. There are two delocalized π systems at right angles to each other, each made up of a *2p* orbital on the C atoms overlapping side-by-side with a *3p* orbital on each S atom. As in CO_2, eight electrons occupy the delocalized π orbitals, four in bonding and four in non-bonding orbitals.

7.37 Determine the Lewis structures using the usual procedures. The NCN^{2-} anion has $2(5) + 4 + 2 = 16$ valence electrons, making it isoelectronic with CO_2:

$$\ddot{N} = C = \ddot{N}$$

The Lewis structure and MO description are the same as that for CO_2. There is no lone pair on the inner C atom, giving SN = 2, *sp* hybridization, and a linear geometry. There are two sets of three-centre delocalized π orbitals:

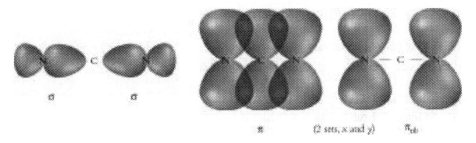

7.39 Determine the Lewis structure using the usual procedures. The molecule has $3(5) + 1 = 16$ valence electrons. Begin by drawing the bonding framework and placing lone pairs on the outer atoms:

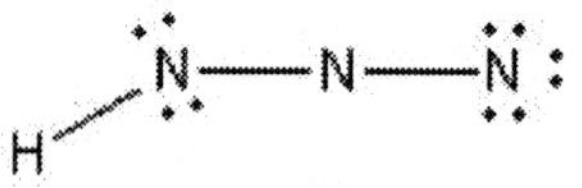

This uses all the valence electrons. Form double and triple bonds to complete the octet on the inner N atom:

The N atom bonded to H has SN = 3 and 4 in the two structures, so it has bent geometry with an angle between 109° and 120° (the experimental value is 112° 39'). The best hybridization is sp^2, leaving one p orbital free to form a π bond. The other inner N atom has SN = 2, sp hybridization, and bond angles of 180°. There are two π networks, one delocalized over all three N atoms and the other localized between the outer N atom and its adjacent inner N atom.

7.41 Conjugation stabilizes a molecule whenever there are more than two adjacent atoms involved in π-bond formation, thereby leading to delocalized π orbitals. Of the molecules whose structures appear in Focus 7-1: Chemistry and Life of your textbook, only xanthin has a delocalized π system, to which all 26 atoms that are part of the double-bond system contribute.

7.43 We can determine hybridizations directly from line structures: each carbon atom with no double bonds, and each inner oxygen atom, has SN = 4, and sp^3 hybrids can be used to describe the bonding. A C atom with one double bond has SN = 3 and sp^2 hybridization:

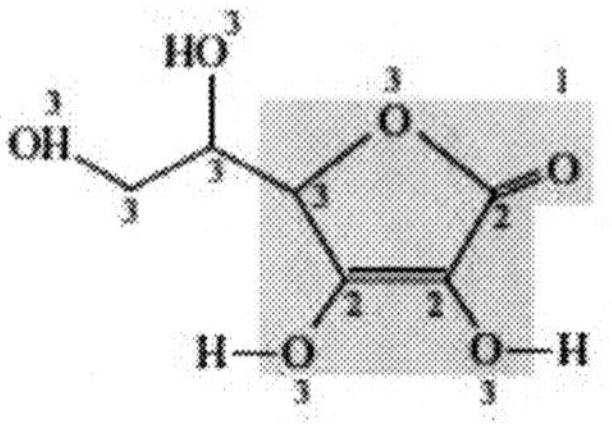

(a) Hybridization is as shown on the line drawing ($3 = sp^3$, $2 = sp^2$).

(b and c) The Lewis structure shows two π bonds on adjacent atoms, so as in butadiene there are four electrons in the delocalized π system.

(d) The four atoms with double bonds and the four additional atoms bonded directly to them all lie in the same plane.

7.45 Determine the Lewis structure using the usual procedures: the molecule has 3(4) + 2(6) = 24 valence electrons. Begin by drawing the bonding framework and adding in lone pairs of electrons to the outer atoms. Use the lone pairs to form double bonds to complete the octets on the inner carbons:

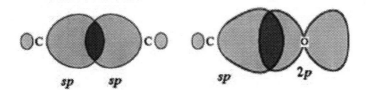

Each carbon atom has SN = 2, so the molecule is linear and the σ-bond network can be described using *sp* hybrids. There are two *sp–sp* bonds between carbon atoms and two *sp–2p* bonds between carbon and oxygen atoms:

If the molecular axis is the *z*-axis, a p_x and a p_y orbital on each atom overlap side-by-side, giving two sets of delocalized π orbitals, each extending over all five atoms:

7.47 A description of bonding in a molecule starts with its Lewis structure, from which steric number, geometry, hybridization, and extension of π bonds can be deduced: ClO_4^- has $7 + 4(6) + 1 = 32$ valence electrons.

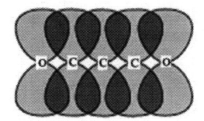

The provisional Lewis structure has $FC_{Cl} = 7 - 4 = +3$, so three electron pairs are shifted to make double bonds:

$SN_{Cl} = 4$, indicating that the molecule has tetrahedral geometry and the σ bond framework can be described using sp^3 hybrids on Cl overlapping with $2p$ orbitals from O. The resonance structures signal that the π system is extended over all five atoms between the *p* orbitals on the O atoms and the *d* orbitals of the Cl atom, and there are three bonding π orbitals occupied by six electrons.

7.49 Doped semiconductors are *p* type if the dopant has fewer valence electrons than the bulk element and *n* type if the dopant has more valence electrons than the bulk element: (a and c) GaP and CdSe are stoichiometric compounds, and hence are undoped semiconductors. (b) InSb (Groups 13 and 15) doped with Te (Group 16) is *n* type.

7.51 The bonding of any metal can be described by constructing bands from the valence orbitals and then distributing the valence electrons into these bands. Iron and potassium are both Row 4 elements with similar band structures, but whereas each potassium atom contributes just one valence electron to the valence (bonding) band of the solid, each iron atom contributes eight electrons. Thus, iron has much greater bonding, making it harder and giving it a higher melting point than potassium.

7.53 The appearance of a band-gap diagram depends on the type of doping in the semiconductor. GaAs doped with Zn is a *p*-type semiconductor, meaning that it has vacancies in its valence band:

p-type semiconductor

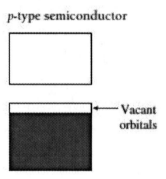

← Vacant orbitals

7.55 Describe bonding and geometry starting with a Lewis structure and a count of bonds and non-bonding pairs around inner atoms. The provisional structure of each compound has a positive formal charge on the Row 3 sulphur atom, which is reduced to zero by making double bonds to each oxygen atom:

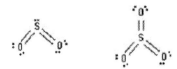

The sulphur atom in each compound has SN = 3, so sp^2 hybrids overlapping with oxygen 2*p* orbitals describe the σ bonds: two in SO_2 and three in SO_3. SO_2 is bent, like O_3, and SO_3 is trigonal planar, like NO_3^-. The Lewis structures indicate the presence of two π bonds in SO_2 and three π bonds in SO_3. All the π orbitals extend over the entire molecule. Because sulphur is a third-row element, its 3*d* orbitals contribute to the extended π-bonding orbitals, which form from side-by-side overlap of oxygen 2*p* orbitals with sulphur 3*p* and 3*d* orbitals.

7.57 The band gap of a semiconductor is related to the frequency of light that can promote an electron through the equation $VE_{gap} = E_{photon} = h\nu$

Divide by Avogadro's number to convert the band-gap energy in kilojoules per mole to the energy of one photon:

$$E_{photon} = \frac{(34.7 \text{ kJ/mol})(10^3 \text{ J/kJ})}{6.022 \times 10^{23} \text{ photons/mol}} = 5.76 \times 10^{-20} \text{ J}$$

$$\nu = \frac{E_{photon}}{h} = \frac{5.76 \times 10^{-20} \text{ J}}{6.626 \times 10^{-34} \text{ J s}} = 8.70 \times 10^{13} \text{ s}^{-1}$$

This frequency is in the infrared region of the spectrum.

7.59 To describe bonding, use the Lewis structure to determine steric numbers and hybridization. Then build orbital pictures accordingly. The π system in acrolein is identical to that in methyl methacrylate, which is described in Example 7-12 in your textbook. The outer oxygen atom uses $2p$ atomic orbitals, and the three double-bonded carbons are sp^2 hybridized. These assignments lead to the following inventory of σ bonds: four sp^2–$1s$ C–H σ bonds, two sp^2–sp^2 C–C σ bonds, and one sp^2–$2p$ C–O σ bond. This leaves one p orbital on each C atom, plus one from the oxygen atom, to form a delocalized π network extending over four atoms. That network has two bonding π orbitals, each with a pair of electrons. To complete the valence electron inventory, there is one $2s$ and one $2p$ lone pair on the outer O atom.

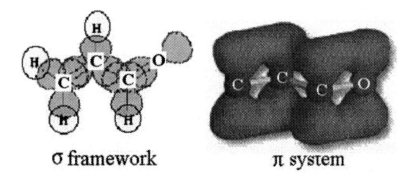

σ framework   π system

7.61 The formula of cinnamic acid reveals that it contains C, H, and O atoms. From the line structure below, we see that there is one outer O atom, whose bonding can be described using $2p$ orbitals. All other C and O atoms are inner. Those with only single bonds have SN = 4, sp^3 hybridization, and tetrahedral geometry. Those with double bonds have

Cinnamic acid
$C_9H_8O_2$

SN = 3, sp^2 hybridization, and trigonal planar geometry:

(a) There are lone pairs only on the two O atoms.

(b and c) *A*: sp^2 and atomic *p*, 120°; *B*: 2*p*, no angle (outer); *C*: sp^3, <109.5°; *D*: sp^2 and atomic *p*, 120°; *E*: atomic 1*s*, no angle (outer).

(d) Cinnamic acid has five π bonds.

7.63 A molecule is paramagnetic if it has unpaired electrons. For diatomic molecules, use MO diagrams to place the valence electrons and determine whether or not they are all paired:

CO	N_2^+	O_2^+	CN^-
10 e's	9 e's	11 e's	10 e's

(a) Diamagnetic; (b) paramagnetic; (c) paramagnetic; and (d) diamagnetic.

7.65 To determine which orbitals atoms use in a molecule, we must determine steric numbers from either the line structure or the Lewis structure. Any inner C, N, or O atom with no double bonds has SN = 4; with one double bond, an inner C or N atom has SN = 3. Remember that outer atoms (including H) always use atomic orbitals:

C and N use sp^3   *all C's use sp^2*

7.67 Band-gap diagrams have valence bands and conduction bands, which can be filled, empty, or partially filled. A metallic conductor has no gap between the bands. An *n*-type semiconductor has a full valence band and a partially filled conduction band:

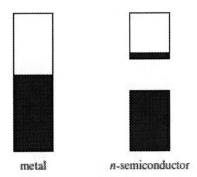

metal   *n*-semiconductor

7.69 A description of bonding and geometry starts with determination of the Lewis structure. $(CH_3)_3C^+$ has $4(4) + 9(1) - 1 = 24$ electrons. All the electrons are required to complete the bonding framework, so there are no lone pairs to shift and the provisional structure is correct even though the central carbon atom has only six electrons associated with it (as this suggests, carbocations are highly reactive species):

The central carbon has SN = 3, trigonal planar geometry, and its bonding orbitals can be described using sp^2 hybrid orbitals. The methyl carbons have SN = 4, tetrahedral geometry, and their bonding orbitals can be described using sp^3 hybrid orbitals. The vacant *p* orbital on the central C atom makes this cation highly reactive.

7.71 As Figure 7-50 shows, an electrical potential applied to a material distorts the bands. As Figure 7-52 shows, a semiconductor has a relatively small band gap between its filled valence band and empty conduction band. The distortion of the bands allows electrons to "hop" from valence to conduction band without changing energy, leading to electron mobility and conduction:

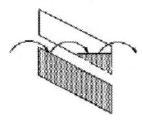

7.73 We obtain the charges on the ions from the chemical formulas: KO_2 contains one K^+ per O_2 unit, so superoxide has one negative charge; K_2O_2 contains two K^+ per O_2 unit, so peroxide has two negative charges. These ions are second-row diatomic species whose electron configurations can be written using the orbital energy-level diagrams in Figure 7-35 of your textbook ($Z > 7$). Superoxide has one more π electron than O_2, and peroxide has two more:

O_2: $(\sigma_s)^2 (\sigma_s{*})^2 (\pi_x)^2 (\pi_y)^2 (\sigma_p)^2 (\pi{*}_x)^1 (\pi{*}_y)^1$ Bond order $= \dfrac{1}{2}(8-4) = 2$

O_2^-: $(\sigma_s)^2 (\sigma_s{*})^2 (\pi_x)^2 (\pi_y)^2 (\sigma_p)^2 (\pi{*}_x)^2 (\pi{*}_y)^1$ Bond order $= \dfrac{1}{2}(8-5) = 1.5$

O_2^{2-}: $(\sigma_s)^2 (\sigma_s{*})^2 (\pi_x)^2 (\pi_y)^2 (\sigma_p)^2 (\pi{*}_x)^2 (\pi{*}_y)^2$ Bond order $= \dfrac{1}{2}(8-6) = 1$

From the bond orders, we see that O_2 has the strongest, shortest bond and O_2^{2-} the longest, weakest bond. From the number of unpaired electrons, we see that O_2 and O_2^- are both magnetic, with O_2 showing the largest magnetism.

7.75 A description of bonding in a molecule starts with its Lewis structure, from which steric number, geometry, hybridization, and extension of π bonds can be deduced. ClO_3^- has $7 + 3(6) + 1 = 26$ valence electrons. Draw the bonding framework, add three lone pairs to each outer oxygen atom and place the two remaining electrons on the inner Cl atom:

The provisional Lewis structure has $FC_{Cl} = 7 - 5 = +2$, so two electron pairs are shifted to make double bonds:

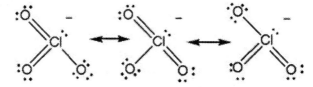

$SN_{Cl} = 4$, indicating that the molecule has tetrahedral geometry and the σ bond framework can be described using sp^3 hybrids on Cl overlapping with $2p$ orbitals from O. The resonance structures signal that the π system is extended over all four atoms, and there are two bonding π orbitals occupied by four electrons.

ClO_2^- has $7 + 2(6) + 1 = 20$ valence electrons. The provisional Lewis structure has two single bonds and two lone pairs on Cl, giving $FC_{Cl} = 7 - 2 - 2(2) = 1$, so one electron pair is shifted to make a double bond:

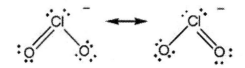

$SN_{Cl} = 4$, indicating that the electron pair geometry is tetrahedral, the molecule is bent, and the σ bond framework can be described using sp^3 hybrids on Cl overlapping with $2p$ orbitals from O. The resonance structures signal that the π system is extended over all three atoms, with one bonding π orbital occupied by two electrons.

7.77 Bond order measures the amount of electron sharing, based on Lewis structures or orbital configurations. Because bond order is based on models, it cannot be directly measured by experiments. The more electrons there are that are shared between atoms, however, the stronger and shorter the bond generally will be, so both bond energy and bond length are correlated with bond order. These bond properties can be measured experimentally, so we obtain confirmation of the bond orders predicted by bonding models by seeing whether experimental bond strengths and lengths match predicted bond orders. A higher bond order should be accompanied by increased bond strength and decreased bond length.

7.79 A description of bonding always starts with the Lewis structure for the molecule. Here is the Lewis structure for acrylonitrile. The steric numbers of the inner atoms are shown:

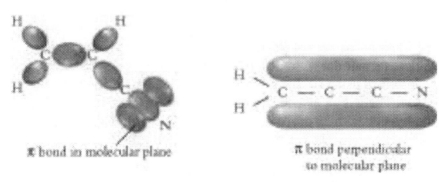

(a) The carbon atoms with SN = 3 have σ bonding that can be described using sp^2 hybrid orbitals, and the σ bonding of the carbon atom with SN = 2 can be described using sp hybrids.

(b) The entire molecule lies in the same plane. The σ bond network falls within this plane, and there is a localized π bond in this plane. Perpendicular to the molecular plane is an extended π network:

(c) There is an extended π network that includes the three C atoms and the N atom. It has four π electrons in two bonding orbitals that span the four atoms and have their electron density perpendicular to the molecular plane.

7.81 The bonding in second-row diatomic molecules can be described using the orbital energy-level diagrams shown in Figure 7-35 of your textbook. CN has nine valence electrons and the following orbital energy-level diagram for $Z \leq 7$:

There are seven bonding electrons and two antibonding electrons, for a net of five, or bond order 2.5. The least stable occupied orbital is the σ_p:

7.83 A description of bonding and geometry starts with determination of the Lewis structure. ClO_2 has $7 + 2(6) = 19$ valence electrons. The bond framework requires two pairs; six additional electrons are placed on each O atom, leaving three lone electrons on the Cl atom. $FC_{Cl} = 7 - 2 - 3 = +2$, so shift two pairs to make double bonds:

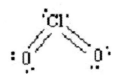

The Cl atom has SN = 4 (remember that a lone electron requires an orbital, just as an electron pair does), so its electron pair geometry is tetrahedral; the σ bonds can be described using sp^3 hybrids from Cl, and the molecule has a bond angle near 109.5°. There is an extended π system formed from *d* orbitals on Cl overlapping side-by-side with $2p$ orbitals from the two O atoms. This molecule is considered unusual because it has an odd number of electrons.

7.85 The MO diagrams for these two diatomic molecules, which appear at right, show that they are very different in stabilities. Whereas dilithium resembles dihydrogen, except that it uses $2s$ valence orbitals rather than $1s$ valence orbitals, diberyllium is like dihelium. Thus, we expect that diatomic molecules of any Group 2 element will not exist, and in fact diberyllium has never been prepared.

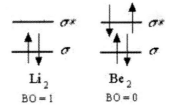

7.87 The bonding in H–*X* molecules can be represented by overlap between the 1*s* orbital of the H atom and the *n*p orbital of the *X* atom. As *n* increases, the *p* orbital becomes larger but more diffuse because the electron is spread out over a larger volume. Consequently, electron density decreases and the amount of overlap with the 1*s* orbital decreases:

In these sketches, the shading represents the electron density within the orbital. Lighter shading represents lower electron density, hence a weaker bond.

7.89 The differences in behaviour between *n* = 2 elements and their *n* = 3 counterparts from the same column of the periodic table can be explained by the difference in side-by-side π overlap for *n* = 2 and *n* = 3 orbitals. Multiple bonds involving π overlap are strong for N and O, but π overlap is very slight for P and S. Consequently, P–P and S–S diatomic molecules are minimally stabilized by π bonding, and these elements are more stable in chains with single bonds.

7.91 The standard procedure gives a Lewis structure from which hybridization, geometry, and π bonding are identified. Ketene has 2 C (4 e⁻) + 1 O (6 e⁻) + 2 H (1 e⁻) = 16 valence electrons, eight of which are used for the four bonds in the framework. Add six electrons around the outer O atom, and then place the remaining pairs on one C atom. Complete the octet around the other C atom by shifting electron pairs to form two double bonds in the final Lewis structure:

The steric numbers of the two C atoms differ, as indicated on the Lewis structure. All atoms lie in a common plane, with a σ bond framework described as shown in the sketch. The C atom with SN = 3 has trigonal planar geometry and 120° angles. The C atom with SN = 2 has linear geometry and 180° angles. There is a two-centre π bond in the plane of the molecule and an extended π bond perpendicular to the molecular plane:

7.93 Pyrrole has 26 valence electrons, resulting in the Lewis structure shown below. The N atom has SN = 4, which predicts tetrahedral geometry and sp^3 hybridization. In contrast to this, the planar ring of the molecule is evidence that the N atom is sp^2 hybridized, with trigonal planar geometry matching that of the C atoms. This allows side-by-side overlap of the remaining *p* orbital on the N atom with *p* orbitals on adjacent C atoms. The result is a five-orbital π set: two bonding, one non-bonding, and two antibonding. The Lewis structure is consistent with this composite picture, provided that the non-bonding pair of electrons is in a delocalized π_{nb} orbital rather than in a localized sp^3 hybrid orbital.

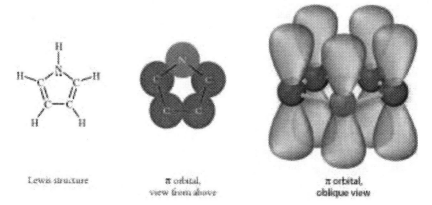

Lewis structure   π orbital,   π orbital,
                                     view from above   oblique view

7.95 We can determine steric numbers, hybridizations, and hence geometries directly from line structures (remember that C–H bonds and lone pairs are not explicitly shown in line structures). Each C or N atom with no double bonds, and each inner O atom, has SN = 4, and sp^3 hybrids can describe the bonding. C, N, and O with one double bond have SN = 3 and sp^2 hybridization. C and N with two double bonds have SN = 2 and *sp* hybridization. The steric numbers of all inner atoms are designated on the line structure below:

(a) The bonding for the two N atoms labelled 4 in the line structure can be described using *sp*³ hybrids. The N atom labelled 3 in the line structure uses *sp*² hybrids, the N atom labelled 2 in the line structure uses *sp* hybrids, and the terminal N atom uses a 2*p* orbital for its **σ** bond.

(b) The ring O atom has SN = 4, tetrahedral electron pair geometry, and a bond angle around 109°.

(c) The six C atoms labelled 4 have bonding that can be described using *sp*³ hybrids.

(d) The number of **π** bonds equals the number of double bonds in the line structure: five. Each of these bonds forms from side-by-side overlap of 2*p* atomic orbitals. An extended **π** system extends over the three adjacent N atoms. One 2*p* orbital from the N atom with SN = 2 overlaps with 2*p* orbitals on each neighbouring N atom.

7.97 There is only one isomer of 1,2-dichloroethane because the C–C bond is a **σ** bond only, and the molecule can freely rotate without disrupting the amount of overlap. When rotation occurs about the C=C double bond in 1,2-dichloroethylene, the side-by-side overlap of the **π** bond is disrupted:

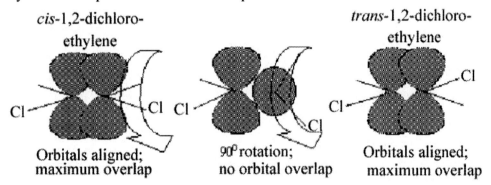

cis-1,2-dichloro-ethylene

Orbitals aligned; maximum overlap

90° rotation; no orbital overlap

trans-1,2-dichloro-ethylene

Orbitals aligned; maximum overlap

As the molecule rotates about its C=C bond, the double bond breaks, leaving only a single bond. Thus, we can estimate the energy required to convert the isomers from the difference between a C=C and a C–C bond (see Table 3-2 in your textbook): 615 − 345 = 270 kJ/mol.

7.99 A description of bonding always begins with a Lewis structure. Trigonal planar O_4 is isoelectronic with NO_3^- and CO_3^{2-} and has the same Lewis structure:

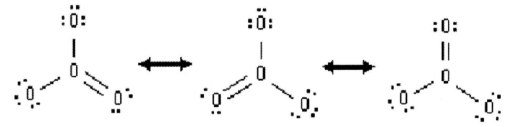

Average bond energies allow us to estimate the energy change associated with the reaction:

$$2O_2 \rightarrow O_4 \quad \Delta E_{reaction} \cong \left[2(495)\right] - \left[2(145) + 1(495)\right] = 990 - 785 = +205 \text{ kJ}$$

The positive energy change explains why the O_4 molecule is less stable than two O_2 molecules. Although the resonance structures signal an extended π system distributed over all four O atoms that would stabilize the molecule relative to this calculation, resonance energies are on the order of 50 kJ/mol, which is not enough to counteract the large positive energy change given by the bond energy calculation.

Another approach to stability is to look at formal charges in the Lewis structure: the inner O atom has FC = +2, an unfavourable situation for this highly electronegative atom. (Notice that in the stable nitrate and carbonate anions, the formal charge on the inner atom is smaller: +1 on N and 0 on C.) Thus, both energy estimates and formal charge considerations predict that O_4 is not stable.

Chapter 8 Effects of Intermolecular Forces

Solutions to Problems in Chapter 8

8.1 The condensation temperature of a substance reflects the strength of its interatomic attractions. Xe has the highest condensation temperature, indicating that it has the greatest interatomic attractions. Ar, with the lowest condensation temperature, has the least interatomic attractions. At 140 K, Xe is in a condensed phase because the kinetic energy of the atoms is not enough to overcome the interatomic attractions. Kr and Ar, however, are gases because the kinetic energy of their atoms is sufficient to overcome their interatomic attractions. At 100K, only argon atoms have kinetic energies sufficient to overcome its interatomic attractions, so it is the only gas. Xe and Kr are both in condensed phases.

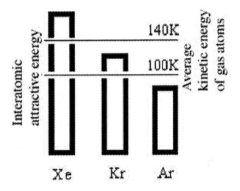

8.3 For any given substance, intermolecular attractions are constant. Their relative importance depends on two features: how close together molecules are, on average; and how much kinetic energy of motion molecules have, on average. For any given substance, molecular volume is constant, so its relative importance depends on how close together molecules are.

(a) Molecules are farther apart, making intermolecular attractions less significant, and molecular size is less significant.

(b) Molecules are closer together, so intermolecular attractions and molecular sizes are more significant.

(c) Increasing the temperature at constant pressure leads to a volume increase, so molecules are farther apart, making intermolecular attractions and molecular size less significant.

8.5 Pictures of atomic arrangements should show a high degree of order for a solid, close packing but some disorder for a liquid, and large distances between atoms for a gas:

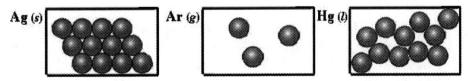

8.7 To determine deviation from ideal behaviour, calculate the pressure using the ideal gas equation and compare the calculated result with the experimental value:

$$p_{ideal} = \frac{nRT}{V} = \frac{(1.00 \text{ mol})(0.08314 \text{ L bar mol}^{-1} \text{ K}^{-1})(40.0 + 273.15 \text{ K})}{1.20 \text{ L}} = 21.7 \text{ bar}$$

$$\% \text{ Deviation} = (100\%)\left(\frac{p_{actual} - p_{ideal}}{p_{ideal}}\right) = (100\%)\left(\frac{19.7 \text{ bar} - 21.7 \text{ bar}}{21.7 \text{ bar}}\right) = -9.2\%$$

8.9 To calculate pressure using the van der Waals equation, we must know *n, V, T, a,* and *b*:

$$p = \left(\frac{nRT}{V - nb}\right) - \left(\frac{n^2 a}{V^2}\right)$$

Table 2-3 in your textbook lists values for Cl_2: $a = 657.9 \text{ kPa L}^2 \text{ mol}^{-2}$, $b = 0.0562$ L/mol

$$n_{Cl_2} = \left(\frac{m}{M}\right) = 1.25 \text{ kg}\left(\frac{10^3 \text{g}}{1 \text{ kg}}\right)\left(\frac{1 \text{ mol}}{70.906 \text{ g}}\right) = 17.63 \text{ mol}$$

$$(V - nb) = 15.0 \text{ L} - \left(17.63 \text{ mol}\left(\frac{0.0562 \text{ L}}{1 \text{ mol}}\right)\right) = 14.01 \text{ L}$$

$$p = \frac{(17.63 \text{ mol})(8.314 \text{ L kPa mol}^{-1} \text{ K}^{-1})(295 \text{ K})}{14.01 \text{ L}} - \frac{(17.63 \text{ mol})^2 (657.9 \text{ kPa L}^2 \text{ mol}^{-2})}{(15.0 \text{ L})^2}$$

$$p = 3086 \text{ kPa} - 909 \text{ kPa} = 2177 \text{ kPa} = 21.8 \text{ bar}$$

$$p_{ideal} = \frac{nRT}{V} = \frac{(17.63 \text{ mol})(0.08314 \text{ L bar mol}^{-1} \text{ K}^{-1})(295 \text{ K})}{15.0 \text{ L}} = 28.8 \text{ bar}$$

$$\% \text{ Deviation} = (100\%)\left(\frac{p_{actual} - p_{ideal}}{p_{ideal}}\right) = (100\%)\left(\frac{21.8 \text{ bar} - 28.8 \text{ bar}}{28.8 \text{ bar}}\right) = -24\%$$

8.11 Ease of liquefaction depends on the magnitude of intermolecular attractions: the larger the attractions, the easier the substance is to liquefy. All these molecules are symmetrical (tetrahedral), so the ranking depends entirely on dispersion forces. Ease of liquefaction will increase with molecular size: CH_4 (hardest to liquefy) < CF_4 < CCl_4 (easiest to liquefy).

8.13 Polarizability increases with the size of the highest occupied orbitals. Draw the molecules with larger orbitals (CCl_4) to show greater distortion than for the molecules with smaller orbitals (CH_4). These molecules are close to spherical in the absence of polarization:

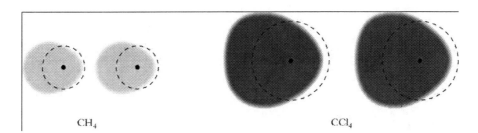

CH$_4$ CCl$_4$

8.15 Boiling point depends on the magnitude of intermolecular attractions: the larger the attractions, the higher the boiling point. The size of intermolecular attractions depends on the amount of dispersion forces (size of molecule and size of orbitals), the presence of dipolar forces, and hydrogen bonding. Among these substances, ethanol forms hydrogen bonds, so it has the highest boiling point. Propane, being smaller than *n*-pentane, has the lowest boiling point:

Propane (lowest bp) < *n*-pentane < ethanol (highest bp).

8.17 Hydrogen bonding capability requires the presence of an electronegative atom with lone pairs (O, F, or N) and an H–*X* bond to a highly electronegative atom (*X* = O, N, or F): (b and d) hydrogen bonding; (a and c) no hydrogen bonding.

8.19 Hydrogen bonding capability requires the presence of an electronegative atom with lone pairs (O, F, or N) and an H–*X* bond to a highly electronegative atom (*X* = O, N, or F):

(a) Ammonia has one N atom with a lone pair that can form a hydrogen bond to any H atom of another ammonia molecule:

(b) Two types of hydrogen bonds are possible, between an N atom of NH$_3$ and an H atom of H$_2$O, and between an O atom of H$_2$O and an H atom of NH$_3$.

$$H \quad\quad H \quad\quad\quad\quad H \quad\quad H$$
$$:\ddot{O}—H\text{-----}:N—H \quad\quad :N—H\text{-----}:\ddot{O}—H$$
$$H \quad\quad\quad\quad\quad H$$

8.21 The viscosities of a set of similar substances increases with increasing chain length, because of increased dispersion forces (stronger intermolecular attractions make it harder for molecules to slide by one another) and increased "tangling" (longer chains are more entangled). Thus, the order of increasing viscosity is pentane (C_5) < gasoline (C_8) < fuel oil (C_{12}).

8.23 An oxide surface has the ability to form hydrogen bonds to water, so the meniscus of water in aluminum tubing will be concave, just like inside glass tubing. The non-metallic nature of the oxide layer means that there will not be strong liquid-surface forces for mercury inside aluminum tubing, so the meniscus will be convex, just like inside glass tubing.

8.25 Paper towels contain cellulose (polysaccharide), which forms good hydrogen-bonding interactions with water, so aqueous solutions wet paper towels well and are absorbed effectively. There are no strong intermolecular interactions between an oil and the fibres of paper towels, so salad dressing is not absorbed effectively.

8.27 The vapour pressure of a substance at any given temperature is determined by the strength of intermolecular forces: the stronger the forces, the lower the vapour pressure. (a) Benzene has a higher vapour pressure, because chlorobenzene has a polar C–Cl bond that gives it a dipole moment and generates dipolar intermolecular forces; (b) hexane has a higher vapour pressure, because there are hydrogen bonding interactions for 1-hexanol that increase its intermolecular forces; (c) heptane has a higher vapour pressure, because octane, being the larger molecule, has stronger dispersion forces.

8.29 The position of an element in the periodic table, the chemical formula, and a knowledge of polyatomic ions all help in identifying types of solids: Sn, a metal, is a metallic solid; S_8 has a specific molecular formula, so it is a molecular solid; Se is not a metal, so it is a network solid; SiO_2 is described in your textbook as a network solid; and Na_2SO_4 contains Na^+ cations and SO_4^{2-} polyatomic anions, so it is an ionic solid.

8.31 Consult your textbook for information about the different types of solids.

(a) The bonding in metals comes from extended networks of delocalized electrons, whereas the bonding in network solids includes many individual covalent bonds.

(b) Metals conduct electricity, are malleable and ductile, and are shiny. Network solids are non-conductors or semiconductors, are brittle, and often have dull appearances.

8.33 The position of an element in the periodic table, the chemical formula, and knowledge of polyatomic ions all help in identifying types of solids:

(a) Br_2 is a discrete neutral molecule. It forms a molecular solid.

(b) KBr contains cations and anions. It forms an ionic solid.

(c) Ba is an alkaline earth metal. It forms a metallic solid.

(d) SiO_2 is described in your textbook as a network solid.

(e) CO_2 is a covalent molecule. It forms a molecular solid.

8.35 A body-centred cubic unit cell has 1/8 of an atom at each corner plus a whole atom at the centre, for a total of two atoms. The mass of the unit cell is therefore:

$$mass = 2\left(\frac{183.84\,g\,/\,mol}{6.02\times10^{23}\,mol^{-1}}\right) = 6.11\times10^{-22}\,g$$

8.37 The volume of the unit cell is V:

$$V = l^3 = (352.4\times10^{-12}\,m)^3 = (352.4\times10^{-10}\,cm)^3 = 4.38\times10^{-23}\,cm^3$$

The mass of the unit cell is the mass of four nickel atoms:

$$mass = 4\left(\frac{58.69\,g\,/\,mol}{6.02\times10^{23}\,mol^{-1}}\right) = 3.90\times10^{-22}\,g$$

The density is therefore:

$$\rho = \frac{mass}{volume} = \frac{3.90\times10^{-22}\,g}{4.38\times10^{-23}\,cm^3} = 8.90\,g\,/\,cm^3$$

8.39 The mass of the unit cell is the mass of four gold atoms:

$$mass = 4\left(\frac{196.97\,g\,/\,mol}{6.02\times10^{23}\,mol^{-1}}\right) = 1.31\times10^{-21}\,g$$

The volume of the unit cell is:

$$V = \frac{mass}{density} = \frac{1.31\times10^{-21}\,g}{19.3\,g\,/\,cm^3} = 6.79\times10^{-23}\,cm^3$$

The edge length of the unit cell is:

$$l = V^{1/3} = (6.79\times10^{-23}\,cm^3)^{1/3} = 4.08\times10^{-8}\,cm$$

In an FCC structure, the diagonal across a face is four times the atomic radius:

$$l^2 + l^2 = (4r)^2$$
$$2l^2 = 16r^2$$
$$l^2 = 8r^2$$
$$r = \left(\frac{l^2}{8}\right)^{1/2} = \left(\frac{(4.08 \times 10^{-8}\,\text{cm})^2}{8}\right)^{1/2} = 1.44 \times 10^{-8}\,\text{cm} = 144\,\text{pm}$$

8.41 The density of a solid depends on the nature of the elements it contains and on how compact the bonding network is. A structure with a more open bonding pattern has a lower density than one with a more compact bonding pattern. Graphite uses sp^2 hybrids and has an extended *p*-bonding network, making it planar with a relatively open structure between successive planes. Diamond uses sp^3 hybrids and has a compact network of bonds:

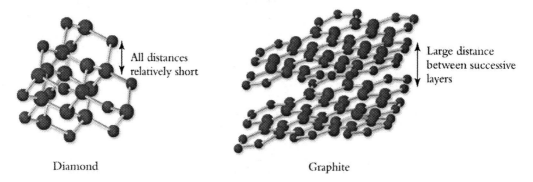

Diamond   Graphite

8.43 To determine a chemical formula from a unit cell, count the number of atoms of each element contained within the unit cell. Atoms on a face count $\frac{1}{2}$ (each is part of two unit cells), those on an edge count $\frac{1}{4}$ (each is part of four unit cells), and those at a corner count $\frac{1}{8}$ (each is part of eight unit cells). There is one Ti atom within the cell.

Ca: $\frac{1}{8}$ (8 corner atoms) = 1 atom

O: $\frac{1}{2}$ (6 face atoms) = 3 atoms

The chemical formula is $CaTiO_3$.

8.45 The problem states that carborundum is diamond-like. Diamond is composed entirely of tetrahedral carbon atoms (SN = 4), and the empirical formula of carborundum, SiC, indicates that alternate C atoms should be replaced with Si atoms. Each C atom bonds to four Si atoms and each Si atom bonds to four C atoms:

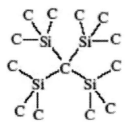

8.47 A unit cell has to be a shape that can be used over and over to build the entire pattern. The upper screen shows a unit cell containing two complete fish, and the lower screen shows a unit cell that contains no complete fish, but notice that it contains $\frac{1}{2}$ (2 edge) + $\frac{1}{4}$ (4 corner) = 2 fish overall, just like the other unit cell:

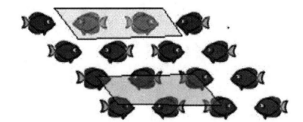

8.49 First determine the number of sulphide ions in the unit cell. A face-centred ionic cube has six face ions and eight corner ions, giving $\frac{1}{2}$ (6) + $\frac{1}{8}$ (8) = 4 sulphide ions.

The sulphide anion has a charge of –2 and the lithium cation has a charge of +1, so we need twice as many lithium cations as sulphide anions in the unit cell. There are eight lithium cations in the interior of the unit cell.

8.51 The heats of phase changes indicate the strengths of intermolecular forces; the larger the intermolecular forces, the larger the heat of the phase change.

(a) Methane has a lower heat of vapourization than ethane because, being a smaller molecule, it has smaller dispersion forces.

(b) Ethanol has a significantly higher heat of vapourization than diethyl ether because of strong hydrogen bonding between ethanol molecules.

(c) Argon (18 electrons) has a higher heat of fusion than methane (10 electrons) because it has a higher polarizability due to its larger number of electrons.

8.53 A phase diagram is a map of the stable phases at different p and T, and phase boundary lines meet at the triple point and cross the horizontal line corresponding to $p = 1.01$ bar at the normal freezing and boiling points:

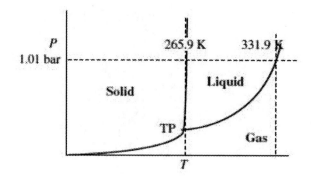

8.55 Phase diagrams provide "road maps," allowing us to determine what phase changes occur as temperature and pressure vary in particular ways:

(a) At $T = 400$ K, $p = 1.00$ bar, Br_2 is a gas. As it cools under a pressure of 1.00 bar, it liquefies at 331.9 K and solidifies at 265.9 K.

(b) At $T = 265.8$ K, $p = 1.00 \times 10^{-3}$ bar, Br_2 is a gas. Compressing it at this temperature causes it to liquefy at about $p = 6 \times 10^{-2}$ bar and solidify at about 0.5 bar.

(c) $p = 2.00 \times 10^{-2}$ bar is below the triple point of Br_2. Heating solid Br_2 at this pressure from 250 to 400 K causes sublimation to the vapour at about 265 K.

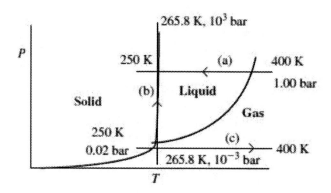

8.57 Summarize the data and the process in a flow chart:

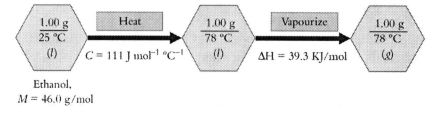

The heat required is $nC\Delta T$ for the heating and $n\Delta H$ for the vapourization.

The amount of ethanol is $n = \dfrac{1.00\ \text{g}}{46.0\ \text{g/mol}} = 2.17 \times 10^{-2}\ \text{mol}$

Substitute the values and do the calculation:

$q = (2.17 \times 10^{-2}\ \text{mol})[(111\ \text{J mol}^{-1}\ ^{\circ}\text{C}^{-1})(78 - 25\ ^{\circ}\text{C}) + (39.3\ \text{kJ/mol})(10^3\ \text{J/kJ})] = 980.\ \text{J}$

8.59 All substances are subject to dispersion forces. In addition, look for polarity and hydrogen-bonding ability:

(a) NH_3 is polar and forms hydrogen bonds in addition to its dispersion interactions.

(b) $CHCl_3$ is a polar molecule (has dipole-dipole intermolecular forces of attraction) in addition to having dispersion interactions.

(c) CCl_4 is symmetrical (tetrahedral), so it has only dispersion interactions.

(d) CO_2 is symmetrical (linear), so it has only dispersion interactions.

8.61 Face-centred cubic crystals are close-packed, each atom having 12 nearest neighbours. In body-centred cubic crystals, each atom has eight nearest neighbours. Thus, the face-centred cubic structure contains more atoms in a given volume and is denser.

8.63 All substances are subject to dispersion forces. In addition, look for polarity and hydrogen-bonding ability:

(a) CH_3OH can hydrogen bond, so it boils at a higher temperature than CH_3OCH_3.

(b) SiO_2 is a network solid with covalent bonds that must break for it to boil, so it has a higher boiling point than SO_2.

(c) HF forms strong hydrogen bonds, so it boils at a higher temperature than HCl.

(d) I_2 has larger orbitals, giving it higher polarizability, larger dispersion forces, and a higher boiling point than Br_2.

8.65 Consult the phase diagram for N_2 in Figure 8-37 of your textbook to determine the temperatures and pressures at which phase changes occur:

(a) A constant temperature process is a vertical line on the phase diagram. $T = 70$ K is in between the triple point (0.126 bar, 63 K) and normal boiling point (1.01 bar, 77 K) of nitrogen. At 70 K, 1.01 bar, nitrogen is liquid. When the pressure falls below about 0.5 bar, the liquid boils, and at 70 K, 0.1 bar, the substance is gaseous.

(b) A constant temperature process is a vertical line on the phase diagram. $T = 298$ K is within the gaseous region at all pressures, so compression from 1.00 bar to 50.0 bar leaves the substance gaseous.

(c) A constant pressure process is a horizontal line on the phase diagram, and 1.01

bar represents "normal" pressure, shown on the figure as a dotted line. When the temperature of the gas reaches 77 K, the gas liquefies, and when the temperature reaches 63 K, the liquid solidifies. At 50 K, N_2 is solid.

8.67 When two substances with otherwise similar structures have different boiling points, look for differences in polarity and/or hydrogen-bonding behaviour.

The Lewis structures of the two isomers, shown with arrows representing bond polarities, indicate why the isomers have different boiling points:

cis
dipole moment
bp = 60 °C

trans
no dipole moment
bp = 47 °C

In the *trans* isomer, the two polar bonds oppose each other and cancel, making this a non-polar compound. In the *cis* isomer, the two polar bonds add, making this a polar compound. Thus, *cis* 1,2-dichloroethylene has the higher boiling point.

8.69 Deviations from ideal gas behaviour can be predicted based on the strengths of intermolecular interactions. The larger the intermolecular interactions, the greater will be the deviations from ideality. Both pairs of substances are non-polar, so dispersion interactions dominate their intermolecular forces. The substance with more electrons will have larger dispersion forces and deviate more from ideality: Cl_2.

8.71 The pressure scale of Figure 8-38 is in kbar, so a pressure of 1.01 bar (atmospheric pressure) essentially lies along the *x*-axis. There are four phase transitions as silica is heated. At room temperature, the stable phase is α quartz, which converts to β quartz at a temperature of around 600 °C. At around 850 °C, the solid becomes tridymite, and just below 1500 °C, this converts to cristobalite. At around 1700 °C, cristobalite melts to give liquid silica.

8.73 Refer to Figures 8-24 and 8-27 in your textbook for views of body-centred cubic and hexagonal close-packed crystals. The more closely packed form will be stable at higher pressures, so the body-centred cubic structure is the low temperature, low pressure form.

8.75 Solids are molecular, metallic, network, or ionic: (a) ZrO_2 (mp = 2677 °C) is a network solid, as indicated by its high melting point; (b) Zr is a transition metal, so the element is a metallic solid; and (c) $ZrSiO_4$ contains the silicate anion, so this solid is ionic.

8.77 Phase diagrams provide "road maps," allowing us to determine what phase changes occur as temperature and pressure vary in particular ways. (a) For rhombic sulphur to melt, it must be heated above 153 °C at a pressure that is greater than 1439 bar.

(b) For rhombic sulphur to change to monoclinic sulphur, it must be heated to a temperature that depends on the pressure, ranging from 95.3 °C at 5.2×10^{-6} bar to 153 °C at 1439 bar. (c) Rhombic sulphur only sublimes when the pressure is less than 5.2×10^{-6} bar.

8.79 (a) There is a rhenium atom at each corner of the unit cell cube but nowhere else, so the rhenium lattice is simple cubic. (b) Each corner of a cube is shared by eight cubes, and there are eight corners, so there are $\frac{1}{8}(8) = 1$ Re atom per unit cell. Each edge of a cube is shared by four cubes, and there are 12 edges, so there are $\frac{1}{4}(12) = 3$ O atoms per unit cell. The chemical formula is ReO_3.

8.81 The melting point of a metal is determined by the strength of interatomic forces. The smaller those forces, the lower the melting point of the metal. The low melting points of Hg, Cs, and Ga indicate that these metals have relatively small interatomic forces.

8.83 (a) When sulphur is heated at any constant pressure between 3.2×10^{-5} bar and 1439 bar, there is a temperature at which each phase is stable. Below 95.3 °C, the rhombic phase is stable. At a pressure-dependent temperature between 95.3 °C and 153 °C, the monoclinic phase becomes stable. At a pressure-dependent temperature between 115.2 °C and 153 °C, monoclinic sulphur melts and the liquid becomes stable. At a still higher temperature that depends strongly on the pressure, the vapour phase is stable.

(b) At any temperature between 115.2 °C and 153 °C, there is a pressure at which each phase is stable. Below 3.2×10^{-5} bar, the vapour is stable. Above this pressure, the liquid is stable up to a pressure that depends on the exact temperature, but at higher pressure, the monoclinic solid phase is stable. At extremely high pressure, the rhombic phase becomes stable.

8.85 The magnitudes of dispersion forces depend on how extended the valence electrons are. In a molecule, valence electrons are spread over a larger volume because they are shared between atoms. Sketches of the electron clouds of H_2 and He illustrate this feature:

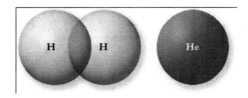

The more extended valence electron cloud of H_2 leads to stronger dispersion forces than for He, giving H_2 the higher boiling point. Likewise for Ne and CH_4. In CH_4, the valence electrons are spread over a larger volume because they are shared between atoms, giving CH_4 the higher boiling point.

8.87 See Figure 8-25 in your textbook for an example of the face-centred cube. Here is a rendering of the unit cell. The ion in the centre of the front face of the unit cell, shown in black, can be used to determine how many nearest neighbours each ion has. The thick lines connect to four ions in the same face, one in the centre of the unit cell and one in front of the front face (ion not shown). Thus, each ion has six nearest neighbours of opposite charge:

8.89 To answer these questions, we need to make reasonable extrapolations of the phase diagram for aluminum silicate. (a) The phase diagram extends to 900 °C without showing a liquid phase, so we can say that the melting point is higher than 900 °C; how much higher is not known. (b) The only stable solid phase of aluminum silicate at high temperature is sillimanite, so at all pressures from 0–10 kbar, the liquid will solidify to this form of the solid. (c) The kyanite–sillimanite phase boundary is moving toward higher temperature as the pressure increases, so there may be a sufficiently high pressure where this is the phase that solidifies from liquid aluminum silicate. (d) The stable phase at highest pressure is the densest phase; this is kyanite. The stable phase at lowest pressure is the least dense; this is andalusite.

8.91 This problem asks us to construct a graph that summarizes how the temperature changes as a liquid sample is cooled to below its freezing point. As energy is removed from the sample, the temperature will drop until the liquid begins to freeze. Then, the temperature will remain fixed until all the water has frozen. Finally, the temperature of the ice will decrease until the final temperature is reached:

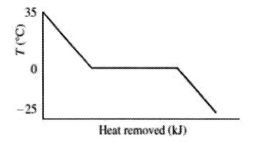

8.93 (a)

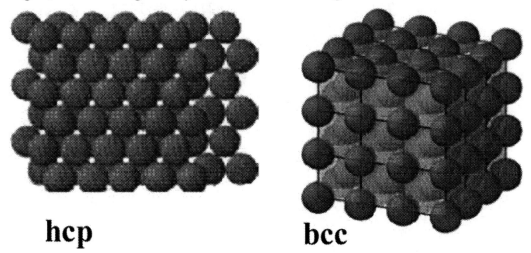

(b) The fraction of paired molecules decreases as temperature rises, because energy added to the system breaks some of the hydrogen bonds.

8.95 Hexagonal closed-packed (hcp) and body-centred cubic (bcc) structures are similar in that both contain ABAB alternating layers of spheres. They differ in that the hcp structure contains hexagonally packed layers, whereas the bcc contains cubic packed layers. Each sphere in an hcp structure has 12 nearest neighbours, whereas each sphere in a bcc structure has eight nearest neighbours. The hcp structure is somewhat more compact than the cubic structure. The geometry of hcp contains 60° angles, whereas the geometry of bcc contains 90° angles:

hcp   **bcc**

8.97 To determine the mass of methane required, first determine the heat required to melt the ice and heat the resulting liquid to the boiling point. The calculation can be done in two steps: melting the ice, and then heating from 0.0 °C to 100.0 °C. The total heat required is the sum of the heats for the two steps: $q = n\,\Delta H_{fus} + nC\Delta T$

$$n_{water} = \frac{m}{M} = \frac{75.0\ \text{g}}{18.02\ \text{g/mol}} = 4.162\ \text{mol}$$

$$q_{water} = (4.162\ \text{mol})(6.01\ \text{kJ/mol})$$
$$+ (4.162\ \text{mol})(75.291\ \text{J mol}^{-1}\ °\text{C}^{-1})(10^{-3}\ \text{kJ/J})(100.0\ °\text{C}) = 56.3\ \text{kJ}$$

$$q_{methane} = -q_{water} = -56.3\ \text{kJ}$$

Next, calculate the heat of combustion of methane using Equation 3-12 and data from Appendix D. We need the balanced combustion reaction.

In a combustion reaction, the products are CO_2 (g) and H_2O (g):

$$CH_4(g) + O_2(g) \rightarrow CO_2(g) + H_2O(g) \quad \text{(unbalanced)}$$

Follow standard procedures to balance the equation. C is balanced. Give H_2O a coefficient of 2 to balance H:

$$CH_4(g) + O_2(g) \rightarrow CO_2(g) + 2\,H_2O(g)$$
$$1\,C + 4\,H + 2\,O \rightarrow 1\,C + 4\,H + 4\,O$$

Give O_2 a coefficient of 2 to balance O:

$$CH_4(g) + 2\,O_2(g) \rightarrow CO_2(g) + 2\,H_2O(g)$$

Calculate the standard enthalpy change from Equation 3-12 using standard enthalpies of formation, which can be found in Appendix D of your textbook:

$$\Delta H^\circ_{\text{reaction}} = \Sigma \text{coeff}_{\text{products}} \Delta H^\circ_{\text{f, products}} - \Sigma \text{coeff}_{\text{reactants}} \Delta H^\circ_{\text{f, reactants}}$$

$$\Delta H^\circ_{\text{combustion}} = \left[2\text{ mol}(-241.83\text{ kJ/mol}) + 1(-393.5\text{ kJ/mol}) \right]$$

$$- \left[1\text{ mol}(-74.6\text{ kJ/mol} + 2\text{ mol}(0\text{ kJ/mol}) \right] = -803\text{ kJ}$$

Use the molar mass of methane to determine the mass required:

$$m = -56.3\text{ kJ} \left(\frac{1\text{ mol}}{-803\text{ kJ}} \right) \left(\frac{16.043\text{ g}}{1\text{ mol}} \right) = 1.12\text{ g}$$

8.99 To prepare an accurate heating curve, we need to determine the heats required to melt the bromine, heat it to boiling, and vapourize it. Then we use the heating rate, 25 J/s, to determine how long each process takes:

$$n_{\text{Br}_2} = \frac{50.0\text{ g}}{159.8\text{ g/mol}} = 0.313\text{ mol}$$

$$\Delta H_{\text{melt}} = n\,\Delta H_{\text{fusion}} = (0.313\text{ mol})(10.8\text{ kJ/mol}) = 3.38\text{ kJ}$$

$$\Delta H_{\text{vap}} = n\,\Delta H_{\text{vap}} = (0.313\text{ mol})(30.5\text{ kJ/mol}) = 9.54\text{ kJ}$$

$$\Delta T = (353\text{ K} - 266\text{ K}) = 87\text{ K} = 87\text{ °C}$$

$$\Delta H_{\text{heating}} = nC\Delta T = (0.313\text{ mol})(76\text{ J mol}^{-1}\text{ °C}^{-1})(87\text{ °C})(10^{-3}\text{ kJ/J}) = 2.07\text{ kJ}$$

The heating rate is $(25\text{ J/s})(10^{-3}\text{ kJ/J}) = 2.5 \times 10^{-2}\text{ kJ/s}$.

Divide each amount of heat by the heating rate to obtain the time required for each process:

$$t_{\text{melt}} = \frac{3.38\text{ kJ}}{2.5 \times 10^{-2}\text{ kJ/s}} = 135\text{ s}$$

$$t_{\text{vap}} = \frac{9.54\text{ kJ}}{2.5 \times 10^{-2}\text{ kJ}} = 382\text{ s}$$

$$t_{\text{heating}} = \frac{2.07\text{ kJ}}{2.5 \times 10^{-2}\text{ kJ/s}} = 92\text{ s}$$

Now we can construct the heating curve. The temperature remains fixed at 266 K for 135 s, as the bromine melts. Then the temperature increases linearly for 92 s, until the liquid boils at 353 K. Then the temperature remains fixed at 353 K for 382 s, as the bromine vapourizes:

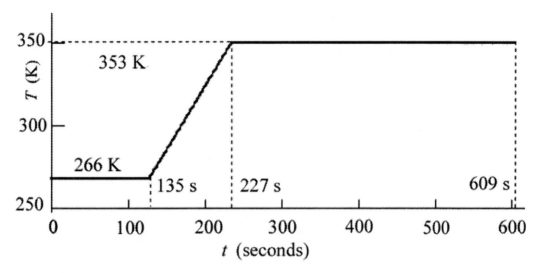

8.101 The unit cell has Cl⁻ ions at each corner, slightly pushed outwards by the Cs⁺ ion at the centre of the unit cell. A diagonal through the body of the unit cell therefore has a length equal to the diameter (= 2 x radius) of a cesium ion plus twice the diameter) of a chloride ion, i.e. 2(169) + 2(181) = 700 pm. Referring to the diagram below, the unit cell is a cube with edge length l. The diagonal d on the bottom can be found by Pythagoras' theorem:

$d^2 = 700^2 - l^2$

but also, $d^2 = l^2 + l^2$

Thus, $l^2 + l^2 = 700^2 - l^2$

$3l^2 = 700^2$

solving, l = 404.1 pm.

Now we can calculate the volume of the unit cell:

V = l³ = (404.1 x 10⁻¹⁰ cm)³ = 6.60 x 10⁻²³ cm³

The mass of the unit cell is that of one chloride ion plus one cesium ion:

$$m = \frac{35.45 \text{ g mol}^{-1}}{6.02 \times 10^{23} \text{mol}^{-1}} + \frac{132.91 \text{ g mol}^{-1}}{6.02 \times 10^{23} \text{mol}^{-1}} = 2.80 \times 10^{-22} \text{g}$$

and the density is therefore:

$$\rho = \frac{m}{V} = \frac{2.80 \times 10^{-22}\,g}{6.60 \times 10^{-23}\,cm^3} = 4.24\ g\ cm^{-3}$$

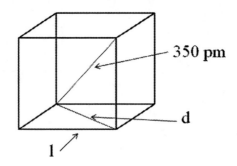

Chapter 9 Properties of Solutions

Solutions to Problems in Chapter 9

9.1 The solvent is the component that determines the phase of the solution; usually it is the component present in the largest amount. Any other components are solutes. (a) Water (normally a liquid) is the solvent and carbon dioxide (normally a gas) is the primary solute; (b) water (normally a liquid) is the solvent and ethanol (normally a liquid) is the primary solute; (c) molecular nitrogen (normally a gas) is the solvent, molecular oxygen (normally a gas) and water (normally a liquid) are the primary solutes.

9.3 Mole fractions are determined by calculating the number of moles of each component of the solution and then using the equation for *X*:

$$X_A = \frac{n_A}{n_{total}}$$

For CH_3OH: $n = \dfrac{14.5 \text{ g}}{32.04 \text{ g/mol}} = 0.4526$ mol

For H_2O: $n = \dfrac{101 \text{ g}}{18.01 \text{ g/mol}} = 5.608$ mol

$n_{total} = 5.608 + 0.4526 = 6.061$ mol

$X_{methanol} = \dfrac{0.4526 \text{ mol}}{6.061 \text{ mol}} = 0.0747$ $X_{water} = 1 - 0.0747 = 0.925$

9.5 To determine the concentrations that are asked for, we need to know the mass and number of moles of each component. Convert volume into mass using the density equation:

$m = \rho V = (0.818 \text{ g/mL})(85.0 \text{ mL}) = 69.53$ g

A table is a useful way to summarize the information that we need:

Substance	Formula	*M*	*m*	*n*
Acetone	C_3H_6O	58.08 g/mol	69.53 g	1.197 mol
Maleic acid	$C_4H_4O_4$	116.07 g/mol	1.521 g	1.31×10^{-2} mol

Use this information to calculate each concentration:

Molality: $b = \dfrac{n_{\text{solute}}}{m_{\text{solvent}} \text{ (in kg)}} = \dfrac{1.31 \times 10^{-2} \text{ mol}}{(69.53 \text{ g})(10^{-3} \text{ kg/g})} = 1.88 \times 10^{-1} \text{ mol/kg}$

Mole fraction: $X = \dfrac{n_{\text{solute}}}{n_{\text{total}}} = \dfrac{1.31 \times 10^{-2} \text{ mol}}{(1.197 + 0.0131 \text{ mol})} = 1.08 \times 10^{-2}$

Mass percent: $\% = (100\%) \dfrac{m_{\text{solute}}}{m_{\text{total}}} = (100\%) \dfrac{1.521 \text{ g}}{(1.521 + 69.53 \text{ g})} = 2.14\%$

9.7 To determine the various concentration values for a solution whose mass percent is known, take exactly 100 g of the solution as a convenient amount to work with. Determine the masses of each substance in this amount of solution, convert to moles, and then apply the appropriate equations:

For H_2O_2: $n = \dfrac{30. \text{ g}}{34.01 \text{ g/mol}} = 0.88 \text{ mol}$

For H_2O: $n = \dfrac{70. \text{ g}}{18.01 \text{ g/mol}} = 3.89 \text{ mol}$

$n_{\text{total}} = 0.88 + 3.89 = 4.77 \text{ mol}$

$X_{\text{hydrogen peroxide}} = \dfrac{0.88 \text{ mol}}{4.77 \text{ mol}} = 0.18$ $X_{\text{water}} = 1 - 0.18 = 0.82$

To determine the molarity, we need the volume of this amount of solution:

$\rho = \dfrac{m}{V}$ so $V = \dfrac{m}{\rho} = \dfrac{100 \text{ g}}{1.11 \text{ g/mL}} = 90.1 \text{ mL}$

$c = \dfrac{n}{V} = \dfrac{0.88 \text{ mol}}{(90.1 \text{ mL})(10^{-3} \text{ L/mL})} = 9.7 \text{ M}$

To determine the molality, we need the mass of solvent in kg:

$b = \dfrac{n_{\text{solute}}}{m_{\text{solvent}}} = \dfrac{0.88 \text{ mol}}{(70 \text{ g})(10^{-3} \text{ kg/g})} = 13 \text{ mol/kg}$

9.9 To determine other concentration measures for a solution whose molarity is known, work with exactly 1 L of solution and use the density to determine the mass of the solution and of its solvent.

$\rho = \dfrac{m}{V}$ $m = \rho V = (0.9811 \text{ g/mL})(1000 \text{ mL/L})(1 \text{ L}) = 981.1 \text{ g}$

Find the mass of ammonia in 1 L of solution from the molarity and molar mass:

$$m_{\text{ammonia}} = nM = (2.30 \text{ mol/L})(1 \text{ L})(17.03 \text{ g/mol}) = 39.17 \text{ g}$$

$$m_{\text{water}} = 981.1 \text{ g} - 39.17 \text{ g} = 941.9 \text{ g}$$

$$n_{\text{water}} = \frac{941.9 \text{ g}}{18.01 \text{ g/mol}} = 52.30 \text{ mol}$$

$$X_{\text{ammonia}} = \frac{2.30 \text{ mol}}{(2.30 + 52.30) \text{ mol}} = 0.042 \qquad X_{\text{water}} = 1 - 0.042 = 0.958$$

$$\text{Mass fraction ammonia} = \frac{\text{mass}}{\text{total mass}} = \frac{39.17 \text{ g}}{981.1 \text{ g}} = 0.0399$$

$$b = \frac{n_{\text{solute}}}{m_{\text{solvent}}} = \frac{2.30 \text{ mol}}{(941.9 \text{ g})(10^{-3} \text{ kg/g})} = 2.44 \text{ mol/kg}$$

9.11 Work with moles to determine the amounts of solute and solvent needed when mole fraction is the target.

First determine the number of moles of methanol:

$$n_{\text{methanol}} = \frac{25.0 \text{ g}}{32.04 \text{ g/mol}} = 0.780 \text{ mol}$$

(a) Substitute into the mole fraction expression to determine moles of water:

$$X_A = \frac{n_{\text{methanol}}}{n_{\text{methanol}} + n_{\text{water}}} = \frac{0.780 \text{ mol}}{0.780 \text{ mol} + n_{\text{water}}} = 0.105$$

$$0.780 \text{ mol} = 0.105 (0.780 \text{ mol} + n_{\text{water}})$$

Divide both sides by 0.105: $7.43 \text{ mol} = 0.780 \text{ mol} + n_{\text{water}}$

$$n_{\text{water}} = 7.43 \text{ mol} - 0.780 \text{ mol} = 6.65 \text{ mol}$$

$$m_{\text{water}} = (6.65 \text{ mol})(18.01 \text{ g/mol}) = 120. \text{ g}$$

(b) Use the results of part (a) and the equation for molality to calculate *b*:

$$b = \frac{n_{\text{solute}}}{m_{\text{solvent}}} = \frac{0.780 \text{ mol}}{(120. \text{ g})(10^{-3} \text{ kg/g})} = 6.50 \text{ mol/kg}$$

9.13 The immiscibility of liquids can be explained using the concept that "like dissolves like." A molecule of salad dressing consists almost entirely of hydrocarbon segments, which have dispersion-type intermolecular forces but low polarity and no hydrogen-bonding capability. A solution of acetic acid in water has high polarity and a large hydrogen-bonding capability. Thus, these two liquids are not like each other and do not mix. In terms of balance of forces, hydrogen bonds would have to be broken for the two liquids to mix, making mixing energetically unfavourable.

173

9.15 The types of solids that dissolve in a solvent can be predicted using the concept that "like dissolves like." Liquid ammonia is similar to liquid water: It is a polar molecule with a strong hydrogen-bonding capability. Thus, liquid ammonia should dissolve the same sorts of substances as water: ionic salts and polar organic molecules such as alcohols.

9.17 Miscible pairs are similar in the types of intermolecular forces that they experience. H_2O has strong hydrogen bonding, and so does CH_3OH, so this pair is miscible. C_8H_{18} and CCl_4 both have only dispersion forces, so this pair is miscible.

All the other possible pairs match unlike substances, so none of the other pairs is miscible.

9.19 To match a solute to its most appropriate solvent, use the "like dissolves like" principle:

Solute	I_2	NaCl	Au	Paraffin
Solvent	CCl_4	Water	Hg	*n*-Octane

CCl_4 is most like I_2; NaCl is a salt, so it only dissolves in water. Au, a metal, requires another metal, Hg; and paraffin, a hydrocarbon, is best matched by *n*-octane, another hydrocarbon.

9.21 Copper atoms and nickel atoms are nearly the same size, so this alloy will be substitutional. Your picture should show nine Cu atoms for every one Ni atom. Copper forms face-centred cubes, which are most easily shown as hexagonal arrays:

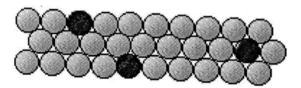

9.23 Methane differs from water in two major respects: it is non-polar, and it does not form hydrogen bonds. Consequently, methane is a much poorer solvent than water. The lack of hydrogen-bonding capability makes it very unlikely that methane could support life, because the complex interactions needed for life processes depend heavily on hydrogen-bonding interactions.

9.25 Your picture should look similar to the one in Figure 9-5 in your textbook, with K^+ cations in place of Na^+ cations:

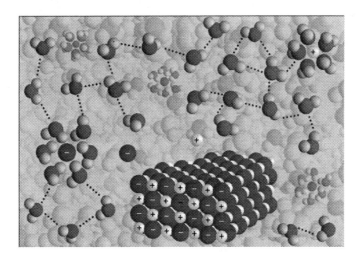

9.27 In order for solute molecules to escape from a solution, they must overcome the intermolecular forces of attraction holding them in solution. To accomplish this requires energy, which is supplied by the solvent. Thus, the total energy of the remaining solution decreases, and the temperature falls.

9.29 Table 9-2 indicates that the solubility of O_2 in water decreases as the temperature rises from 0 °C to 25 °C. Use Henry's law to calculate the concentration at each temperature. Note that although equilibrium constants are dimensionless, each concentration must be in standard units (M for solutes, bar for gases):

$$\frac{[O_{2(aq)}]_{eq}}{(p_{O_2})_{eq}} = K_H \qquad [O_{2(aq)}]_{eq} = (p_{O_2})_{eq} K_H$$

The atmosphere is 21% O_2, so the partial pressure of O_2 is 0.20 bar:

At 0 °C, $[O_{2(aq)}]_{eq}$= (0.20 bar)(2.5 × 10^{-3}) = 5.0 × 10^{-4} M

At 25 °C, $[O_{2(aq)}]_{eq}$= (0.20 bar)(1.3 × 10^{-3}) = 2.6 × 10^{-4} M

The percentage change is: $(100\ \%)\dfrac{(5.0 \times 10^{-4}\ \text{M}) - (2.6 \times 10^{-4}\ \text{M})}{5.0 \times 10^{-4}\ \text{M}} = 48\%$

9.31 Use the information about equilibrium pressure to determine the Henry's law constant. Then use the Henry's law constant to determine the equilibrium pressure for any other concentration. Note that although equilibrium constants are dimensionless, each concentration must be in standard units (M for solutes, bar for gases):

$$[\text{gas }(aq)]_{\text{eq}} = K_{\text{H}}(p_{\text{gas}})_{\text{eq}} \qquad \text{so} \qquad K_{\text{H}} = \frac{[\text{gas }(aq)]_{\text{eq}}}{(p_{\text{gas}})_{\text{eq}}}$$

$$K_{\text{H}} = \frac{0.500 \text{ M}}{(6.8 \text{ Torr})(1 \text{ bar}/760 \text{ Torr})} = 55.88 \text{ M/bar}$$

$$(p_{\text{gas}})_{\text{eq}} = \frac{[\text{gas }(aq)]_{\text{eq}}}{K_{\text{H}}} = \left(\frac{2.5 \text{ M}}{55.88 \text{ M/bar}}\right)(760 \text{ Torr/bar}) = 34 \text{ Torr}$$

$$(p_{\text{gas}})_{\text{eq}} = 0.84 \text{ kPa} = 8.4 \times 10^{-3} \text{ bar}$$

$$K_{\text{H}} = \frac{0.5 \text{ M}}{8.4 \times 10^{-3} \text{ bar}} = 59.53 \text{ M/bar}$$

$$(p_{\text{gas}})_{\text{eq}} = \frac{2.5 \text{ M}}{59.53 \text{ M/bar}} = 0.042 \text{ bar} = 4.2 \text{ kPa}$$

9.33

$$p_{\text{total}} = X_{\text{water}}\, p_{\text{vap,water}} + X_{\text{acetone}}\, p_{\text{vap,acetone}}$$
$$= (0.900)(3.17 \text{ kPa}) + (0.100)(30.17 \text{ kPa})$$
$$= 5.87 \text{ kPa}$$

9.35

$$p_{\text{total}} = X_{\text{water}}\, p_{\text{vap,water}} + X_{\text{ethanol}}\, p_{\text{vap,ethanol}}$$
$$= (0.800)(52.6 \text{ kPa}) + (0.200)(119.8 \text{ kPa})$$
$$= 66.0 \text{ kPa}$$

The mole fraction of ethanol in the distillate will be the same as that in the gas phase from which it is condensed:

$$X_{\text{water}} = \frac{p_{\text{water}}}{p_{\text{total}}} = \frac{(0.800)(52.6 \text{ kPa})}{66.0 \text{ kPa}} = 0.637$$

$$X_{\text{ethanol}} = \frac{p_{\text{ethanol}}}{p_{\text{total}}} = \frac{(0.200)(119.8 \text{ kPa})}{66.0 \text{ kPa}} = 0.363$$

9.37 The vapour pressure of a solvent above a solution is described by Raoult's law, Equation 9-3 in your textbook:

$$p_{\text{vap,solution}} = X_{\text{solvent}}\, p_{\text{vap,pure solvent}}$$

In addition to the vapour pressure of pure solvent (101 bar, as we are at the boiling point of water), we need the mole fraction of the solvent. To calculate this, we must know the number of moles of solute and solvent:

For ethylene glycol : $n = \dfrac{65.0\text{g}}{62.07 \text{ g/mol}} = 1.047 \text{ mol}$

For water : $m = (0.500 \text{ L})(1.00 \text{ g/mL})(1000 \text{ mL/L}) = 500. \text{ g}$

$$n = \dfrac{500. \text{ g}}{18.01 \text{ g/mol}} = 27.76 \text{ mol}$$

$n_{\text{total}} = 1.047 + 27.76 = 28.81 \text{ mol}$ $\qquad X_{\text{water}} = \dfrac{27.76 \text{ mol}}{28.81 \text{ mol}} = 0.964$

$p_{\text{vap,solution}} = X_{\text{solvent}}\, p_{\text{vap,pure solvent}} = (0.964)(760 \text{ Torr}) = 732 \text{ Torr}$

$p_{\text{vap,solution}} = (0.964)(1.01 \text{ bar}) = 0.974 \text{ bar}$

9.39 Freezing points are calculated using Equation 9-5 from your textbook: $\Delta T_f = iK_f b$. For water, K_f is 1.858 °C kg/mol. For sucrose, $i = 1$. First, the molality of the solution must be calculated:

$$b = \dfrac{n_{\text{solute}}}{m_{\text{solvent}} \text{ (in kg)}}$$

The solution contains 12.50 g of sucrose and 155 mL of solvent (water). The density of water is 1.00 g/mL:

$$n_{\text{solute}} = \dfrac{12.50 \text{ g}}{342.3 \text{ g/mol}} = 3.652 \times 10^{-2} \text{ mol}$$

$$m_{\text{solvent}} = 155 \text{ mL}\left(\dfrac{1.00 \text{ g}}{1 \text{ mL}}\right)\left(\dfrac{10^{-3} \text{ kg}}{1 \text{ g}}\right) = 0.155 \text{ kg}$$

$$b = \dfrac{0.03652 \text{ mol}}{0.155 \text{ kg}} = 0.2356 \text{ mol/kg}$$

$\Delta T_f = iK_f b = 1((1.858 \text{ °C kg/mol})(0.2356 \text{ mol/kg})) = 0.438 \text{ °C}$

This is the freezing point depression, so the new freezing point is

$T_f = 0.000 - 0.438 = -0.438 \text{ °C}$.

9.41 Boiling points can be calculated for solutions of non-volatile solutes, using the boiling point elevation constant. $\Delta T_b = iK_b b$. For water, K_b is 0.512 °C kg/mol. First, the molality of the solution must be calculated:

$$b = \dfrac{n_{\text{solute}}}{m_{\text{solvent}} \text{ (in kg)}}$$

The solution contains 12.50 g of sucrose and 155 mL of solvent (water). The density of water is 1.00 g/mL. For sucrose, $i = 1$:

$$n_{solute} = 12.50 \text{ g}\left(\frac{1 \text{ mol}}{342.3 \text{ g}}\right) = 3.652 \times 10^{-2} \text{ mol}$$

$$m_{solvent} = 155 \text{ mL}\left(\frac{1.00 \text{ g}}{1 \text{ mL}}\right)\left(\frac{10^{-3} \text{ kg}}{1 \text{ g}}\right) = 0.155 \text{ kg}$$

$$b = \frac{0.03652 \text{ mol}}{0.155 \text{ kg}} = 0.2356 \text{ mol/kg}$$

$$\Delta T_b = iK_b b = 1\left(\frac{0.512 \text{ °C kg}}{1 \text{ mol}}\right)\left(\frac{0.2356 \text{ mol}}{1 \text{ kg}}\right) = 0.121 \text{ °C}$$

This is the boiling point elevation, so the new boiling point is

$T_b = 100.000 + 0.121 = 100.121 \text{ °C}$

9.43 The boiling point elevation of an aqueous solution depends on the *total* molality of solute particles. For an ionic salt, we must take into account the cations and anions, remembering that the net effect will be slightly diminished by ion pairing. The solution with the smallest total molality will have the lowest boiling point.

The total molalities in these solutions are:

Sucrose, 0.20 M ($i = 1$); NaCl, 0.40 M ($i = 2$); sucrose, 0.40 M (($i = 1$); MgCl$_2$, 0.75 M ($i = 3$);

The order of boiling points of these solutions is therefore as follows:

0.20 M sucrose (lowest) < 0.20 M NaCl < 0.40 M sucrose < 0.25 M MgCl$_2$ (highest).

(Note that the slight ion pairing in 0.20 M NaCl gives this solution a smaller boiling point elevation than 0.20 M sucrose.)

9.45 (a) Molar masses can be calculated from osmotic pressures using Equation 9-7 in your textbook.

$$V = (1.00 \times 10^2 \text{ mL})\left(\frac{10^{-3} \text{ L}}{1 \text{ mL}}\right)$$

$$M = \frac{mRT}{\Pi V} = \frac{(1.00 \text{ g})(0.08314 \text{ L bar mol}^{-1} \text{ K}^{-1})(25 + 273 \text{ K})}{(1.38 \text{ bar})(0.100 \text{ L})} = 180 \text{ g/mol}$$

(b) To determine the molecular formula, start from the general chemical formula of a sugar, C$_x$(H$_2$O)$_y$, with $y = x$, $x - 1$, or $x - 2$. When $y = x = 1$, $M = 30$ g/mol.

Dividing the actual molar mass by this molar mass per unit gives

$$\frac{180 \text{ g/mol}}{30 \text{ g/mol}} = 6.$$

Thus, $y = x = 6$ and the formula is $C_6H_{12}O_6$.

9.47 The freezing point of a solution allows us to calculate the solution's molality, but to calculate the osmotic pressure, we need the solution's molarity. To convert from molality to molarity requires knowledge of the density of the solution. Assume pantothenic acid does not dissociate in water, i.e. $i = 1$:

$$\Delta T_f = iK_f b \qquad\qquad b = \frac{\Delta T_f}{iK_f}$$

$$b = \frac{0.00 \text{ °C} - (-0.65\square\text{C})}{1(1.858\square\text{C kg/mol})} = 0.350 \text{ mol/kg}$$

To convert from molality, consider a solution containing 1 kg of solvent. It contains 0.350 mole of pantothenic acid ($M = 205.3$ g/mol):

$$m = (0.350 \text{ mol})(205.3 \text{ g/mol}) = 71.8 \text{ g}.$$

The total mass of this solution is 1071.8 g, from which we can calculate its volume:

$$V = \frac{m}{\rho} = \frac{1071.8 \text{ g}}{(1.015 \text{ g/mL})(10^3 \text{ mL/L})} = 1.056 \text{ L}$$

Now we can calculate the molarity:

$$c = \frac{n}{V} = \frac{0.350 \text{ mol}}{1.056 \text{ L}} = 0.331 \text{ M}$$

Knowing the molarity allows us to determine the osmotic pressure of this solution:

$$\Pi = cRT = (0.331 \text{ mol/L})(8.314 \text{ L kPa mol}^{-1} \text{ K}^{-1})(18 + 273 \text{ K}) = 800 \text{ kPa} = 8.00 \text{ bar}$$

9.49 Both milk and whipped cream are colloidal suspensions, and both have similar compositions. Milk is a suspension of solids (fats) in water, whereas whipped cream is a suspension of a gas (air) in a solid (coagulated fat) that also contains liquid.

9.51 Surfactants must have a highly polar (or ionic) end and a non-polar end. Having no polar end, octadecane will be the worst surfactant. Sodium alkylbenzenesulphonate ($C_{18}H_{29}SO_3Na$) has a longer non-polar portion than decanoic acid, and $C_{18}H_{29}SO_3Na$ has an ionic portion compared with a carboxylate end, making $C_{18}H_{29}SO_3Na$ the best surfactant.

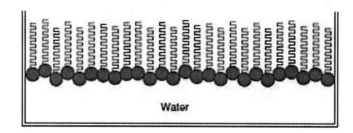

9.53 Surfactants have a polar (or ionic) end and a non-polar end. The polar end of the surfactant molecule associates with a polar substance. Water is polar, so your figure should show the polar heads of sodium stearate in contact with the solvent:

Use this information to calculate each concentration:

9.55 To determine the concentrations that are asked for, we need to know the mass and number of moles of each component and the volume of the solution. Work with exactly 1 L of solution. Round the final results to two significant figures to match the molarity given in the problem.

1 L of solution contains 2.0 mol ammonia. Determine the mass of this solution using the density equation:

$$m = \rho V = (0.787 \text{ g/mL})(1 \text{ L})(10^3 \text{ L/mL}) = 787 \text{ g}$$

The 2.0 mol ammonia has mass $m = nM = (2.0 \text{ mol})(17.03 \text{ g/mol}) = 34.1 \text{ g}$.

The rest of the solution is water: $m_{water} = (787 \text{ g} - 34.1 \text{ g}) = 753 \text{ g}$.

A table is a useful way to summarize the information that we need:

Substance	Formula	M	m	n
Ammonia	NH_3	17.03 g/mol	34.1 g	2.0 mol
Water	H_2O	18.02 g/mol	753 g	41.8 mol

Use this information to calculate each concentration:

Molality: $b = \dfrac{n_{solute}}{m_{solvent} \, (\text{in kg})} = \dfrac{2.0 \text{ mol}}{(753 \text{ g})(10^{-3} \text{ kg/g})} = 2.7 \text{ mol/kg}$

Mole fraction: $X = \dfrac{n_{solute}}{n_{total}} = \dfrac{2.0 \text{ mol}}{(2.0 \text{ mol} + 41.8 \text{ mol})} = 4.6 \times 10^{-2}$

Mass percent: $\% = (100\%)\dfrac{m_{solute}}{m_{total}} = (100\%)\dfrac{34.1 \text{g}}{(34.1 + 753 \text{ g})} = 4.3\%$

9.57 Hydrophilicity is conveyed by the presence of ionic, polar, or hydrogen-bonding groups. BHT contains only one OH group, so it is highly hydrophobic. Thus, it is stored in fatty tissues (which are non-polar) rather than being excreted from the body. BHT would be dangerous to organisms if the molecule was toxic.

9.59 Molar masses can be calculated from freezing points using Equation 9-5 in your textbook. First calculate the molality of the solution from its freezing point. Assume Vitamin C does not dissociate in water ($i = 1$):

$$b = \dfrac{\Delta T_f}{iK_f} = \dfrac{2.33 \text{ °C}}{1.858 \text{ °C kg/mol}} = 1.254 \text{ mol/kg.}$$

Next, rearrange the definition for molality to determine mole of solute:

$$b = \dfrac{n_{solute}}{m_{solvent} \, (\text{in kg})} \qquad n_{solute} = m_{solvent} b$$

$$n_{solute} = (1.00 \times 10^2 \text{g})\left(\dfrac{10^{-3} \text{ kg}}{1 \text{ g}}\right)\left(\dfrac{1.254 \text{ mol}}{1 \text{ kg}}\right) = 0.1254 \text{ mol}$$

Finally, calculate M using $n = \dfrac{m}{M}$: $\qquad M = \dfrac{m}{n} = \dfrac{22.0 \text{ g}}{0.1254 \text{ mol}} = 175 \text{ g/mol}$

9.61 Cyclohexane is a non-polar solvent, so it will dissolve non-polar substances best. Among these compounds, KCl is ionic and has the lowest solubility, whereas C_3H_8 is non-polar and dissolves readily. Polar C_2H_5OH lies in between.

KCl (lowest solubility) < C_2H_5OH < C_3H_8 (highest solubility)

9.63 Differential solubilities can be determined using the concept that "like dissolves like." Here, benzene is a non-polar liquid that dissolves in non-polar liquids. Water is highly polar and CCl_4 is non-polar, so these two liquids are immiscible and benzene preferentially dissolves in CCl_4.

9.65 The notion of "like dissolves like" can be extended to the wetting behaviour of one substance for another. A pipette wall is "dirty" if water beads on it, meaning that a hydrophobic substance is on the wall.

(a) Grease is "dirty." It is hydrophobic, containing long hydrocarbon chains, so water will bead on a greasy buret wall.

(b) Mg^{2+} ions are not "dirty." They will dissolve in water, so they might contaminate the solution in the buret.

(c) Acetone is not "dirty." It can hydrogen bond to water, so it is miscible with water. Again, the presence of acetone might contaminate the solution in the buret.

(d) SiO_2 is not "dirty." Silica, like glass, contains many polar Si–O bonds to which water molecules can form hydrogen bonds, wetting the surface.

9.67 The molal freezing point constant for a liquid is related to temperature and concentration through the freezing point depression equation, Equation 9-5 in your textbook. Note that naphthalene does not dissociate (hence $i = 1$):

$$\Delta T_f = iK_f b \qquad\qquad K_f = \frac{\Delta T_f}{ib}$$

(a) Use the data provided in the problem to evaluate the molality and the temperature change:

$\Delta T_f = 5.50\ ^\circ C - 5.03\ ^\circ C = 0.47\ ^\circ C$

$$n_{\text{naphthalene}} = \frac{1.28\ \text{g}}{128\ \text{g/mol}} = 1.00 \times 10^{-2}\ \text{mol}$$

$$m_{\text{benzene}} = V\rho = (125\ \text{mL})(0.88\ \text{g/mL})(10^{-3}\ \text{kg/g}) = 0.11\ \text{kg}$$

$$b = \frac{1.00 \times 10^{-2}\ \text{mol}}{0.11\ \text{kg}} = 9.1 \times 10^{-2}\ \text{mol/kg}$$

$$K_f = \frac{\Delta T_f}{ib} = \frac{0.47\ ^\circ C}{1(9.1 \times 10^{-2}\ \text{mol/kg})} = 5.16\ ^\circ C\ \text{kg/mol (round to 5.2 as a final result)}$$

(b) Use the result of part (a) to determine the molality of the solution of the unknown compound. From the molality, determine the number of moles, and then use the mass to find the molar mass of the unknown compound.

$\Delta T_f = 5.50\ ^\circ C - 5.24\ ^\circ C = 0.26\ ^\circ C$

$$b = \frac{\Delta T_f}{K_f} = \left(\frac{0.26\ ^\circ C}{5.16\ ^\circ C\ \text{kg/mol}}\right) = 0.0504\ \text{mol/kg}$$

$$b = \frac{n_{\text{solute}}}{m_{\text{solvent}}\ (\text{in kg})} \quad \text{so}\quad n_{\text{solute}} = m_{\text{solvent}} b$$

$$n_{\text{solute}} = (25.0\ \text{mL})(0.88\ \text{g/mL})(10^{-3}\ \text{kg/g})(0.0504\ \text{mol/kg}) = 1.11 \times 10^{-3}\ \text{mol}$$

Finally, calculate M: $M = \dfrac{m}{n} = \dfrac{0.125\ \text{g}}{1.11 \times 10^{-3}\ \text{mol}} = 1.1 \times 10^{2}\ \text{g/mol}$

9.69 If water were linear rather than bent, the two polar O–H bonds would oppose each other and the molecule would be non-polar like CO_2. Non-polar water molecules could not associate with ions through ion–dipole interactions, so linear water would not be a good solvent for salts.

9.71 A surfactant has a polar (or ionic) end and a non-polar end. The polar end of the surfactant molecule associates with a polar substance, whereas the non-polar end of a surfactant molecule associates with a non-polar substance. A bilayer membrane between two aqueous layers will have its polar "heads" associated with the aqueous layers and its non-polar "tails" associated with one another, as the drawing shows:

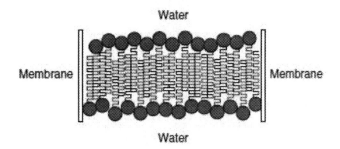

9.73 Fish blood is isotonic with seawater, meaning that the overall concentrations of the two are equal. The molality of seawater can be calculated from its freezing point. To calculate the osmotic pressure, assume that molality and molarity differ negligibly for sufficiently dilute aqueous solutions:

$$\Delta T_f = 0\ °C - (-1.96\ °C) = 1.96\ °C$$

$$b = \frac{\Delta T_f}{iK_f} = 1.96\ °C\left(\frac{1\ mol}{1(1.858\ °C\ kg)}\right) = 1.055\ mol/kg \quad c \cong b = 1.055\ mol/L$$

$$\Pi = cRT = (1.055\ mol/L)(0.08314\ L\ bar\ mol^{-1}\ K^{-1})(273+15\ K) = 25.3\ bar$$

(Note that the value of c calculated is the total concentration of ions.)

9.75 Use the boiling point elevation equation, Equation 9-6 in your textbook, to determine the molality of a solution that boils at 145 °C. For sugar, $i = 1$:

$$\Delta T_b = iK_b b \qquad so \qquad b = \frac{\Delta T_b}{iK_b}$$

$$b = \frac{(145\ °C - 100\ °C)}{1(1.858\ °C\ kg/mol)} = 24.2\ mol/kg$$

The question asks for the mass ratio of sugar to water needed to raise the boiling point by this much. We need to convert the molality into a mass ratio. Convert moles to mass in kilograms:

$$m_{sucrose} = nM = (24.2\ mol)(342\ g/mol)(10^{-3}\ kg/g) = 8.27\ kg$$

To have a boiling point of 145 °C, the solution must contain 8.3 kg of sugar for every 1 kg of water; in other words, it is mostly sugar.

9.77 Seawater has a higher osmotic pressure than fresh water, so the effect of sea water ingested into the body is to leach water from the insides of cells into the fluids that are contaminated with seawater. Water molecules cross the cell membranes, but proteins and ions do not. This causes the cells to collapse.

9.79 (a) Use the volume equation to calculate the volume of a single colloidal particle:

$$V = \frac{4}{3}\pi r^3 = \frac{4\pi}{3}\left[\left(\frac{15 \text{ nm}}{2}\right)\left(\frac{10^{-7} \text{ cm}}{1 \text{ nm}}\right)\right]^3 \left(\frac{10^{-3} \text{ L}}{1 \text{ cm}^3}\right) = 1.8 \times 10^{-21} \text{ L}$$

(b) Use the result of part (a), the density of gold, and its molar mass to determine the number of atoms in one particle:

$$\rho = \frac{m}{V} \qquad m = \rho V$$

$$m = (19.3 \text{ g/mL})(1.8 \times 10^{-21} \text{ L})(10^3 \text{ mL/L}) = 3.4 \times 10^{-17} \text{ g}$$

$$n = \frac{m}{M} = \frac{3.4 \times 10^{-17} \text{ g}}{197 \text{ g/mol}} = 1.7 \times 10^{-19} \text{ mol}$$

$$\# = nN_A = (1.7 \times 10^{-19} \text{ mol})(6.02 \times 10^{23} \text{ mol}^{-1}) = 1.0 \times 10^5 \text{ atoms}$$

(c) Use the mass of a single particle and the mass in 1 L of suspension to determine how many particles are in 1 L:

$$\# = \frac{0.10 \text{ g/L}}{3.4 \times 10^{-17} \text{ g/particle}} = 2.9 \times 10^{15} \text{ particles}$$

(d) Use the surface area equation to calculate the surface area of a single particle, and then multiply by the number of particles to get the total surface area:

$$A = 4\pi r^2 = (4)(\pi)(7.5 \text{ nm})^2 (10^{-9} \text{ m/nm})^2 = 7.1 \times 10^{-16} \text{ m}^2$$

Total area = $(7.1 \times 10^{-16} \text{ m}^2/\text{particle})(2.9 \times 10^{15} \text{ particles}) = 2.1 \text{ m}^2$

(e) Use mass and density to determine the volume of the cube. Then find the dimensions of the cube from its volume, and use the dimensions to calculate the surface area:

$$\rho = \frac{m}{V} \qquad V = \frac{m}{\rho} = \frac{0.10 \text{ g}}{19.3 \text{ g/cm}^3} = 5.2 \times 10^{-3} \text{ cm}^3$$

For a cube, $V = l^3$ \qquad Cube dimension $= l = \sqrt[3]{V} = 0.17 \text{ cm}$

Area per face $= (0.17 \text{ cm})^2 = 3.0 \times 10^{-2} \text{ cm}^2$

$$\text{Total area} = 6(\text{area per face}) = 6(3.0 \times 10^{-2} \text{ cm}^2) = 0.18 \text{ cm}^2$$

9.81 Begin this problem by using Henry's law, Equation 9-2 in your textbook, to calculate the concentration of dinitrogen in blood under the conditions described:

$$\left[\text{gas}(aq) \right]_{eq} = K_H \left(p_{gas} \right)_{eq} = \left(3.8 \times 10^{-4} \text{ M/bar} \right) \left(4.05 \text{ bar} \right) \left(\frac{78\%}{100\%} \right) = 1.2 \times 10^{-3} \text{M}$$

If a diver ascends rapidly to the surface, the new concentration of dinitrogen in the blood is the equilibrium value at 1.01 bar:

$$\left[\text{gas}(aq) \right]_{eq} = K_H \left(p_{gas} \right)_{eq} = \left(3.8 \times 10^{-4} \text{ M/bar} \right) \left(1.01 \text{ bar} \right) \left(\frac{78\%}{100\%} \right) = 0.3 \times 10^{-3} \text{M}$$

The difference, 0.9×10^{-3} mol/L, escapes into the gas phase, and we can calculate the volume of the released gas using the ideal gas equation:

$$V = \frac{nRT}{p} = \frac{(0.9 \times 10^{-3} \text{mol/L})(1 \text{ L})(0.08314 \text{ L bar mol}^{-1} \text{ K}^{-1})(37 + 273 \text{ K})}{(1.01 \text{ bar})}$$

$$V = 0.023 \text{ L}$$

We can convert this volume into mm^3, using 1 L = 1000 mL = 1000 cm^3:

$$V = (0.023 \text{ L})(1000 \text{ cm}^3/\text{L})(10 \text{ mm/cm})^3 = 2.3 \times 10^4 \text{ mm}^3$$

Use the equation for the volume of a sphere to calculate the volume of a bubble that is 1 mm in diameter (0.5 mm in radius):

$$V = \frac{4}{3}\pi r^3 = \frac{4\pi(0.50 \text{ mm})^3}{3} = 0.52 \text{ mm}^3$$

To find the number of bubbles, divide the total volume of gas by the volume per bubble:

$$\text{\# of bubbles} = \frac{2.3 \times 10^4 \text{ mm}^3}{0.52 \text{ mm}^3} = 4.4 \times 10^4 \text{ bubbles}$$

9.83 The freezing point depression is $\Delta T_f = iK_f b$. For MEA, i = 1. Thus, $K_f = \dfrac{\Delta T_f}{b}$. The given concentrations are mass % and must be converted to molalities. As an example, 2.489 % MEA can be converted to molality as follows:

2.489 mass % means 2.489 g MEA in 100 g of solution.

The water must therefore weigh 100 − 2.489 = 97.511 g.

$$n_{MEA} = \frac{2.489 \, g}{61.08 \, g \, mol^{-1}} = 0.0407 \, mol \, MEA$$

$$b = \frac{0.04075 \, mol}{0.097511 \, kg} = 0.4179 \, m$$

Performing the same calculation at each concentration of MEA, and dividing ΔT_f by b gives a calculated K_f at each concentration:

Monoethanolamine Concentration (mass %)	Monoethanolamine Concentration, b (m)	Freezing Point (K)	ΔT_f	Calculated K_f
2.489	0.42	272.45	0.7	1.68
5.035	0.87	271.46	1.69	1.95
7.359	1.30	270.70	2.45	1.88
9.192	1.66	269.92	3.23	1.95
14.119	2.69	267.81	5.34	1.98
19.845	4.05	264.79	8.36	2.06
21.863	4.58	263.89	9.26	2.02
24.666	5.36	261.62	11.53	2.15
29.921	6.99	258.11	15.04	2.15
30.649	7.24	256.88	16.27	2.25

The apparent value of K_f increases with the concentration of MEA because increased hydrogen bonding in the solution lowers the vapour pressure (and hence the freezing point) of the solution more than if there had been no hydrogen bonding.

Chapter 10 Organic Chemistry—Structure

Solutions to Problems in Chapter 10

10.1

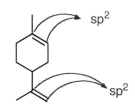

All other carbon atoms sp^3

10.3 (a) 2-Methylpentane

   (b) 2,3-Dimethylbutane

   (c) 2-Methylpentane

   (d) 2,4-Dimethylpentane

   (e) 3-Methylpentane

10.5

a)   b)   c)   d)   e)

10.7 (a) *sp* hybridization

   (b) sp^3 hybridization

   (c) sp^2 hybridization

10.9   2,2-dimethylbutane

10.11 2,3-dimethylpentane

10.13

   (a)   2: 1,1-dimethylcyclopropane and 1,2-dimethylcyclopropane
   (b)   3: 1,1-dimethylcyclobutane, 1,2-dimethylcyclobutane, and 1,3-dimethylcyclobutane

10.15 (a) *o*-Bromonitrobenzene

(b) *p*-Dicyclohexylbenzene

(c) *m*-Dinitrobenzene

10.17

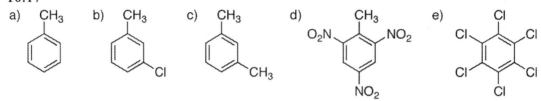

10.19 Benzene is a planar molecule where all three "double" bonds are conjugated such that the π electrons of these double bonds are delocalized around the ring. Two resonance structures can be drawn for benzene. Any situation in which resonance structures can be drawn for a molecule increases the stability of the molecule over similar structures for which no resonance is possible.

10.21 Two

10.23

CH₃

CH₃   *para*-methyltoluene (*p*-methyltoluene)

10.25

Br

Br   *para* disubstitution

10.27

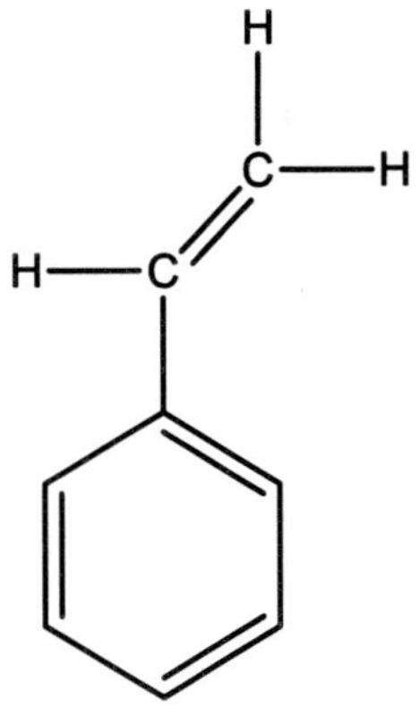

10.29

(a)  Primary 1°

(b) 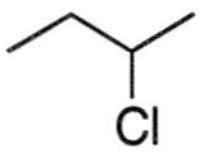 Secondary 2°

(c) 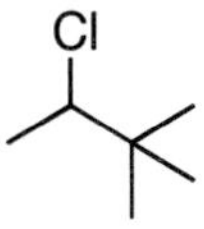 Secondary 2°

(d) 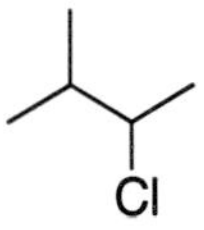 Secondary 2°

(e) 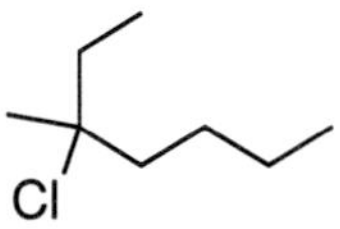 Tertiary 3°

10.31

(a) 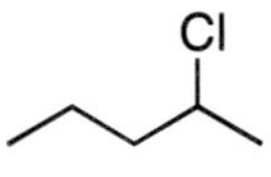 2-Chloropentane

(b) 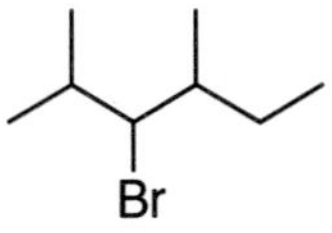 3-Bromo-2,4-dimethylhexane

(c) 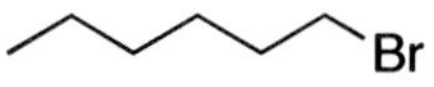 1-Bromohexane (or *n*-hexyl bromide also acceptable)

(d) 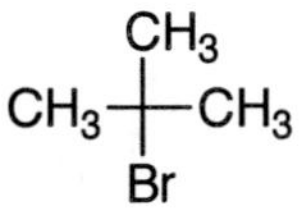 *tert*-Butyl bromide

(e) 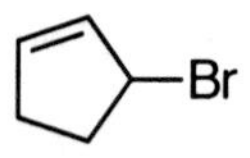 3-Bromocyclopentene

10.33

(a) 2-Butanol

(b) Cyclohexylmethanol

(c) Cyclohexanol

(d) 1-Methylcyclohexanol

(e) 2,2-Dimethyl-3-pentanol

10.35

a) 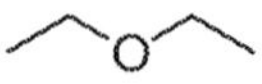   b) 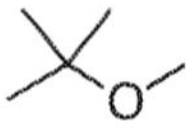   c) 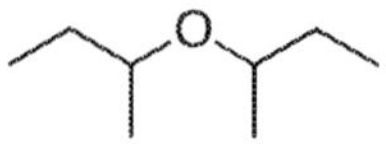   d) 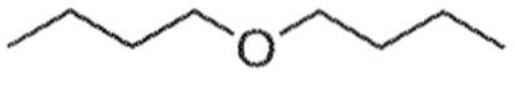   e)

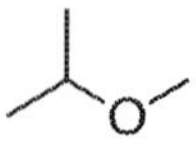

10.37

(a) 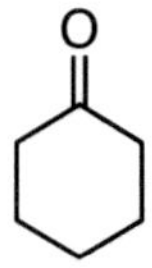 Ketone

(b) 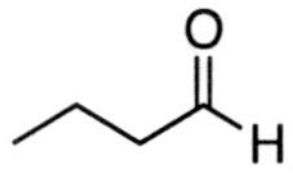 Aldehyde

(c) 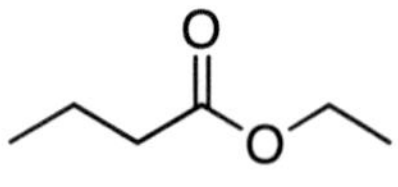 Ester

(d) 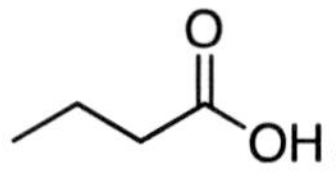 Carboxylic Acid

(e) 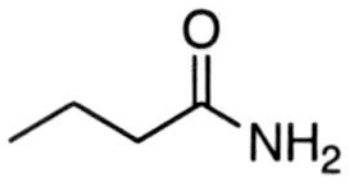 Amide

10.39

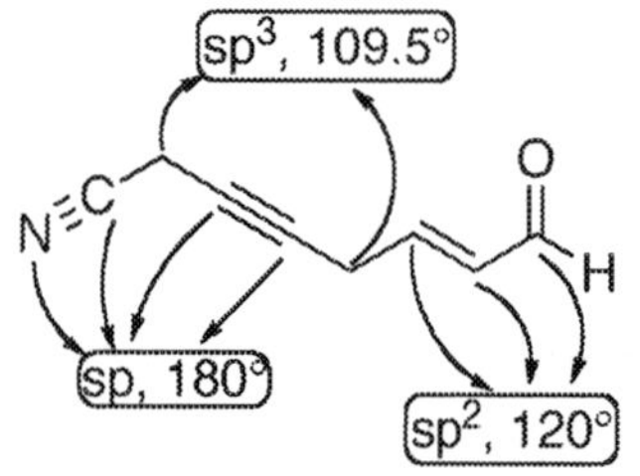

sp³, 109.5°

sp, 180°

sp², 120°

10.41

(a)

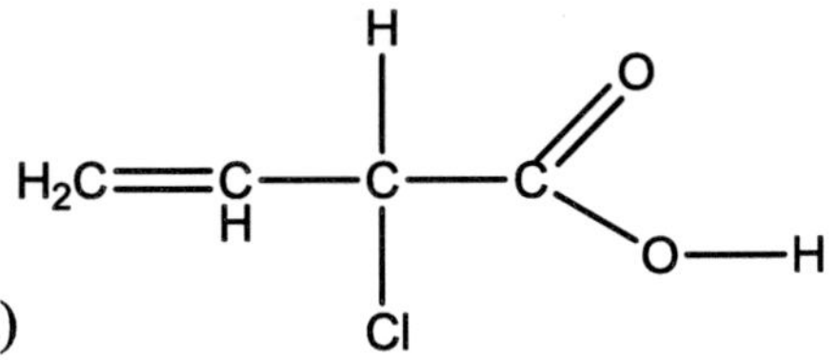

(b)

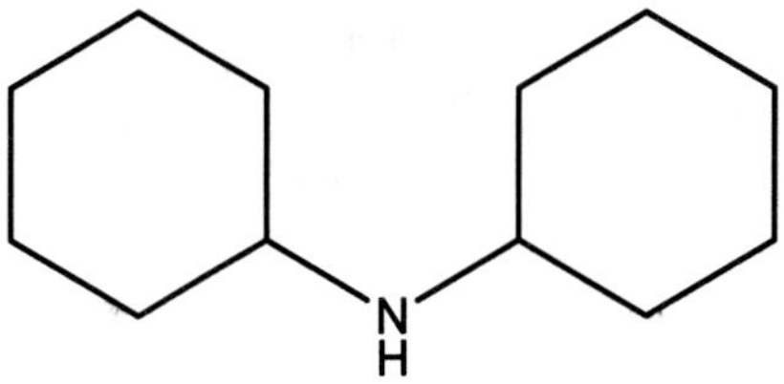

(c)

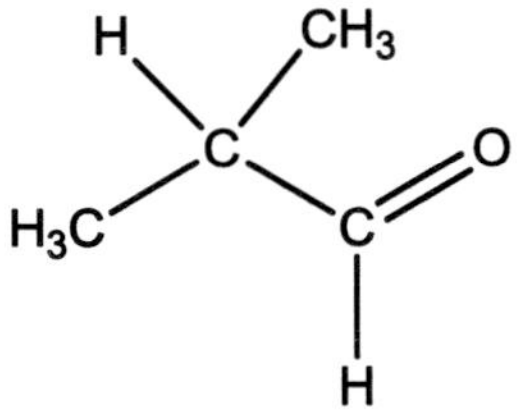

10.43 (a) No *E/Z* isomerization

(b) No *E/Z* isomerization

(c)  *E*-2-pentene 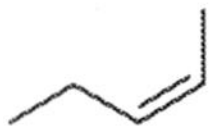   *Z*-2-pentene

(d) No *E/Z* isomerization

(e) No *E/Z* isomerization

10.45 (a) Achiral

(b) Achiral

(c) Chiral

(d) Chiral

(e) Chiral

10.47

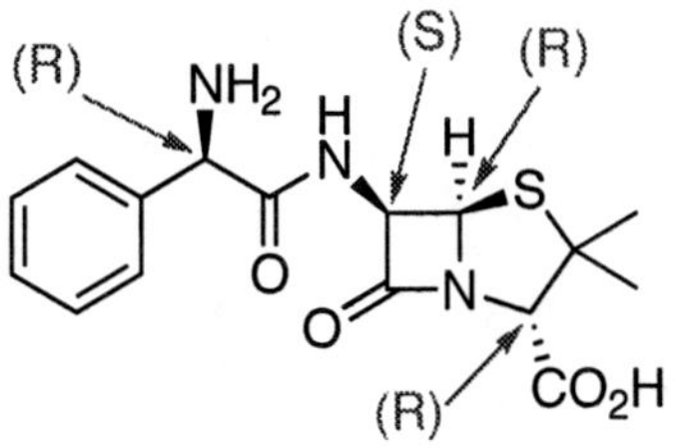

Amoxicillin

10.49

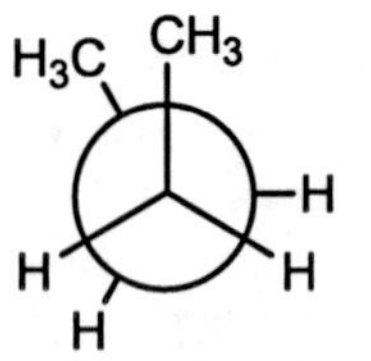

10.51 (a) 4; (b) 1

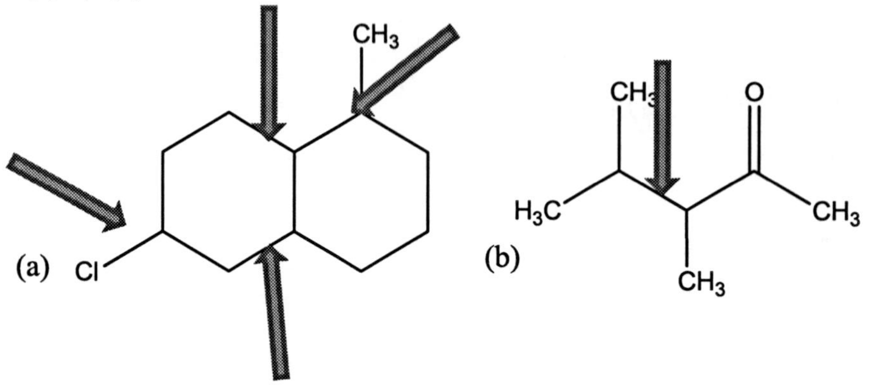

(a) Cl   (b)

10.53 This is (*R*)-carvone. No, only the double bond in the ring could be labelled as Z. The other double bond has two hydrogens on one end, and so *cis*/*trans* and *E*/*Z* notations would be meaningless here.

10.55 1-(*R*), 2-(*S*). They are enantiomers (i.e., non-superimposable mirror images).

10.57 (a) *S*; (b) Not chiral!; (c) *S*

10.59 Because of the non-polar nature of C–H bonds, the only significant intermolecular forces in hydrocarbons are dispersion forces. Despite the weak nature of dispersion forces in hexane, there is a long, large, and somewhat polarizable electron cloud that results in enough intermolecular interactions that this molecule is a liquid at room temperature.

10.61 Ethanol (CH_3CH_2–OH) has an alcohol functional group, and with a hydrogen bonded directly to an oxygen atom is capable of hydrogen bonding to water solvent molecules, resulting in strong intermolecular forces between ethanol and water molecules. Dimethylether (CH_3–O–CH_3) has no hydrogen bonding, and as a result the major intermolecular forces are dipole–dipole, which in this case are weaker than hydrogen bonding.

10.63 The carbonyl group of acetone would have a dipole, as shown below. The $\delta+$ end of the dipole is centred on the carbon atom, whereas the $\delta-$ end is centred on the oxygen atom. Attack by a nucleophile (such as the OH⁻ ion) would occur at the positive end of the dipole, as shown below.

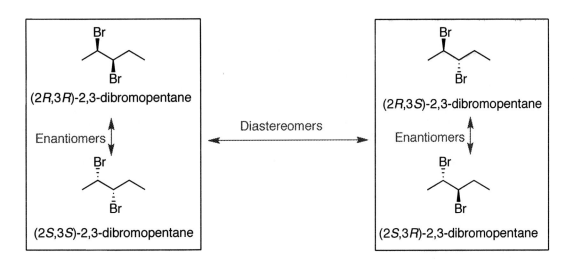

10.65 The two stereoisomers (2*R*,3*R*)-2,3-dibromopentane and (2*S*,3*S*)-2,3-dibromopentane are enantiomers of each other. The two stereoisomers (2*R*,3*S*)-2,3-dibromopentane and (2*S*,3*R*)-2,3-dibromopentane are also enantiomers of each other. The (2*S*,3*S*)-2,3-dibromopentane and the (2*R*,3*R*)-2,3-dibromopentane stereoisomers are diastereomers of the (2*R*,3*S*)-2,3-dibromopentane and the (2*S*,3*R*)-2,3-dibromopentane. This is illustrated below.

10.67 3-Ethyl-1,1-diethylcyclohexane

10.69 3-Ethyl-5-methylheptane

10.71 (a) and (b) are both aromatic molecules. (c) does not have alternating double and single bonds for all the double bonds involved in the system. The alternating double and single bonds in (a) and (b) allow electrons to delocalize around the system leading to an alternate resonance structure and hence increased stability.

10.73 2-Amino-4-chlorophenol

10.75 Secondary alkyl halide (chloride), ether, secondary amine, alkene, ketone

10.77

10.79

10.81

10.83 The fact that more energy is released when *Z*-2-butene is hydrogenated must mean that there was more energy stored in this double bond than in *E*-2-butene. This leads to the conclusion that *E*-alkenes are more stable (i.e., lower in energy) than *Z*-alkenes. The reason for this is that in *Z*-2-butene, the two methyl groups that are on the same side of the double bond can interact with each other, raising the energy of the molecule. In *E*-2-butene, this unfavourable interaction is removed as the two methyl groups are as far apart as possible, lowering the overall energy of the system.

10.85 2-Bromopentane

10.87 Cyclohexane

10.89 Enantiomers

10.91 Racemate or racemic mixture

10.93 The cycloheptatrienyl cation is an aromatic system, the same as benzene. This is demonstrated by the fact that all of the bond lengths are of equal length and the molecule is planar owing to the presence of the sp^2-hybridized positively charged carbon and the sp^2-hybridized carbons of the double bonds.

10.95 The large sterically bulky size of the *tert*-butyl group would lead to unfavourable interactions with the rest of the atoms in the cyclohexane ring, whereas in the chair configuration, that is not observed. In particular, the two hydrogen atoms on carbon atoms 3 and 5 that stick "up" in the same direction as the *tert*-butyl group would lead to unfavourable interactions. These steric interactions are completely eliminated when the *tert*-butyl group points away from the ring in the favoured conformation.

10.97 The two esters formed in this reaction are diastereomers; that is, stereoisomers that are not related by a mirror image. It should be possible to separate these two molecules by exploiting physical property differences between the two. For example, they will have different boiling points and it may be possible to separate them by distillation. Other physical separation techniques, such as recrystallization, can also in theory separate these molecules.

10.99 You would first choose a base with a known chirality centre; call this (*S*)-Base (or you could choose (*R*)-Base as long as it is chirally pure). Next you would allow the (*R*)/(*S*) mixture of acids to react with the (*S*)-Base to form two diastereomeric salts, (*R*)-Acid/(*S*)-Base and (*S*)-Acid/(*S*)-Base. These salts now have different physical properties and can be separated by recrystallization. Once the salts have been separated by recrystallization, we need to convert them back to the acid by acidification and then extraction of the more organically soluble acid from the inorganic side products. In this manner, we can isolate pure (*R*) and pure (*S*) acid.

10.101

(i) Distillation would be the easiest way to separate the desired molecule (b) from the side products. Molecule (a) would distill first, followed by (b) and then (c).

(ii)

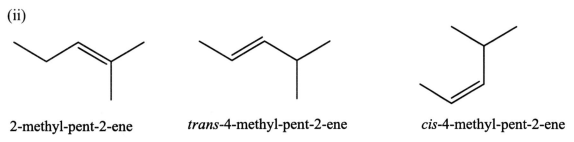

2-methyl-pent-2-ene   *trans*-4-methyl-pent-2-ene   *cis*-4-methyl-pent-2-ene

10.103 (a) 8

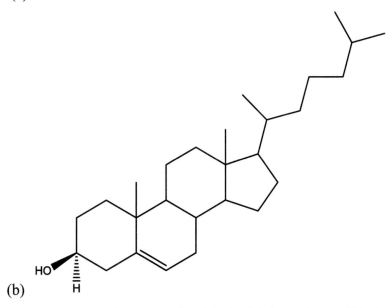

(b)

(c) Only one configuration for this double bond is possible inside this cyclic molecular structure. The opposite configuration to that shown would lead to a too highly strained ring system with bond lengths too far removed from those available, which would therefore not exist.

Chapter 11 Organic Chemistry—Reactions

Solutions to Problems in Chapter 11

11.1 (a) Elimination; (b) addition; (c) substitution; (d) substitution

11.3 (a) Addition; (b) addition; (c) none, this is a rearrangement, not an addition, elimination, or substitution. To confirm this, count all the atoms in the starting material and the product. You will find that nothing is added, eliminated, or substituted; (d) substitution.

11.5 (a) Addition; (b) addition; (c) addition

11.7 The nucleophilic centres are indicated by arrows:

11.9 Molecule (a) is not nucleophilic at all. It is actually an electrophile called a Lewis acid. The nucleophilic centres are indicated by arrows:

11.11 Electrophilic centres are indicated by arrows:

(a)   (b)

(c)   (d)

11.13 Electrophilic centres are indicated by arrows:

(a)   (b)   (c)   (d)

Note: Look at (b) and (c) and note the difference in the positions of P and N in the periodic table. P can accept a pair of electrons to expand its octet, whereas N cannot. Although N can accept the pair of electrons from the C—N bond in (c), it cannot accept a pair of electrons from an external nucleophile, and thus, in a strict sense, the N in molecule (c) is an electrophilic centre.

11.15 The electrophilic centre of the electrophile is shown by an arrow:

(a)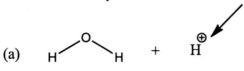

(b)

(c)

(d)

11.17 (a) A pair of π electrons from the benzene structure move to the positive carbon of the carbocation to form a new bond to a carbon at one end of the "double" bond (remember Kekulé structures!!). Within the ring, the carbon at the other end has a positive charge (because negative electron density left this centre to form the new bond):

(b) A pair of π electrons from the double bond move to the positive proton to form a bond from one end of the alkene, and the carbon at the other end of the double bond has a positive charge (electron density left this centre to form the new bond):

(c) A lone pair of electrons on the oxygen atom of the water molecule move to the positive centre of the carbocation (the carbon bearing the positive charge), forming a bond between oxygen and the carbon atom. As the oxygen has increased its normally allowable number of bonds to other atoms beyond two, it is positively charged:

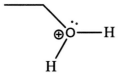

11.19 (a) Electrons move from the negative bromide ion to the positive carbocation centre to form a bond:

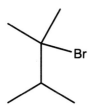

(b) Electrons move from the negative chloride ion to the positive carbocation centre to form a bond:

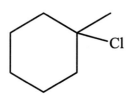

(c) Electrons move from the negative oxygen to the positive carbocation centre to form a bond:

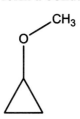

11.21 (a) The nucleophile is the alkoxide, the substrate is the alkyl bromide, the product is the ether, and the leaving group is bromide.

(b) The nucleophile is the thiolate (the sulphur equivalent of an alkoxide), the substrate is the alkyl bromide, the product is the thiolether (the sulphur equivalent of an ether), and the leaving group is bromide.

(c) The nucleophile is the carbanion, the substrate is the alkyl bromide, the product is the alkane (heptane), and the leaving group is bromide.

11.23 (a) The nucleophile is azide ion (the N_3^- ion), the substrate is the alkyl iodide, the product is the alkyl azide (RN_3), and the leaving group is the iodide ion.

(b) Both the nucleophile and the substrate are present on the same molecule here; the nucleophile is the alkoxide (-O- end) of the molecule, the substrate is the entire alkoxy bromide molecule, the product is called a pyran (a six-membered ring with an oxygen), and the leaving group is the bromide ion. Note that these types of reactions where substrate, nucleophile, and leaving group are present in the same molecule are common for a type of reaction called a cyclization, where the product is a cyclic system.

(c) The nucleophile is hydride ion (H^- ion), the substrate is the ketone (acetone), the product is the alkoxide, and there is no leaving group in this example. The reason for this is that the nucleophile attacked at a multiply bonded centre, and the "leaving group" in this case was a pair of electrons from the π bond of the C=O bond moving to the electronegative oxygen of the ketone. This is a common type of reaction and occurs at carbonyl carbons quite frequently with various nucleophiles.

11.25 Good leaving groups are conjugate bases of strong acids (like Cl^- and Br^-); that is, weak bases. Poor leaving groups are conjugate bases of weak acids (like OH^-); that is, strong bases.

(a) Br^- = good leaving group

(b) $CH_3–O^-$ = poor leaving group

(c) NH_2^- = poor leaving group

(d) Cl^- = good leaving group

(e) CH_3COO^- (acetate anion) = good leaving group

11.27 (a)

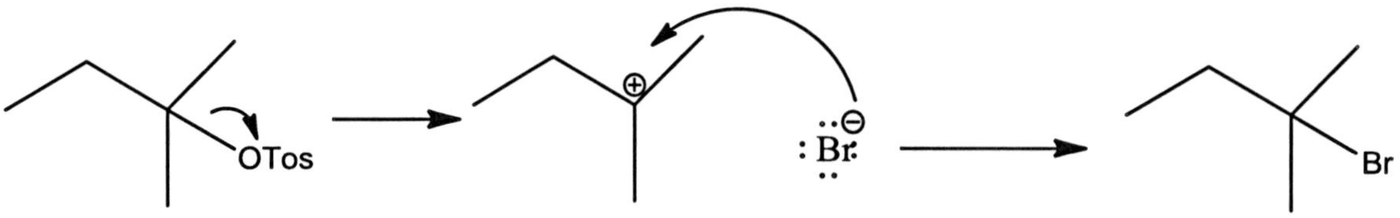

(b)

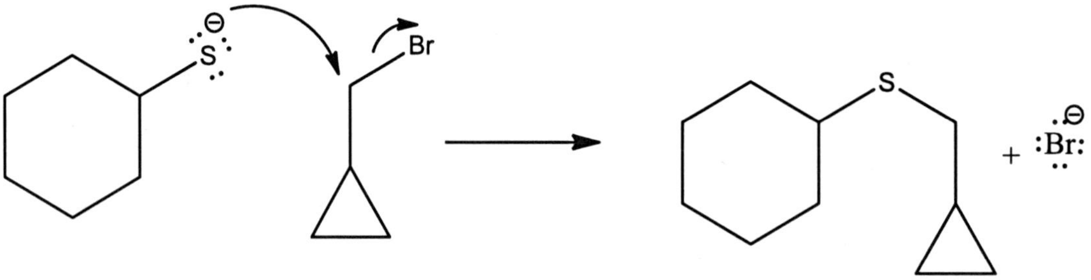

11.29 (a)

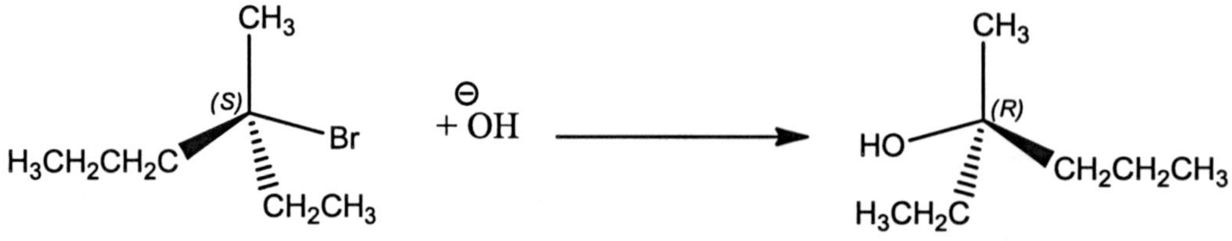

(b)

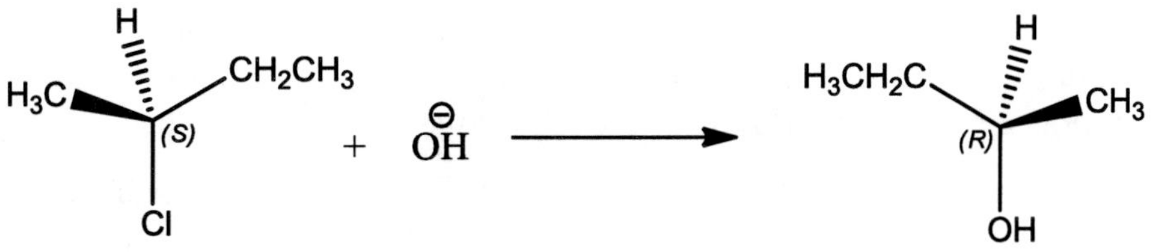

((R)-2-Butanol is formed)

11.31 (a)

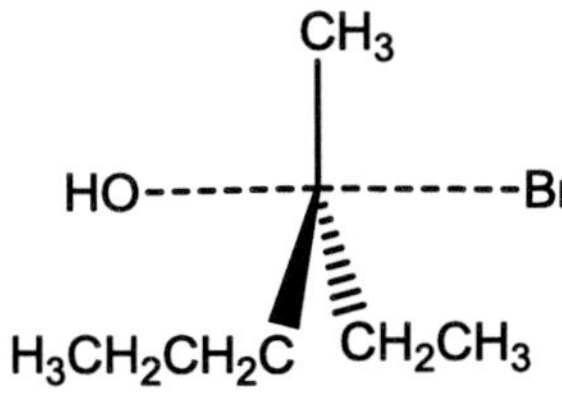

(b)

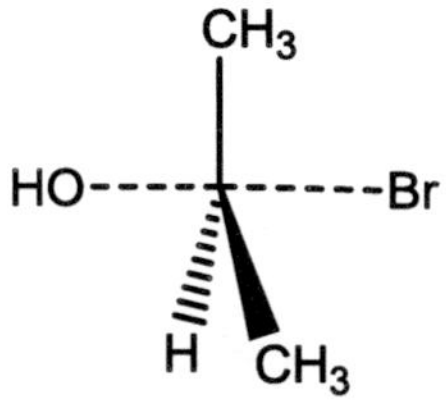

Molecule (b) will be faster as it is less sterically hindered.

11.33

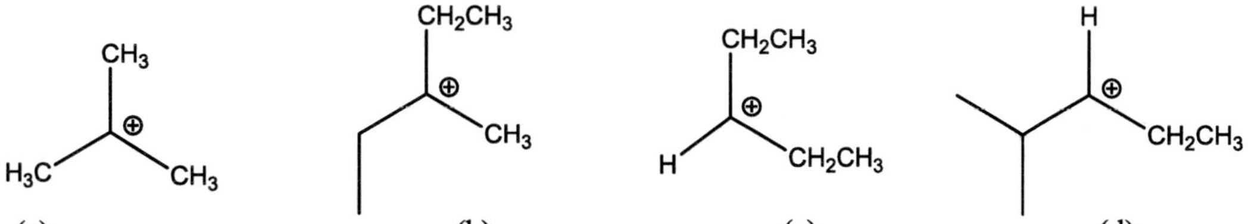

(a)   (b)   (c)   (d)

11.35 (a)

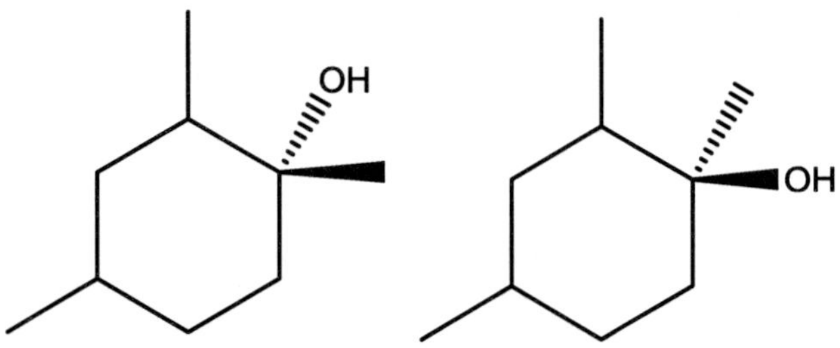

(b)

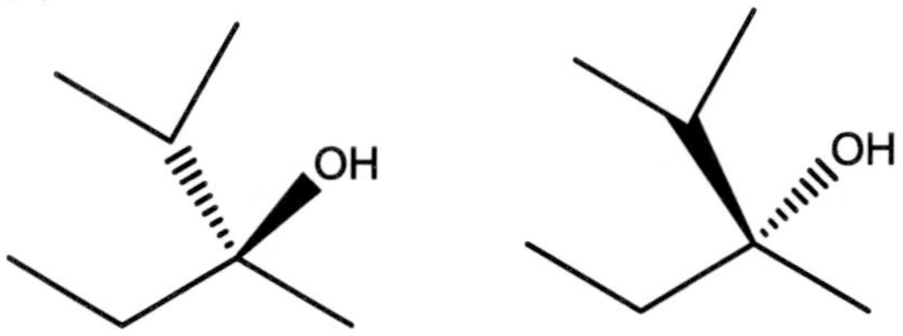

11.37 (a) The ethoxide ion is the nucleophile. The primary alkyl bromide is the substrate. The substitution results in an ether, and the leaving group is the bromide ion. Since the substrate is a primary alkyl halide, the mechanism would be S_N2.

(b) The EtO⁻ ion is the nucleophile. The tertiary alkyl iodide is the substrate. The substitution results in an ether (R-O-R'), and the leaving group is the iodide ion. Since the substrate is a tertiary alkyl halide, the mechanism would be S_N1.

11.39

(a) CH_3-Cl S_N2

(b) 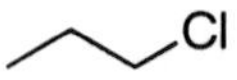 S_N2

(c) 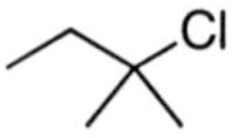 S_N1

(d) 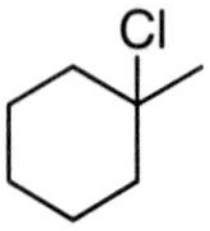 S_N1

(e) 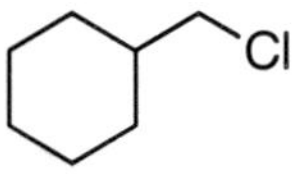 S_N2

11.41

(a) 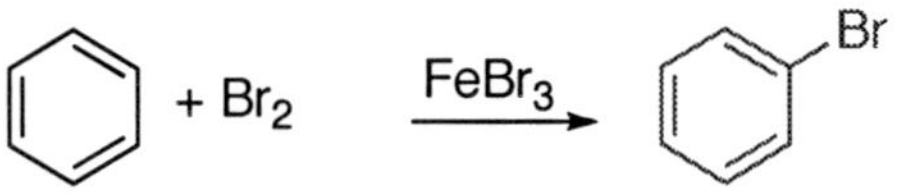 + Br_2 FeBr₃ →

(b) 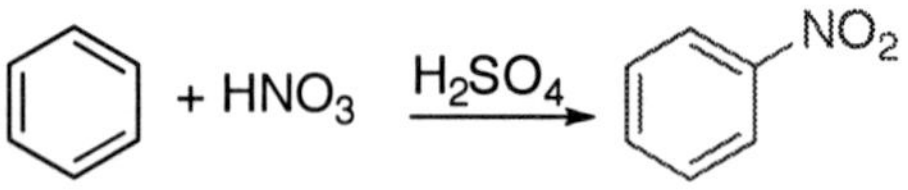 + HNO_3 H_2SO_4 →

(c) 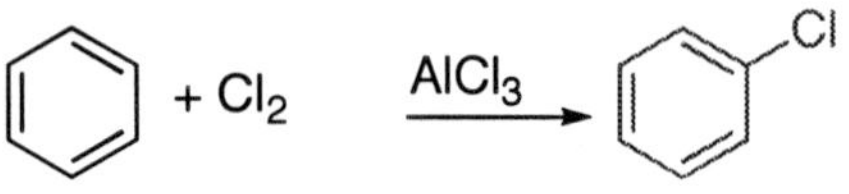 + Cl_2 $AlCl_3$ →

(d) 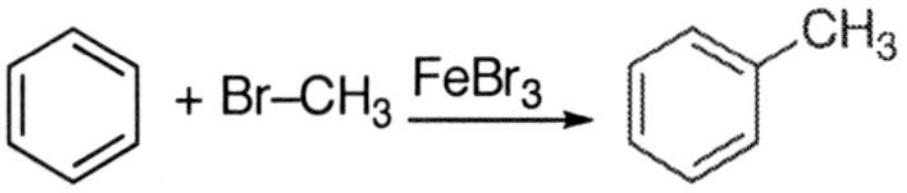 + $Br–CH_3$ FeBr₃ →

(e) 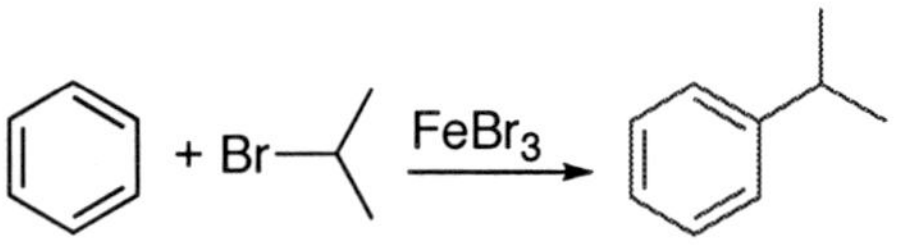 + Br— → FeBr₃ →

11.43 (a)

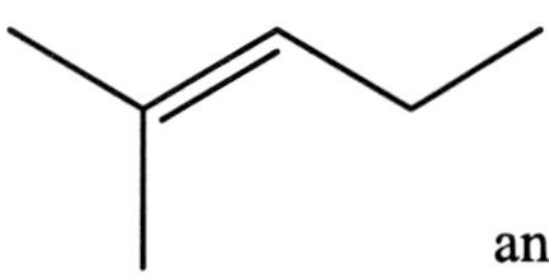

and

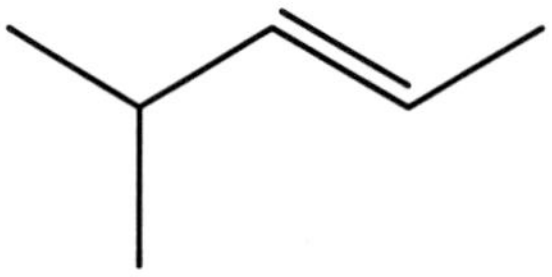

The structure on the left is the more highly substituted alkene and will therefore be the major product.

(b)

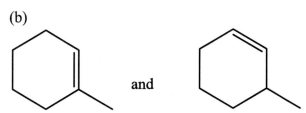

and

The structure on the left is the more highly substituted alkene and will therefore be the major product.

(c)

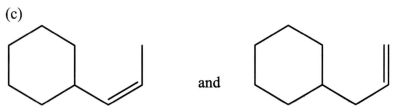

and

The structure on the left is the more highly substituted alkene and will therefore be the major product.

11.45 (a) The mechanism will be E2, as the alkyl halide is a primary alkyl halide. The product will be

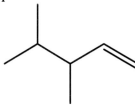

(b) The mechanism will be E1, as the alkyl halide is a tertiary alkyl halide. Two products could be formed:

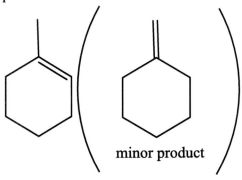

minor product

11.47

(a) Br E2

(b) E1

(c)   E1

(d)   E2

(e)   E1

11.49

(a)   Br $\diagup\!\!\!\!\triangle$ + NaOCH$_2$CH$_3$ ⟶

(b)   Br + NaOCH$_2$CH$_3$ ⟶

(c)   Br + NaOCH$_2$CH$_3$ ⟶

(d)
$$\underset{\underset{CH_3}{|}}{\overset{\overset{CH_3}{|}}{H-C-CH_2-Br}} + NaOCH_2CH_3 \longrightarrow \underset{H_3C}{\overset{H_3C}{>}}C=CH_2$$

(e)
$$\underset{\underset{CH_3}{|}}{\overset{\overset{CH_3}{|}}{Br-C-CH_3}} + NaOCH_2CH_3 \longrightarrow \underset{H_3C}{\overset{H_3C}{>}}C=CH_2$$

11.51 (a)

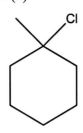

1-Chloro-1-methylcyclohexane

(b)

2-Chloro-2-methylpropane (tertiary butyl chloride)

11.53 (a)

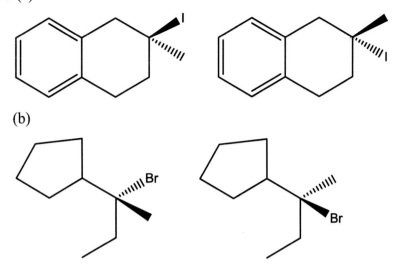

(b)

11.55 Heterogeneous refers to the fact that the catalyst is a different phase of matter
(solid Pt in this case) than the reagent (gaseous H_2 in this case) (and also usually a
liquid solution of alkene). The role of the catalyst is to lower the energy required for
the reaction to proceed without taking part in the overall reaction. In hydrogenation,
the Pt metal serves to activate the $H_{2(g)}$ by "stretching" the H–H bond.

11.57

(a)

(b)

(c)

(d)

(e)

11.59

(a)

(b)

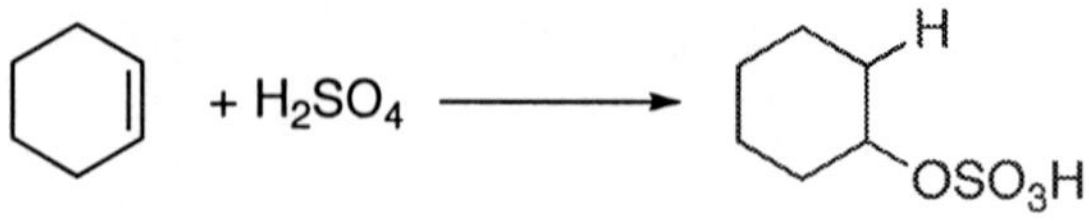

(c)

(d)

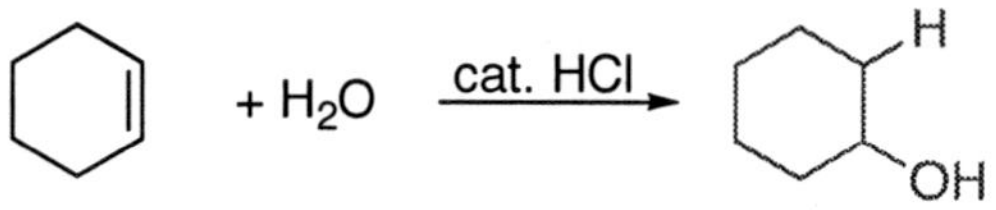

(e)

11.61

(a)

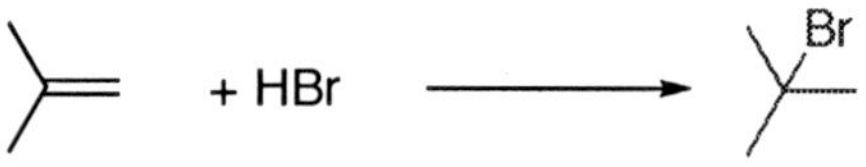

(b)

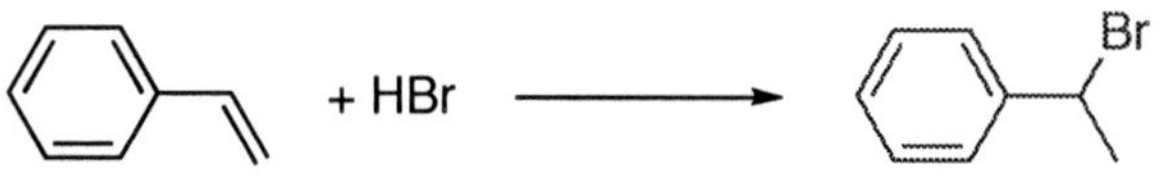

(c)

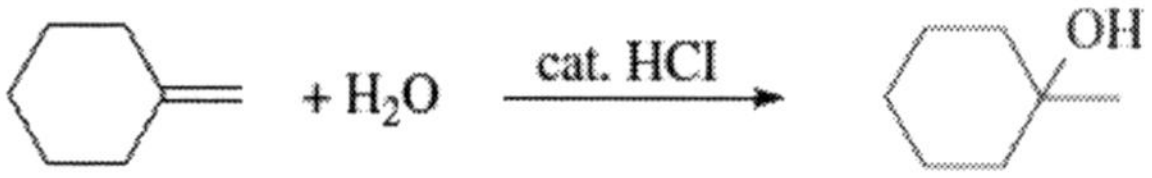

(d)

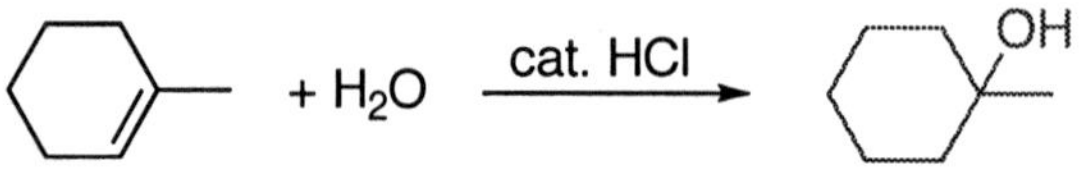

(e)

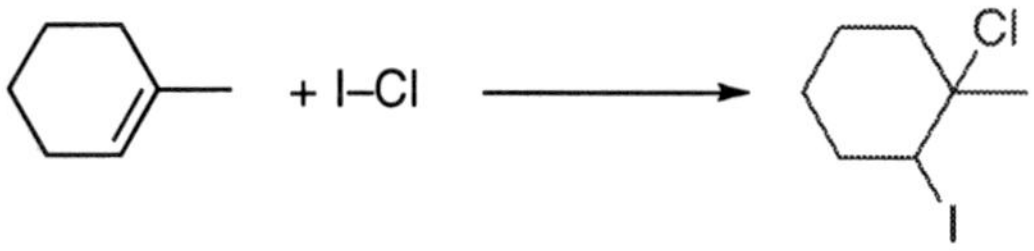

11.63 Products (a), (c), and (d) are chiral.

11.65

(a)

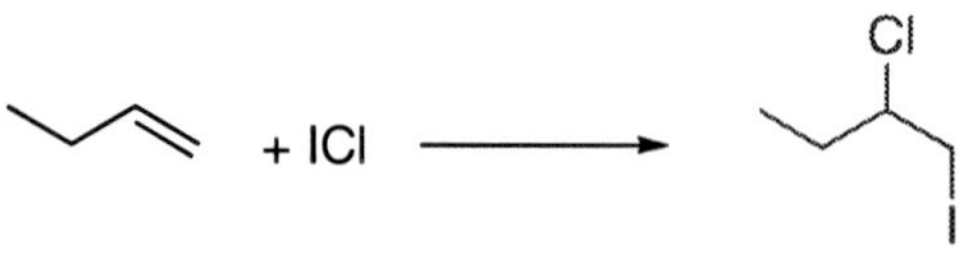

(b)

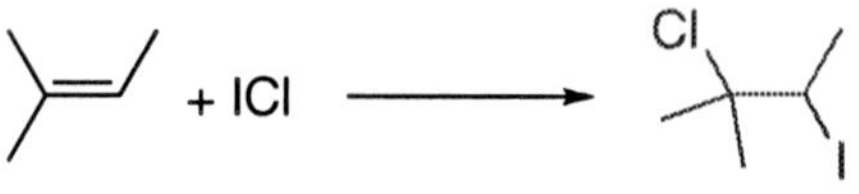

(c) + ICl ⟶

(d)  + ICl ⟶

(e) 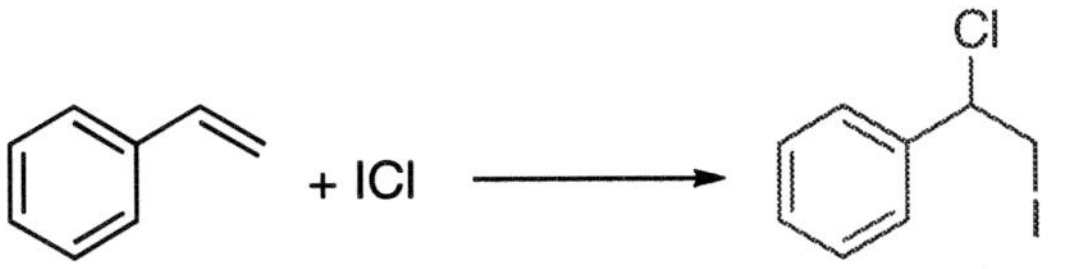 + ICl ⟶

11.67 All of the products in 11.65 are chiral.

11.69

(a)

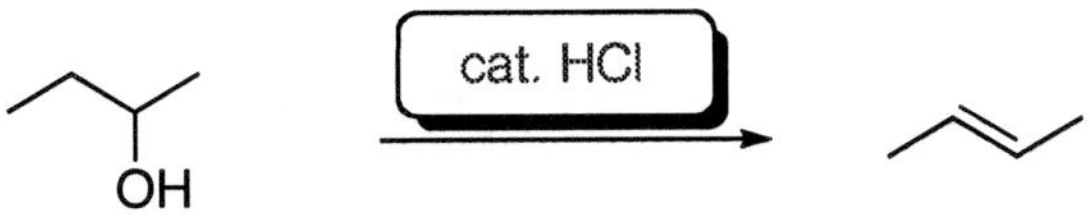

(b)

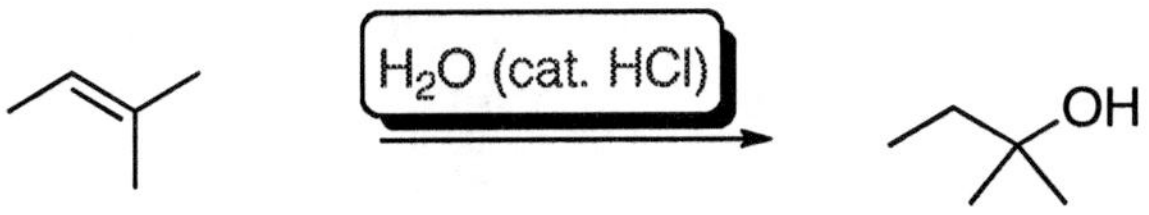

(c)

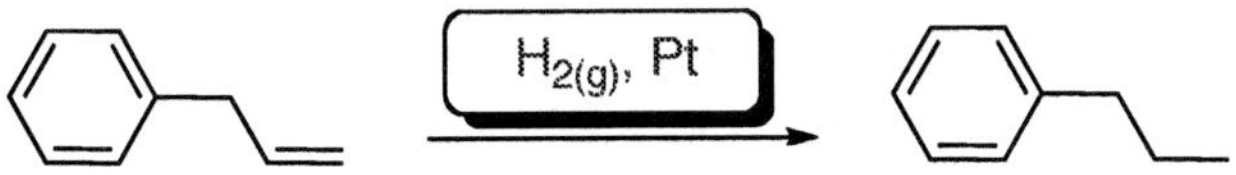

(d)

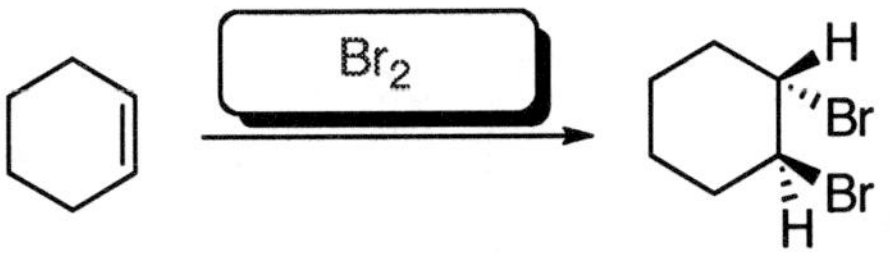

11.71

(a)

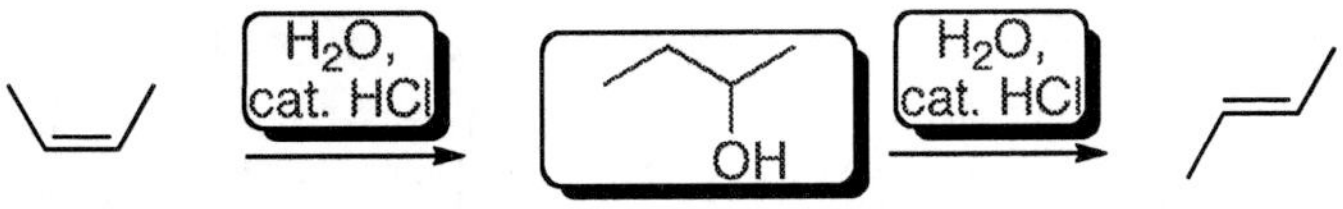

(b)

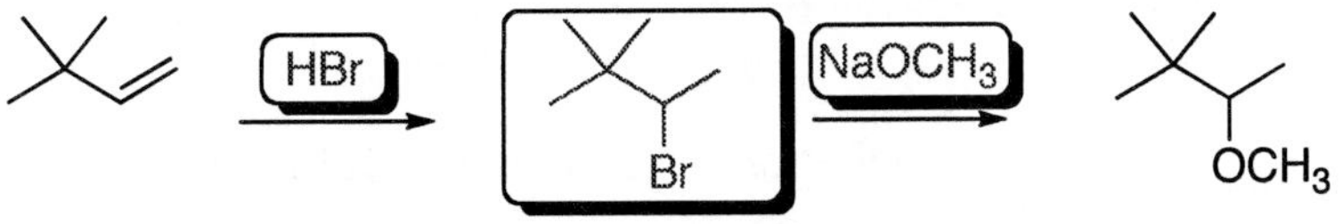

(c)

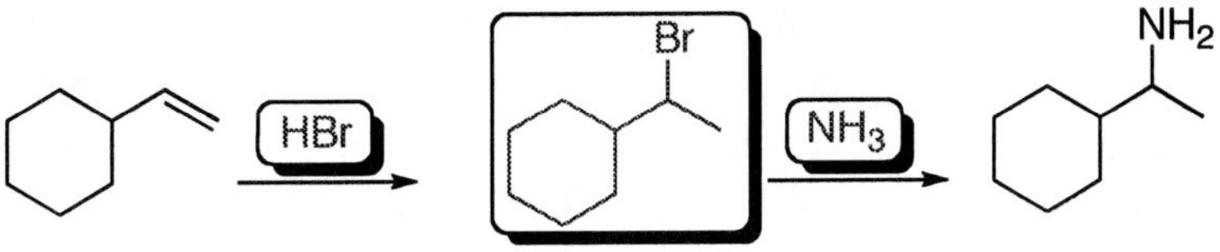

(d)

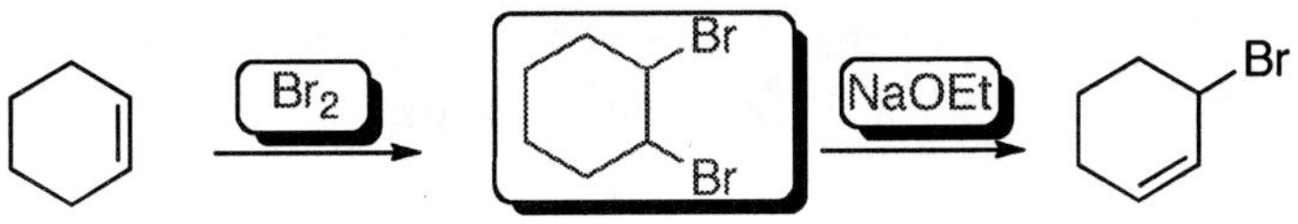

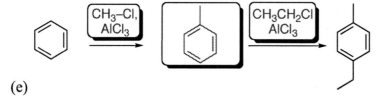

(e)

11.73 In the absence of acid, the leaving group on *tert*-butanol would be a hydroxide ion, which is the conjugate base of a weak acid (H_2O). Hydroxide ion is, therefore, a very poor leaving group. By adding acid, the first step in the mechanism is to protonate the oxygen atom of the alcohol. In this case then, the leaving group is a molecule of water, the conjugate base of a strong acid (H_3O^+). This makes water a very good leaving group.

11.75 This molecule is unreactive under S_N2 conditions owing to the size of the trifluoromethyl groups preventing backside attack. Under S_N1 conditions, the key carbocation intermediate is strongly disfavoured by the very electronegative fluorine atoms. These electronegative atoms would exert a strong pull on the carbon–chlorine bond and make it much less likely that this bond would break to generate the carbocation.

11.77 The critical step in the S_N1 reaction is the formation of the carbocation intermediate. This carbocation is formally an sp^2-hybridized carbon atom that is planar, with 120° bond angles and an empty p orbital. The incoming nucleophile has no preference as to which side of the carbocation to attack to form the new covalent bond. In this example, the result is an equimolar mixture of the (*S*)-1-bromo-1-phenylethane, the result of attack from the "top" of the carbocation, and (*R*)-1-bromo-1-phenylethane from attack at the "bottom" of the carbocation.

11.79 The *tert*-butyl chloride reacts faster, because under S_N1 conditions the rate-determining step is formation of the carbocation. Tertiary carbocations are more stable than primary carbocations and this is demonstrated by this example. It is much easier to form a tertiary than a primary carbocation. If you picture each C–H bond on the adjacent methyl groups as a "cloud" of electrons (and this is a pretty accurate picture), then each bond is able to donate a little of this cloud into the empty p orbital on the carbocation through a process called hyperconjugation. Branched alkyl groups have a stronger electron donation effect than straight-chain alkyl groups. It is this slight donation of each of the CH bonds in the alkyl branches that accounts for the extra stability of the tertiary carbocation.

11.81 In the E1 reaction, there is a carbocation intermediate that then loses a proton from an adjacent carbon to form the alkene product. The results in the reaction shown in Problem 11.81 show that the *E* alkene is the favoured product. This suggests that *E* alkenes are in some way more stable than the *Z* alkene. This is in fact the case. One of the contributing factors is the physical size (steric bulk) of the aromatic benzene ring and the methyl group present on either side of the double bond. The configuration of the *E* alkene allows these two substituents to be as far away from each other as possible. In the case of the *Z* alkene, the benzene ring and the methyl group will interact with each other. The *E* alkene minimizes this unfavourable interaction, making the overall energy of the molecule lower and resulting in this being the preferred product.

11.83   In general, an E2 reaction can take place on any substrate with a hydrogen atom on a carbon atom adjacent to the leaving group. For the E2 reaction, it makes little difference if the leaving group is on a primary, secondary, or tertiary carbon atom. However, the S_N2 mechanism is favoured by primary and some secondary but never by a tertiary substrate. Moreover, the size of the base has an effect. Large, strong, and bulky bases, such as *tert*-butoxide, favour the elimination pathway. Smaller bases that are derived from weaker acids favour the substitution pathway.

11.85   The electronegativity numbers for hydrogen (2.1) and boron (2.0) mean that the hydrogen atom is more electronegative than the boron atom and that the electrons in the B–H bond are polarized toward the hydrogen atom. This means that in an electrophilic addition, the mechanism occurs as shown below, where the boron actually adds first to the double bond. The second step is the attack of an H⁻ (a hydride ion) on the carbocation. This mechanism agrees with the definition of Markovnikov's rule in Problem 11.84, since the most stable carbocation is still the intermediate in the reaction.

11.87   In this reaction the relative energy difference between starting material and product is the same as in Problem 11.86, and there will be a net release of 63 kJ/mol. The major and significant difference is that this reaction proceeds with the initial formation of a carbocation intermediate. This intermediate will be higher in energy than either the starting material or the product but will have a measurable existence (unlike a transition state). Therefore, the diagram has to take this into account as shown below.

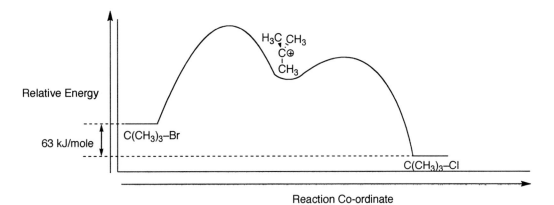

11.89   This trend is explainable based on the electronegativity of chlorine (3.0) compared with hydrogen (2.1). The electronegative chlorine atom will help to stabilize the negative charge that forms on the conjugate base of each of these acids. Each chlorine induces a dipole in the C–Cl bond that is then transferred through the other C–C bonds to the oxygen atom that carries the negative charge. The more chlorine atoms on the acid, the stronger this inductive effect and the more stable the anion, which results in a stronger base. This is shown below.

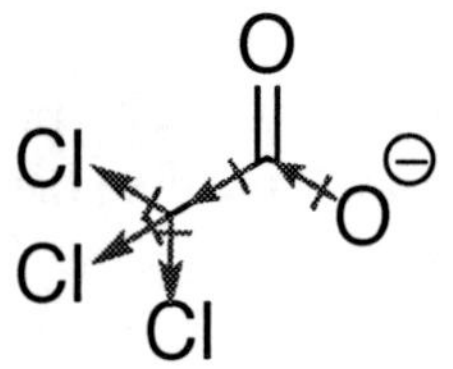

Chapter 12 Spontaneity of Chemical Processes

Solutions to Problems in Chapter 12

12.1 (a) A sand castle represents an organized structure constructed by a person. Waves destroy that structure, dispersing the sand grains.

(b) Two separate liquids represent a constrained arrangement (all molecules of one kind in one container). Upon mixing, the molecules in the liquids are dispersed throughout the container.

(c) Sticks in a bundle are constrained (all aligned in the same direction). When dropped, the sticks lose their alignment and become dispersed.

(d) Water in a puddle is relatively constrained, as it is confined to a small volume. When the water evaporates, the molecules spread over a much larger volume and become more dispersed.

12.3 The molecules in a drop of ink are relatively constrained, because they occupy one particular part of the total volume. As they move about randomly, they spread throughout the volume of the container and become more dispersed.

12.5 (a) The air molecules in a tire are relatively constrained, because they are confined to a specific, small volume. A puncture allows gas molecules to escape from the tire and fill a much larger volume, becoming more dispersed in the process.

(b) Like air in a tire, the fragrant molecules in a perfume bottle are relatively constrained because they are confined to a specific, small volume. When the bottle is open, molecules escape from the confined volume into a much larger space of the room, becoming more dispersed in the process.

12.7 (a) Energy dispersal requires that heat flows from high to low temperature, never in the opposite direction.

(b) Matter dispersal requires that the spontaneous direction be toward greater dispersal, and a "sand castle" represents constrained matter.

12.9 Calculate W by determining how many ways each symbol can be placed. The first X can be placed in any of the nine compartments, the first O in any of the eight empty compartments, so $W = (9)(8) = 72$.

12.11 The entropy change accompanying a constant-temperature process is $\Delta S = \dfrac{q_T}{T}$.

(a) Melting involves absorption of heat:

$$q_{ice} = n\Delta H_{fus} = \left(\frac{13.8\ g}{18.02\ g/mol}\right)\left(\frac{6.01\ kJ}{1\ mol}\right)\left(\frac{10^3\ J}{1\ kJ}\right) = 4.60 \times 10^3\ J$$

$$\Delta S_{ice} = \frac{4.60 \times 10^3\ J}{273.15\ K} = 16.8\ J/K$$

(b) The heat absorbed by the ice cube must be supplied by the pool water:

$$q_{pool} = -q_{ice}$$

$$\Delta S_{pool} = \frac{-4.60 \times 10^3\ J}{27.5 + 273.15\ K} = -15.3\ J/K$$

(c) $\Delta S_{total} = \Delta S_{ice} + \Delta S_{pool} = (16.8\ J/K) + (-15.3\ J/K) = 1.5\ J/K$

12.13 In each of these spontaneous processes, $\Delta S_{total} > 0$. For system and surroundings, the sign of the entropy change will be the same as the sign of q:

(a) The system (water) absorbs heat to boil, so q_{sys} is positive and ΔS_{sys} is positive. The surroundings (stove) release heat to the system, so q_{surr} is negative and ΔS_{surr} is negative.

(b) The system (ice cubes) absorbs heat to melt, so q_{sys} is positive and ΔS_{sys} is positive. The surroundings (table top) release heat to the system, so q_{surr} is negative and ΔS_{surr} is negative.

(c) The system (coffee) absorbs heat, so q_{sys} is positive and ΔS_{sys} is positive. The surroundings (microwave oven) release heat to the system, so q_{surr} is negative and ΔS_{surr} is negative.

12.15 The entropy change accompanying a constant-temperature process is $\Delta S = \dfrac{q_T}{T}$.

Condensation, the reverse of vapourization, involves release of heat:

$$q_{steam} = -n\Delta H_{vap} = -\left(\frac{15.5\ g}{18.02\ g/mol}\right)\left(\frac{40.79\ kJ}{1\ mol}\right)\left(\frac{10^3\ J}{1\ kJ}\right) = -3.509 \times 10^4\ J$$

$$\Delta S_{steam} = \frac{-3.509 \times 10^4\ J}{373.15\ K} = -94.0\ J/K$$

Knowing that ΔS_{total} must be positive because the overall process is spontaneous, we can say without doing further calculations that $\Delta S_{surr} > 94.0\ J/K$.

12.17 The molar entropy change accompanying fusion is $\Delta S_{molar} = \dfrac{\Delta H_{fus}}{T_{fus}}$:

(a) Argon: $\Delta S_{molar} = \left(\dfrac{1.3 \text{ kJ/mol}}{83 \text{ K}}\right)\left(\dfrac{10^3 \text{ J}}{1 \text{ kJ}}\right) = 16 \text{ J mol}^{-1} \text{ K}^{-1}$

(b) Methane: $\Delta S_{molar} = \left(\dfrac{0.84 \text{ kJ/mol}}{90 \text{ K}}\right)\left(\dfrac{10^3 \text{ J}}{1 \text{ kJ}}\right) = 9.3 \text{ J mol}^{-1} \text{ K}^{-1}$

(c) Ethanol: $\Delta S_{molar} = \left(\dfrac{7.61 \text{ kJ/mol}}{156 \text{ K}}\right)\left(\dfrac{10^3 \text{ J}}{1 \text{ kJ}}\right) = 48.8 \text{ J mol}^{-1} \text{ K}^{-1}$

(d) Mercury: $\Delta S_{molar} = \left(\dfrac{23.4 \text{ kJ/mol}}{234 \text{ K}}\right)\left(\dfrac{10^3 \text{ J}}{1 \text{ kJ}}\right) = 1.00 \times 10^2 \text{ J mol}^{-1} \text{ K}^{-1}$

12.19 (a) Both are ionic solutions, but $MgCl_2$ produces 3 moles of ions per mole of substance, whereas NaCl produces 2 moles of ions per mole of substance, so $MgCl_2$ has the larger molar entropy.

(b) Both are similar polar compounds, but HgS has a higher molar mass than HgO, so HgS has the larger molar entropy.

(c) Both are diatomic molecules, but Br_2 (*l*) is liquid and I_2 (*s*) is solid, so Br_2 (*l*) has the larger molar entropy.

12.21 All these substances are small gaseous molecules. Ozone has more entropy than O_2 or H_2 because it has three atoms per molecule whereas the others have only two. O_2 has a higher molar mass than H_2, so it has higher entropy.

12.23 Absolute entropies are tabulated in Appendix D of your textbook. To obtain the entropy per mole of atoms, divide each value by the number of atoms in one particle:

He: $S° = 126.153 \text{ J mol}^{-1} \text{ K}^{-1} \left(\dfrac{1 \text{ mol He}}{1 \text{ mol atom}}\right) = 126.153 \text{ J mol}^{-1} \text{ K}^{-1}$

H_2: $S° = 130.680 \text{ J mol}^{-1} \text{ K}^{-1} \left(\dfrac{1 \text{ mol } H_2}{2 \text{ mol atom}}\right) = 65.340 \text{ J mol}^{-1} \text{ K}^{-1}$

CH_4: $S° = 186.3 \text{ J mol}^{-1} \text{ K}^{-1} \left(\dfrac{1 \text{ mol } CH_4}{5 \text{ mol atom}}\right) = 37.26 \text{ J mol}^{-1} \text{ K}^{-1}$

C_3H_6: $S° = 226.9 \text{ J mol}^{-1} \text{ K}^{-1} \left(\dfrac{1 \text{ mol } C_3H_6}{9 \text{ mol atom}}\right) = 25.21 \text{ J mol}^{-1} \text{ K}^{-1}$

Entropy per mole of atoms decreases as the number of atoms in a species increases, because tying together atoms into a molecule increases the constraints among those atoms (other factors, notably the atomic mass of the atoms, also play important roles).

12.25 Use Equation 11-5 to calculate entropies at pressures different from 1.00 bar:

$$S_{(p \neq 1 \text{ bar})} = S° - R \ln p$$

Take into account amounts different from 1 mole by multiplying by the number of moles.

(a) $S = (2.50 \text{ mol})[154.843 \text{ J mol}^{-1} \text{ K}^{-1} - (8.314 \text{ J mol}^{-1} \text{ K}^{-1}) \ln (0.25 \text{ bar})] = 416 \text{ J/K}$

(b) $S = (0.75 \text{ mol})[238.9 \text{ J mol}^{-1} \text{ K}^{-1} - (8.314 \text{ J mol}^{-1} \text{ K}^{-1}) \ln (2.75 \text{ bar})] = 1.7 \times 10^2$ J/K

(c) Calculate the entropy of each component separately, using $n_i = X_i \, n_{tot}$ and $p_i = X_i \, p_{tot}$:

$$S_{N_2} = (0.78)(0.45 \text{ mol})\{191.61 \text{ J mol}^{-1} \text{ K}^{-1} - (8.314 \text{ J mol}^{-1} \text{ K}^{-1}) \ln [(0.78)(1.00 \text{ bar})]\} = 68 \text{ J/K}$$

$$S_{O_2} = (0.22)(0.45 \text{ mol})\{205.152 \text{ J mol}^{-1} \text{ K}^{-1} - (8.314 \text{ J mol}^{-1} \text{ K}^{-1}) \ln [(0.22)(1.00 \text{ bar})]\} = 22 \text{ J/K}$$

$$S = S_{N_2} + S_{O_2} = 9.0 \times 10^1 \text{ J/K}$$

12.27 Standard entropy changes can be calculated using tabulated values of absolute entropies (Appendix D of your textbook) and Equation 11-6:

$$\Delta S°_{\text{reaction}} = \Sigma \text{coeff}_{\text{products}} \, S°_{\text{products}} - \Sigma \text{coeff}_{\text{reactants}} \, S°_{\text{reactants}}$$

(a) $\Delta S°_{\text{reaction}} = [2 \text{ mol}(192.8 \text{ J mol}^{-1} \text{ K}^{-1})] - [1 \text{mol}(191.61 \text{ J mol}^{-1} \text{ K}^{-1}) + 3 \text{ mol}(130.680 \text{ J mol}^{-1} \text{ K}^{-1})] = -198.1 \text{ J/K}$

(b) $\Delta S°_{\text{reaction}} = [2 \text{ mol}(238.9 \text{ J mol}^{-1} \text{ K}^{-1})] - [3 \text{ mol}(205.152 \text{ J mol}^{-1} \text{ K}^{-1})] = -137.7$ J/K

(c) $\Delta S°_{\text{reaction}} = [1 \text{ mol}(64.8 \text{ J mol}^{-1} \text{ K}^{-1}) + 2 \text{ mol}(37.99 \text{ J mol}^{-1} \text{ K}^{-1})] - [1 \text{ mol}(68.7 \text{ J mol}^{-1} \text{ K}^{-1}) + 2 \text{ mol}(29.9 \text{ J mol}^{-1} \text{ K}^{-1})] = 12.3 \text{ J/K}$

(d) $\Delta S°_{\text{reaction}} = [2 \text{ mol}(213.8 \text{ J mol}^{-1} \text{ K}^{-1}) + 2 \text{ mol}(69.95 \text{ J mol}^{-1} \text{ K}^{-1})] - [1 \text{ mol}(219.3 \text{ J mol}^{-1} \text{ K}^{-1}) + 3 \text{ mol} (205.152 \text{ J mol}^{-1} \text{ K}^{-1})] = -267.3 \text{ J/K}$

12.29 Reactions have negative entropy changes when products are more constrained than reactants.

(a) There is a relatively large negative value because of the reduction in number of moles of gaseous substances: 4 moles of gaseous reactants are converted to 2 moles of gaseous products.

(b) There is a relatively large negative value because of the reduction in number of moles of gaseous substances: 3 moles of gaseous reactants are converted to 2 moles of gaseous products.

(c) This reaction has a near-zero entropy change because all reagents are relatively constrained solid substances.

(d) There is a relatively large negative value because of the reduction in number of moles of gaseous substances: 4 moles of gaseous reactants are converted to 2 moles of gaseous products.

12.31 The sign of $\Delta S°$ for a reaction depends on whether the atoms are more dispersed among the products or the reactants.

(a) A gas is more dispersed than a solid, so $\Delta S°$ is positive.

(b) There are gaseous molecules among the reactants but not in the products, so $\Delta S°$ is negative.

(c) Three molecules of gaseous reactants become two molecules of gaseous products, so there are more constraints and $\Delta S°$ is negative.

12.33 False statements can be made true in various ways; we choose the simplest change that corrects each statement.

(a) $\Delta G_{system} < 0$ for any spontaneous process at constant T and p.

(b) The free energy of a system decreases in any spontaneous process at constant T and p.

(c) $\Delta G = \Delta H - T\Delta S$ at constant T and p.

12.35 Standard free energy changes are calculated from Equation 12-9 using standard free energies of formation, which can be found in Appendix D of your textbook:

$$\Delta G°_{reaction} = \sum coeff_{products} \, \Delta G°_{f, products} - \sum coeff_{reactants} \, \Delta G°_{f, reactants}$$

(a) $\Delta G°_{reaction} = [2 \text{ mol}(-16.4 \text{ kJ/mol})] - [1 \text{ mol}(0 \text{ kJ/mol}) + 3 \text{ mol}(0 \text{ kJ/mol})] = -32.8$ kJ

(b) $\Delta G°_{reaction} = [2 \text{ mol}(163.2 \text{ kJ/mol})] - 3 \text{ mol}(0 \text{ kJ/mol}) = 326.4 \text{ kJ}$

(c) $\Delta G°_{reaction} = [1 \text{ mol}(0) + 2 \text{ mol}(-211.7 \text{ kJ/mol})] - [1 \text{ mol}(-217.3 \text{ kJ/mol}) + 2 \text{ mol}(0 \text{ kJ/mol})] = -206.1 \text{ kJ}$

(d) $\Delta G°_{reaction} = [2 \text{ mol}(-394.4 \text{ kJ/mol}) + 2 \text{ mol}(-237.1 \text{ kJ/mol})] - [1 \text{ mol}(68.4 \text{ kJ/mol}) + 3 \text{ mol}(0 \text{ kJ/mol})] = -1331.4 \text{ kJ}$

12.37 Use Equation 12-11 to estimate the standard free energy change at a temperature different from 298 K: $\Delta G^\circ_{\text{reaction}, T} \cong \Delta H^\circ_{\text{reaction}, 298\text{ K}} - T\Delta S^\circ_{\text{reaction}, 298\text{ K}}$. Standard entropy changes are calculated in Problem 12.27, but standard enthalpy changes need to be calculated from standard enthalpies of formation. $T = 273.15 - 85 = 188$ K:

12.27(b) $\Delta H^\circ_{\text{reaction}} = [2\text{ mol}(142.7\text{ kJ/mol})] - [3\text{ mol}(0\text{ kJ/mol})] = 285.4$ kJ

$$\Delta G^\circ_{\text{reaction}, T} \cong (285.4\text{ kJ}) - (188\text{ K})(-137.7\text{ J/K})\left(\frac{10^{-3}\text{ kJ}}{1\text{ J}}\right) = 311.3\text{ kJ}$$

12.27(c) $\Delta H^\circ_{\text{reaction}} = [1\text{ mol}(0\text{ kJ/mol}) + 2\text{ mol}(-239.7\text{ kJ/mol})] - [1\text{ mol}(-277.4$ kJ/mol$) + 2\text{ mol}(0\text{ kJ/mol})] = -202.0$ kJ

$$\Delta G^\circ_{\text{reaction}, T} \cong (-202.0\text{ kJ}) - (188\text{ K})(12.3\text{ J/K})\left(\frac{10^{-3}\text{ kJ}}{1\text{ J}}\right) = -204.3\text{ kJ}$$

12.39 Use Equation 12-13 to estimate the standard free energy change at pressures different from 1 bar:

$\Delta G_{\text{reaction}} = \Delta G^\circ_{\text{reaction}, T} + RT\ln Q$. $\Delta G^\circ_{\text{reaction}}$ values are calculated in Problem 12.35.

(a) $RT\ln\left(\dfrac{(p_{\text{NH}_3})^2}{(p_{\text{H}_2})^3(p_{\text{N}_2})}\right) = (8.314 \times 10^{-3}\text{ kJ mol}^{-1}\text{ K}^{-1})(298\text{ K})\ln\left(\dfrac{(15\text{ bar})^2}{(15\text{ bar})^3(15\text{ bar})}\right)$
$$= -13.4\text{ kJ}$$

$\Delta G_{\text{reaction}} = (-32.8\text{ kJ}) - (13.4\text{ kJ}) = -46.2\text{ kJ}$

(b) $RT\ln\left(\dfrac{(p_{\text{O}_3})^2}{(p_{\text{O}_2})^3}\right) = (8.314 \times 10^{-3}\text{ kJ mol}^{-1}\text{ K}^{-1})(298\text{ K})\ln\left(\dfrac{(15\text{ bar})^2}{(15\text{ bar})^3}\right) = -6.71\text{ kJ}$

$\Delta G_{\text{reaction}} = 326.4\text{ kJ} - 6.71\text{ kJ} = 319.7\text{ kJ}$

12.41 Use Equations 3-12, 12-6, and 12-9 to calculate standard thermodynamic values:

$$\Delta H^\circ_{\text{reaction}} = \sum \text{coeff}_{\text{products}}\ \Delta H^\circ_{\text{f, products}} - \sum \text{coeff}_{\text{reactants}}\ \Delta H^\circ_{\text{f, reactants}}$$

$$\Delta S^\circ_{\text{reaction}} = \sum \text{coeff}_{\text{products}}\ S^\circ_{\text{products}} - \sum \text{coeff}_{\text{reactants}}\ S^\circ_{\text{reactants}}$$

$$\Delta G^\circ_{\text{reaction}} = \sum \text{coeff}_{\text{products}}\ \Delta G^\circ_{\text{f, products}} - \sum \text{coeff}_{\text{reactants}}\ \Delta G^\circ_{\text{f, reactants}}$$

$\Delta H^\circ_{\text{reaction}} = [1\text{ mol}(-365.6\text{ kJ/mol}) + 1\text{ mol}(-285.83\text{ kJ/mol})] - [2\text{ mol}(-45.9\text{ kJ/mol})$
$+ 2\text{ mol}(0\text{ kJ/mol})] = -559.6\text{ kJ}$

$$\Delta S^{\circ}_{\text{reaction}} = [1 \text{ mol}(151.1 \text{ J mol}^{-1} \text{ K}^{-1}) + 1 \text{ mol}(69.95 \text{ J mol}^{-1} \text{ K}^{-1})] - [2 \text{ mol}(192.8 \text{ J}$$
$$\text{mol}^{-1} \text{ K}^{-1}) + 2 \text{ mol}(205.152 \text{ J mol}^{-1} \text{ K}^{-1})] = -574.9 \text{ J/K}$$

$$\Delta G^{\circ}_{\text{reaction}} = [1 \text{ mol}(-183.9 \text{ kJ/mol}) + 1 \text{ mol}(-237.1 \text{ kJ/mol})] - [2 \text{ mol}(-16.4 \text{ kJ/mol}) +$$
$$2 \text{ mol}(0 \text{ kJ/mol})] = -388.2 \text{ kJ}$$

12.43 (a) The number of molecules of gaseous products is smaller than the number of molecules of gaseous reactants, so the reaction has a negative ΔS°. The reaction is a formation reaction with a positive $\Delta H^{\circ}_{\text{f}}$ (Appendix D). Negative ΔS° combined with positive ΔH° means that ΔG° is positive at all temperatures. Thus, the reaction is not spontaneous at any temperature.

(b) Any gas-phase reaction can be made spontaneous by reducing the partial pressure of products to 0 bar.

12.45 To calculate a vapour pressure, use Equation 12-14:

$$\ln p_{\text{vap}} = -\frac{\Delta H_{\text{vap}}}{RT} + \frac{\Delta S^{\circ}_{\text{vap}}}{R}$$

The vapourization reaction is $4 \text{ P } (s, \text{ white}) \rightarrow P_4 (g)$.

$\Delta H^{\circ} = 58.9 - 4(0) = 58.9 \text{ kJ}$; $\Delta S^{\circ} = 280.0 - 4(41.1) = 115.6 \text{ J/K}$

$$\ln p_{\text{vap}} = -\left(\frac{(58.9 \text{ kJ/mol}) (10^3 \text{ J/kJ})}{(8.314 \text{ J mol}^{-1} \text{ K}^{-1})(298 \text{ K})}\right) + \left(\frac{115.6 \text{ J mol}^{-1} \text{ K}^{-1}}{8.314 \text{ J mol}^{-1} \text{ K}^{-1}}\right) = -9.87$$

$$p_{\text{vap}} = e^{-9.87} = 5.2 \times 10^{-5} \text{ bar}$$

12.47 Note that the "normal boiling point" means the temperature at which octane boils under an applied pressure of 1 atm (i.e., 1.01325 bar). Thus, we let $p_1 = 0.468$ bar, $T_1 = 100 \text{ }^{\circ}\text{C} = 373.15$ K, $p_2 = 1.01325$ bar, $\Delta H_{\text{vap}} = 50.1 \text{ kJ mol}^{-1}$, and solve for T_2:

$$\ln p_2 - \ln p_1 = \frac{\Delta H_{vap}}{R}\left(\frac{1}{T_1} - \frac{1}{T_2}\right)$$

$$\frac{R}{\Delta H_{vap}}\left(\ln p_2 - \ln p_1\right) = \left(\frac{1}{T_1} - \frac{1}{T_2}\right)$$

$$\frac{1}{T_2} = \frac{1}{T_1} - \frac{R}{\Delta H_{vap}}\left(\ln p_2 - \ln p_1\right)$$

$$T_2 = \left(\frac{1}{T_1} - \frac{R}{\Delta H_{vap}}\left(\ln p_2 - \ln p_1\right)\right)^{-1}$$

$$T_2 = \left(\left(\frac{1}{373.15\ \text{K}}\right) - \left(\frac{8.314\ \text{J K}^{-1}\text{mol}^{-1}}{50\ 100\ \text{J}/\text{mol}}\right)\left(\ln 1.01325 - \ln 0.468\right)\right)^{-1}$$

$$= (0.002680\ \text{K}^{-1} - 0.000128\ \text{K}^{-1})^{-1}$$

$$= 391.9\ \text{K}$$

$$= 119\ ^{\circ}\text{C}$$

12.49

$$\Delta H_{vap} = \frac{R\,(\ln p_2 - \ln p_1)}{\dfrac{1}{T_1} - \dfrac{1}{T_2}}$$

$$= \frac{8.314\,\text{J K}^{-1}\text{mol}^{-1}\,(\ln 17.5 - \ln 9.21)}{\left(\dfrac{1}{(10 + 273.15\ \text{K})}\right) - \left(\dfrac{1}{(20 + 273.15\ \text{K})}\right)}$$

$$= 44\ 298\ \text{J}/\text{mol}$$

$$= 44.3\ \text{kJ}/\text{mol}$$

This value is higher than that at the boiling point, because at the boiling point, the water molecules have more kinetic energy and therefore need less additional energy to escape into the gas phase.

12.51 Coupled reactions share a common intermediate that connects the reaction and allows transfer of free energy between reactants and final products. In the case of the seesaw, the first "reaction" would be Child 1 starting on the ground and rising into the air. The second "reaction" would be Child 2 starting in the air and falling to the ground. In this case, Child 2 would be the spontaneous reaction, Child 1 would be non-spontaneous (things tend to fall down, not float up), and so as Child 2 is falling he will cause the first child to rise. The common intermediate is the seesaw board, which acts as a lever transferring energy from one child to the other.

12.53 The reactions that are coupled in this example are the following:

$$\text{ATP} + \text{H}_2\text{O} \rightarrow \text{ADP} + \text{H}_3\text{PO}_4 \qquad \Delta G^{\circ} = -30.6\ \text{kJ}$$

$$C_6H_{12}O_6 \text{ (fructose)} + C_6H_{12}O_6 \text{ (glucose)} \rightarrow C_{12}H_{22}O_{11} + H_2O \qquad \Delta G° = 23.0 \text{ kJ}$$

Coupled reaction:

$$ATP + C_6H_{12}O_6 \text{ (fructose)} + C_6H_{12}O_6 \text{ (glucose)} \rightarrow ADP + C_{12}H_{22}O_{11} + H_3PO_4$$

$$\Delta G° = -30.6 \text{ kJ} + 23.0 \text{ kJ} = -7.6 \text{ kJ}$$

12.55 Spontaneity is determined by free energy, heat flow by enthalpy, and amount of dispersal by entropy: use Equations 3-12, 12-6, and 12-9 to calculate standard thermodynamic values:

$$\Delta H°_{\text{reaction}} = \sum \text{coeff}_{\text{products}} \, \Delta H°_{\text{f, products}} - \sum \text{coeff}_{\text{reactants}} \, \Delta H°_{\text{f, reactants}}$$

$$\Delta S°_{\text{reaction}} = \sum \text{coeff}_{\text{products}} \, S°_{\text{products}} - \sum \text{coeff}_{\text{reactants}} \, S°_{\text{reactants}}$$

$$\Delta G°_{\text{reaction}} = \sum \text{coeff}_{\text{products}} \, \Delta G°_{\text{f, products}} - \sum \text{coeff}_{\text{reactants}} \, \Delta G°_{\text{f, reactants}}$$

(a) $\Delta G°_{\text{reaction}} = [2 \text{ mol}(0 \text{ kJ/mol}) + 3 \text{ mol}(-237.1 \text{ kJ/mol})] - [1 \text{ mol}(-1582.3 \text{ kJ/mol}) + 3 \text{ mol}(0 \text{ kJ/mol})] = 871.0 \text{ kJ}$. The reaction is not spontaneous.

(b) $\Delta H°_{\text{reaction}} = [2 \text{ mol}(0 \text{ kJ/mol}) + 3 \text{ mol}(-285.83 \text{ kJ/mol})] - [1 \text{ mol}(-1675.7 \text{ kJ/mol}) + 3 \text{ mol}(0 \text{ kJ/mol})] = 818.2 \text{ kJ}$. The reaction absorbs heat.

(c) $\Delta S°_{\text{reaction}} = [2 \text{ mol}(28.3 \text{ J mol}^{-1} \text{ K}^{-1}) + 3 \text{ mol}(69.95 \text{ J mol}^{-1} \text{ K}^{-1})] - [1 \text{ mol}(50.9 \text{ J mol}^{-1} \text{ K}^{-1}) + 3 \text{ mol}(130.680 \text{ J mol}^{-1} \text{ K}^{-1})] = -176.5 \text{ J/K}$. Entropy decreases, so the products are less dispersed than the reactants.

12.57 Use Equation 12-6 to calculate the entropy change of a reaction:

$$\Delta S°_{\text{reaction}} = \sum \text{coeff}_{\text{products}} \, S°_{\text{products}} - \sum \text{coeff}_{\text{reactants}} \, S°_{\text{reactants}}$$

(a) $\Delta S°_{\text{reaction}} = 2 \text{ mol}(162 \text{ J mol}^{-1} \text{ K}^{-1}) - [2 \text{ mol}(101 \text{ J mol}^{-1} \text{ K}^{-1}) + 1 \text{ mol}(205.152 \text{ J mol}^{-1} \text{ K}^{-1})] = -83 \text{ J/K}$

(b) $\Delta S°_{\text{reaction}} = [2 \text{ mol}(87.4 \text{ J mol}^{-1} \text{ K}^{-1}) + 6 \text{ mol}(223.1 \text{ J mol}^{-1} \text{ K}^{-1})] - [4 \text{ mol}(142.3 \text{ J mol}^{-1} \text{ K}^{-1}) + 3 \text{ mol}(205.152 \text{ J mol}^{-1} \text{ K}^{-1})] = 328.7 \text{ J/K}$

(c) $\Delta S°_{\text{reaction}} = [6 \text{ mol}(210.8 \text{ J mol}^{-1} \text{ K}^{-1}) + 6 \text{ mol}(69.95 \text{ J mol}^{-1} \text{ K}^{-1})] - [3 \text{ mol}(121.2 \text{ J mol}^{-1} \text{ K}^{-1}) + 4 \text{ mol}(238.9 \text{ J mol}^{-1} \text{ K}^{-1})] = 365.3 \text{ J/K}$

12.59 Use Equation 11-9 to calculate the standard free energy change of a reaction:

$$\Delta G°_{\text{reaction}} = \sum \text{coeff}_{\text{products}} \, \Delta G°_{\text{f, products}} - \sum \text{coeff}_{\text{reactants}} \, \Delta G°_{\text{f, reactants}}$$

(a) $\Delta G^{\circ}_{\text{reaction}} = [2 \text{ mol}(-3 \text{ kJ/mol})] - [2 \text{ mol}(17 \text{ kJ/mol}) + 1 \text{ mol}(0 \text{ kJ/mol})] = -40 \text{ kJ}$

(b) $\Delta G^{\circ}_{\text{reaction}} = [2 \text{ mol}(-742.2 \text{ kJ/mol}) + 6 \text{ mol}(0)] - [4 \text{ mol}(-334.0 \text{ kJ/mol}) + 3 \text{ mol}(0 \text{ kJ/mol})] = -148.4 \text{ kJ}$

(c) $\Delta G^{\circ}_{\text{reaction}} = [6 \text{ mol}(87.6 \text{ kJ/mol}) + 6 \text{ mol}(-237.1 \text{ kJ/mol})] - [3 \text{ mol}(149.3 \text{ kJ/mol}) + 4 \text{ mol}(163.2 \text{ kJ/mol})] = -1997.7 \text{ kJ}$

12.61 To determine ΔG° at a temperature other than 298 K, calculate ΔS° and ΔH° using Equations 12-6 and 3-12, and then use Equation 12-11. The standard entropy changes are calculated in Problem 12.57:

$$\Delta S^{\circ}_{\text{reaction}} = \Sigma \text{coeff}_{\text{products}} \ S^{\circ}_{\text{products}} - \Sigma \text{coeff}_{\text{reactants}} \ S^{\circ}_{\text{reactants}}$$

$$\Delta H^{\circ}_{\text{reaction}} = \Sigma \text{coeff}_{\text{products}} \ \Delta H^{\circ}_{\text{f, products}} - \Sigma \text{coeff}_{\text{reactants}} \ \Delta H^{\circ}_{\text{f, reactants}}$$

Add: $T = 273.15 +$

$$\Delta G^{\circ}_{\text{reaction}, T} \cong \Delta H^{\circ}_{\text{reaction}, 298 \text{ K}} - T\Delta S^{\circ}_{\text{reaction}, 298 \text{ K}}$$

$85 = 358.2 \text{ K}$

(a) $\Delta H^{\circ}_{\text{reaction}} = [2 \text{ mol}(-104 \text{ kJ/mol})] - [2 \text{ mol}(-67 \text{ kJ/mol}) + 1 \text{ mol}(0 \text{ kJ/mol})] = -74 \text{ kJ}$

$$T\Delta S^{\circ}_{\text{reaction}} = (358.2 \text{ K})\left(\frac{-83 \text{ J}}{1 \text{ K}}\right)\left(\frac{10^{-3} \text{ kJ}}{1 \text{ J}}\right) = -29.7 \text{ kJ}$$

$$\Delta G^{\circ}_{\text{reaction}, 358.2 \text{ K}} = (-74 \text{ kJ}) - (-29.7 \text{ kJ}) = -42 \text{ kJ}$$

(b) $\Delta H^{\circ}_{\text{reaction}} = [2 \text{ mol}(-824.2 \text{ kJ/mol}) + 6 \text{ mol}(0 \text{ kJ/mol})] - [4 \text{ mol}(-399.5 \text{ kJ/mol}) + 3 \text{ mol}(0 \text{ kJ/mol})] = -50.4 \text{ kJ}$

$$T\Delta S^{\circ}_{\text{reaction}} = (358.2 \text{ K})\left(\frac{328.7 \text{ J}}{1 \text{ K}}\right)\left(\frac{10^{-3} \text{ kJ}}{1 \text{ J}}\right) = 117.7 \text{ kJ}$$

$$\Delta G^{\circ}_{\text{reaction}, 358.2 \text{ K}} = (-50.4 \text{ kJ}) - (117.7 \text{ kJ}) = -168.1 \text{ kJ}$$

(c) $\Delta H^{\circ}_{\text{reaction}} = [6 \text{ mol}(91.3 \text{ kJ/mol}) + 6 \text{ mol}(-285.83 \text{ kJ/mol})] - [3 \text{ mol}(50.6 \text{ kJ/mol}) + 4 \text{ mol}(142.7 \text{ kJ/mol})] = -1889.8 \text{ kJ}$

$$T\Delta S^{\circ}_{\text{reaction}} = (358.2 \text{ K})\left(\frac{365.3 \text{ J}}{1 \text{ K}}\right)\left(\frac{10^{-3} \text{ kJ}}{1 \text{ J}}\right) = 130.9 \text{ kJ}$$

$$\Delta G^{\circ}_{\text{reaction}, 358.2 \text{ K}} = (-1889.8 \text{ kJ}) - (130.9 \text{ kJ}) = -2020.7 \text{ kJ}$$

12.63 The figure shows a chemical reaction in which a set of diatomic molecules fragment into atoms with no change in volume or temperature.

   (a) There is no volume change, so $w_{sys} = 0$.

   (b) Energy must be supplied to break the chemical bonds, so $q_{sys} > 0$.

   (c) Because $q_{sys} > 0$, $q_{surr} < 0$, so $\Delta S_{surr} < 0$.

12.65 A reaction is thermodynamically feasible if $\Delta G < 0$.

   (a) $\Delta G° = [2\ mol(-73.5\ kJ/mol) + 1\ mol(87.6\ kJ/mol)] - [3\ mol(51.3\ kJ/mol) + 1\ mol(-237.1\ kJ/mol)] = 23.8\ kJ$. This is not thermodynamically feasible at 298 K under standard conditions.

   (b) $\Delta G° = \Delta H° - T\Delta S°$

   $\Delta H° = [2\ mol(-133.9\ kJ/mol) + 1\ mol(91.3\ kJ/mol)] - [3\ mol(33.2\ kJ/mol) + 1\ mol(-285.83\ kJ/mol)] = 9.73\ kJ$

   $\Delta S° = [2\ mol(266.9\ J\ mol^{-1}\ K^{-1}) + 1\ mol(210.8\ J\ mol^{-1}\ K^{-1})] - [3\ mol(240.1\ J\ mol^{-1}\ K^{-1}) + 1\ mol(69.95\ J\ mol^{-1}\ K^{-1})] = -45.65\ J/K$

   $\Delta G° = 9.73\ kJ/mol - (277\ K)(-45.65\ J/K)(10^{-3}\ kJ/J) = 22.4\ kJ$

   This is not thermodynamically feasible at 277 K under standard conditions.

   (c) $\Delta G = \Delta G° + RT\ \ln Q$ $Q = \left[\dfrac{(p_{HNO_3})^2 (p_{NO})}{(p_{NO_2})^3}\right]$

   $\Delta G = 23.8\ kJ + (0.008314\ kJ\ mol^{-1}\ K^{-1})(298\ K)\ \ln\left[\dfrac{(10^{-6})^2(10^{-6})}{(1)^3}\right]$

   $= 23.8\ kJ - 102.69\ kJ = -78.9\ kJ$

   The reaction is thermodynamically feasible under these conditions.

12.67 Energy "transactions" in the body are carried out using the ATP–ADP energy-storing reaction. When a cell needs energy, it "spends" ATP; when fat or carbohydrates ("capital") are consumed, the energy is stored by converting ADP to ATP ("buying" ATP). Like money, ATP is readily transported from place to place and is readily converted into "buying power."

12.69 Spontaneity is determined by free energy, heat flow by enthalpy, and change in order by entropy. Use Equations 3-12, 12-6, and 12-9 to calculate standard thermodynamic values:

$$\Delta H°_{reaction} = \Sigma\ coeff_{products}\ \Delta H°_{f,\ products} - \Sigma\ coeff_{reactants}\ \Delta H°_{f,\ reactants}$$

$$\Delta S^\circ_{\text{reaction}} = \Sigma \text{ coeff}_{\text{products}} \, S^\circ_{\text{products}} - \Sigma \text{ coeff}_{\text{reactants}} \, S^\circ_{\text{reactants}}$$

$$\Delta G^\circ_{\text{reaction}} = \Sigma \text{ coeff}_{\text{products}} \, \Delta G^\circ_{\text{f, products}} - \Sigma \text{ coeff}_{\text{reactants}} \, \Delta G^\circ_{\text{f, reactants}}$$

(a) $\Delta G^\circ_{\text{reaction}}$ = [1 mol(–261.905 kJ/mol) + 1 mol(–131.0 kJ/mol)] –1 mol(–384.1 kJ/mol) = –8.8 kJ. The reaction is spontaneous.

(b) $\Delta H^\circ_{\text{reaction}}$ = [1 mol(–240.3 kJ/mol) + 1 mol(–167.1 kJ/mol)] – 1 mol(–411.2 kJ/mol) = 3.8 kJ. The reaction absorbs heat.

(c) $\Delta S^\circ_{\text{reaction}}$ = [1 mol(58.5 J mol^{-1} K^{-1}) + 1 mol(56.5 J mol^{-1} K^{-1})] – 1 mol(72.1 J mol^{-1} K^{-1}) = 42.9 J/K. Entropy increases, because the ions become dispersed during this reaction.

12.71 Entropy changes for constant-temperature processes can be calculated using Equation 12–2, $\Delta S = \dfrac{q_{\text{T}}}{T}$, and q for water freezing can be found using $q = - n\Delta H_{\text{fus}}$:

$$n_{\text{water}} = \frac{m}{M} = \frac{155 \text{ g}}{18.02 \text{ g/mol}} = 8.60 \text{ mol}$$

$$T_{\text{water}} = 0.0 + 273.15 \text{ K} = 273.2 \text{ K}$$

$$q_{\text{water}} = -8.60 \text{ mol}\left(\frac{6.01 \text{ kJ}}{1 \text{ mol}}\right)\left(\frac{10^3 \text{ J}}{1 \text{ kJ}}\right) = -5.17 \times 10^4 \text{ J}$$

$$T_{\text{surr}} = -20.0 + 273.15 \text{ K} = 253.2 \text{ K}$$

(a) $\Delta S_{\text{water}} = \dfrac{-5.17 \times 10^4 \text{ J}}{273.2 \text{ K}} = -189 \text{ J/K}$

(b) The energy released by the water will be absorbed by the surroundings:

$$q_{\text{surr}} = -q_{\text{water}} = 5.17 \times 10^4 \text{ J}$$

$$\Delta S_{\text{surr}} = \frac{5.17 \times 10^4 \text{ J}}{253.2 \text{ K}} = 204 \text{ J/K}$$

$$\Delta S_{\text{total}} = \Delta S_{\text{water}} + \Delta S_{\text{surr}} = -189 \text{ J/K} + 204 \text{ J/K} = 15 \text{ J/K}$$

(c) Cooling the ice to –20 °C is an irreversible change (because the temperatures of the ice and freezer are not the same, except at the end of the process), for which ΔS_{total} > 0. Thus, the cooling must generate an additional entropy increase for the universe.

12.73 Ammonia is a toxic gas, whereas urea and ammonium nitrate both are relatively non-toxic solids. Thus, the transport and application of ammonia entail significant risks to humans. Even in aqueous solution, ammonia is highly irritating, as anyone who has used strong ammonia as a cleanser knows.

12.75 The standard free energies of combustion are given in your textbook: $\Delta G^\circ_{glucose} =$ −2870 kJ/mol and $\Delta G^\circ_{palmitic\ acid}$ = −9790 kJ/mol. Divide molar quantities by molar masses to obtain free energy per gram:

$$\text{Glucose: } \Delta G_{per\ gram} = \left(\frac{-2870\ kJ}{1\ mol}\right)\left(\frac{1\ mol}{180\ g}\right) = -15.9\ kJ/g$$

$$\text{Palmitic acid: } \Delta G_{per\ gram} = \left(\frac{-9790\ kJ}{1\ mol}\right)\left(\frac{1\ mol}{256\ g}\right) = -38.2\ kJ/g$$

Palmitic acid has the higher energy content per gram.

12.77 (a) Whenever a substance cools, ΔS is negative.

(b) Whenever a gas is compressed (at constant *T*), ΔS is negative.

(c) Whenever two substances mix, matter is dispersed and ΔS is positive.

12.79 (a) Energy is always conserved, so $\Delta E_{total} = 0$.

(b) The teaspoon warms, so $\Delta E_{teaspoon} > 0$.

(c) This is a spontaneous process, so $\Delta S_{total} > 0$.

(d) The water cools, so $\Delta S_{water} < 0$.

(e) The teaspoon warms, so $q_{teaspoon} > 0$.

12.81 Entropy depends on amount, phase, temperature, and concentration. The order is (0.5 mol, liquid, 298 K) < (1 mol, liquid, 298 K) < (1 mol, liquid, 373 K) < (1 mol, gas, 1 bar, 373 K) < (1 mol, gas, 0.1 bar, 373 K).

12.83 Use values from Appendix D of your textbook and Equations 3-12, 12-6, and 12-11 to calculate standard thermodynamic values:

$$\Delta H^\circ_{reaction} = \Sigma\ coeff_{products}\ \Delta H^\circ_{f,\ products} - \Sigma\ coeff_{reactants}\ \Delta H^\circ_{f,\ reactants}$$

$$\Delta S^\circ_{reaction} = \Sigma\ coeff_{products}\ S^\circ_{products} - \Sigma\ coeff_{reactants}\ S^\circ_{reactants}$$

$$\Delta G^\circ_{reaction,\ T} \cong \Delta H^\circ_{reaction,\ 298\ K} - T\Delta S^\circ_{reaction,\ 298\ K}$$

$$\Delta H^\circ_{reaction} = [2\ mol(91.3\ kJ/mol)] - [1\ mol(0\ kJ/mol) + 1\ mol(0\ kJ/mol)] = 182.6\ kJ$$

$$\Delta S^\circ_{reaction} = [2\ mol(210.8\ J\ mol^{-1}\ K^{-1})] - [1\ mol(191.622\ J\ mol^{-1}\ K^{-1}) + 1\ mol(205.152\ J\ mol^{-1}\ K^{-1})] = 24.83\ J/K$$

$$\Delta G^{\circ}_{\text{reaction, 298 K}} = (182.6 \text{ kJ}) - (298 \text{ K})\left(\frac{24.83 \text{ J}}{1 \text{ K}}\right)\left(\frac{10^{-3} \text{ kJ}}{1 \text{ J}}\right) = 175.2 \text{ kJ}$$

To estimate the temperature at which the reaction becomes spontaneous, set $\Delta G^{\circ}_{\text{reaction, } T} = 0$ and solve for T:

$$0 = 182.6 \text{ kJ} - T\left(\frac{24.83 \text{ J}}{1 \text{ K}}\right)\left(\frac{10^{-3} \text{ kJ}}{1 \text{ J}}\right) \qquad T = \frac{182.6 \text{ kJ}}{(24.83 \text{ J/K})(10^{-3} \text{ kJ/J})} = 7354 \text{ K}$$

The temperature in an automobile engine is not anywhere near this high, but pressures are far from standard. In particular, the pressure of the NO product is very low, and this makes the reaction spontaneous.

12.85 Vapourization at the normal boiling point is a constant-temperature process, for which the entropy change can be calculated from Equation 12-2. The change takes place at constant pressure, so $q = \Delta H$:

(a) $\Delta S^{\circ}_{\text{vap}} = \dfrac{\Delta H_{\text{vap}}}{T} = \dfrac{(30.77 \text{ kJ/mol})(10^3 \text{ J/kJ})}{(80.1 + 273.15 \text{ K})} = 87.1 \text{ J/K}$

$\Delta G^{\circ}_{\text{vap}} = \Delta H - T\Delta S = 0$ at the normal boiling point

(b) $\Delta G^{\circ}_{\text{vap}} = \Delta H - T\Delta S$

$\Delta G^{\circ}_{\text{vap}} = (30.77 \text{ kJ/mol}) - (21 + 273.15 \text{ K})(87.1 \text{ J mol}^{-1} \text{ K}^{-1})(10^{-3} \text{ kJ/J})$

$\qquad = 5.16 \text{ kJ/mol at 21 °C}$

(c) Use Equation 12-14 to calculate a vapour pressure:

$$\ln p_{\text{vap}} = -\frac{\Delta H_{\text{vap}}}{RT} + \frac{\Delta S^{\circ}_{\text{vap}}}{R}$$

$$\ln p_{\text{vap}} = -\left(\frac{(30.77 \text{ kJ/mol})(10^3 \text{ J/kJ})}{(8.314 \text{ J mol}^{-1} \text{ K}^{-1})(298 \text{ K})}\right) + \left(\frac{87.1 \text{ J mol}^{-1} \text{ K}^{-1}}{8.314 \text{ J mol}^{-1} \text{ K}^{-1}}\right) = -1.98$$

$$p_{\text{vap}} = e^{-1.98} = 0.14 \text{ bar}$$

12.87 The process that takes place is vapourization of the liquid bromine to form bromine vapour.

(a) At the end of the process, all the bromine is in the gas phase, with the molecules well separated, and the piston has moved back.

(b) The process is reversible, since the temperatures of the system and surroundings are the same and the external and internal pressures both are 1.0 bar (bromine is at its boiling point). Thus, $\Delta S_{\text{total}} = 0$.

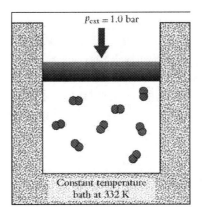

12.89 The attractive forces in CaO are substantially larger than the attractive forces in KCl, so the individual Ca^{2+} and O^{2-} ions cannot vibrate as easily in the crystal. Thus, the amount of vibrational disorder is substantially less in CaO than in KCl.

12.91 Make use of thermodynamic definitions to identify the state functions:

(a) $q_v = \Delta E$; (b) $q_p = \Delta H$; (c) $q_T = T\Delta S$.

12.93 Your molecular picture should show all three phases simultaneously present, with transfers of atoms occurring among all phases:

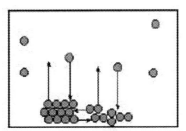

$\Delta G = 0$ for all three processes; when matter becomes more constrained in the system, energy is dispersed into the surroundings, and vice versa.

12.95 Use bond energy, intermolecular forces, and order–disorder to predict signs for ΔH and ΔS:

(a) Bonds form, so energy is released and ΔH is negative. The number of independent particles decreases, so order increases and ΔS is negative.

(b) When a solid melts, energy must be added; q is positive, ΔH is positive, and ΔS is positive.

(c) Combustion reactions release energy, so we expect ΔH to be negative. The number of molecules of gas remains the same during this reaction, so ΔS is expected to be small, but it is not possible to predict whether it is positive or negative without doing actual calculations.

12.97 Standard conditions refer to 1 bar, 298 K, all solutes at 1 M. In a cell, $T = 37\,°C$ (310 K) and no solutes are present at 1 M. In particular, phosphoric acid and phosphate concentrations are significantly lower than 1 M. The elevated temperature makes ΔG more negative in the reaction direction that has a positive ΔS, which is the hydrolysis reaction. The low concentration of free phosphates makes ΔG more negative in the reaction direction that produces free phosphate, which also is the hydrolysis reaction.

12.99 According to your textbook, the complete oxidation of 1 mole of palmitic acid produces 130 moles of ATP. The reactions are

$$\text{ADP} + \text{H}_3\text{PO}_4 \rightarrow \text{ATP} + \text{H}_2\text{O} \qquad\qquad \Delta G° = +30.6 \text{ kJ}$$

$$\text{Palmitic acid: } \text{C}_{15}\text{H}_{31}\text{CO}_2\text{H} + 23\,\text{O}_2 \rightarrow 16\,\text{CO}_2 + 16\,\text{H}_2\text{O} \qquad \Delta G° = -9790 \text{ kJ}$$

Thus, the amount of energy stored is $(130)(30.6 \text{ kJ}) = 3978 \text{ kJ}$ and the efficiency is as follows:

$$\text{Efficiency} = 100\% \left(\frac{\text{energy stored}}{\text{energy released}} \right) = 100\% \left(\frac{3978 \text{ kJ}}{9790 \text{ kJ}} \right) = 40.6\%$$

Metabolism of 1 mole of palmitic acid generates $9790 \text{ kJ} - 3978 \text{ kJ} = 5812 \text{ kJ}$ of free energy. We will assume that all this free energy is converted to heat. On a per-gram basis, this is

$$q_{\text{per gram}} = \left(\frac{5812 \text{ kJ}}{1 \text{ mol}} \right)\left(\frac{1 \text{ mol}}{256.42 \text{ g}} \right) = 22.67 \text{ kJ/g}$$

Evaporation of 75 g of water requires a heat input:

$$q = n\Delta H_{\text{vap}} = \left(\frac{75 \text{ g}}{18.02 \text{ g/mol}} \right)\left(\frac{40.79 \text{ kJ}}{1 \text{ mol}} \right) = 1.7 \times 10^2 \text{ kJ}$$

$$\text{Mass required} = (1.7 \times 10^2 \text{ kJ})\left(\frac{1 \text{ g}}{22.67 \text{ kJ}} \right) = 7.5 \text{ g}$$

12.101 For the first reaction, ATP is consumed and water is produced, so it is likely that energy is released. There are three molecules produced from two, so ΔS is likely to be positive. The process is spontaneous, so ΔG is negative. For the second reaction, light is produced, so energy is released. There is one gaseous reactant and one gaseous product, but the starting material fragments, so ΔS is positive. The process is spontaneous, so ΔG is negative.

12.103 The HCl molecule is heteronuclear, whereas H_2 and Cl_2 both are homonuclear. The homonuclear molecules are more symmetrical than the heteronuclear ones, so they have fewer possible distinguishable orientations in space. Here is an example:

Distinguishable   Indistinguishable

The heteronuclear molecules have more possible orientations, and larger entropy.

12.105 (a) Use Equations 3-12, 12-6, and 12-9 to calculate standard thermodynamic values:

$$\Delta H^\circ_{reaction} = \sum coeff_{products} \ \Delta H^\circ_{f, \ products} - \sum coeff_{reactants} \ \Delta H^\circ_{f, \ reactants}$$

$$\Delta S^\circ_{reaction} = \sum coeff_{products} \ S^\circ_{products} - \sum coeff_{reactants} \ S^\circ_{reactants}$$

$$\Delta G^\circ_{reaction} = \sum coeff_{products} \ \Delta G^\circ_{f, \ products} - \sum coeff_{reactants} \ \Delta G^\circ_{f, \ reactants}$$

$\Delta H^\circ_{reaction} = 2 \ mol(142.7 \ kJ/mol) - 3 \ mol(0 \ kJ/mol) = 285.4 \ kJ$

$\Delta S^\circ_{reaction} = 2 \ mol(238.9 \ J \ mol^{-1} \ K^{-1}) - 3 \ mol(205.152 \ J \ mol^{-1} \ K^{-1}) = -137.7 \ J/K$

$\Delta G^\circ_{reaction} = 2 \ mol(163.2 \ kJ/mol) - 3 \ mol(0 \ kJ/mol) = 326.4 \ kJ$

(b) $\Delta G^\circ = \Delta H^\circ - T\Delta S^\circ$; because $\Delta H^\circ_{reaction}$ is positive and $\Delta S^\circ_{reaction}$ is negative, $\Delta G^\circ_{reaction}$ is positive at all temperatures.

(c) Calculate the pressure of O_3 at which $\Delta G = 0$; the reaction will be spontaneous at any lower pressure:

$$\Delta G = \Delta G^\circ_{reaction} + RT \ \ln\left(\frac{(p_{O_3})^2}{(p_{O_2})^3}\right) = \Delta G^\circ_{reaction} + RT \ \ln\left(\frac{(p_{O_3})^2}{(0.20)^3}\right) = 0$$

$$326.4 \ kJ/mol + \left(8.314 \times 10^{-3} \ kJ \ mol^{-1} \ K^{-1}\right)(298 \ K) \ \ln\left(\frac{(p_{O_3})^2}{(0.20)^3}\right) = 0$$

$$\ln\left(\frac{(p_{O_3})^2}{(0.20)^3}\right) = -\frac{326.4 \ kJ}{\left(8.314 \times 10^{-3} \ kJ \ mol^{-1} \ K^{-1}\right)(298 \ K)} = -131.7$$

$$\frac{(p_{O_3})^2}{(0.20)^3} = 6.4 \times 10^{-58} \qquad p_{O_3} = 2.3 \times 10^{-30} \ bar$$

(d) The calculations show that the production of ozone is not spontaneous, regardless of temperature and pressure. In the upper atmosphere, the reaction is coupled with the absorption of light by O_2 molecules, which deposits energy in the system and breaks O–O bonds. Once O atoms are present, the formation of ozone becomes spontaneous.

12.107 (a)

$$\Delta G^\circ = \Delta H^\circ - T\Delta S^\circ = 5.65 \ kJ/mol - (298 \ K)(28.9 \ J \ mol^{-1} \ K^{-1})\left(\frac{10^{-3} \ kJ}{1 \ J}\right) = -2.96 \ kJ/mol$$

(b) The freezing point is the temperature at which $\Delta G° = 0$:

$$\Delta G° = 0 = (5.65 \text{ kJ/mol}) - T(28.9 \text{ J mol}^{-1} \text{ K}^{-1})\left(\frac{10^{-3} \text{ kJ}}{1 \text{ J}}\right)$$

$$T = \left(\frac{5.65 \text{ kJ/mol}}{28.9 \text{ J mol}^{-1} \text{ K}^{-1}}\right)\left(\frac{10^{3} \text{ J}}{1 \text{ kJ}}\right) = 196 \text{ K}$$

12.109 (a) At sufficiently low temperature, every substance adopts the form that is most highly constrained. The stable form of phosphorus at low temperature is $P_{4\beta}$, so this is the more constrained crystalline structure.
(b) A phase change that occurs when temperature decreases involves the release of heat (consider gaseous water condensing or liquid water freezing). Thus, ΔH for this process is negative. The more constrained form of phosphorus is $P_{4\beta}$, so for this process, ΔS is negative for the system.
(c) The $P_{4\alpha}$ crystal is unstable below 77 K, so it would not have $S° = 0$ at 0 K.

12.111 Figure 12-12 indicates that the entropy increases as the temperature increases, and increases abruptly at each phase change. The data give the entropy of the solid at the melting temperature (83.8 K). As the temperaure is further increased, the argon will melt, then warm further, then vapourize, then warm further, up to the temperature of interest, 500 K. Each of these steps contributes to the entropy at 500 K:
$S°(Ar_{(g), 500 K}) = S°(Ar_{(s), 83.8 K}) + (\Delta S_{fus}) + (\Delta S_{Ar(l)} \text{ from } T_m \text{ to } T_b) + (\Delta S_{vap}) + (\Delta S_{Ar(g)} \text{ from } T_b \text{ to } 500 \text{ K})$
$= 35.5 + 14.1 + 44.6 \ln(87.3/83.8) + 73.6 + 20.8 \ln(500/87.3)$
$= 35.5 + 14.1 + 1.8 + 73.6 + 36.3$
$= 161.3 \text{ J K}^{-1} \text{ mol}^{-1}$
The largest contributor to the absolute entropy of argon gas at 500 K is the entropy of vapourization, since a low entropy liquid is converted into a high entropy gas.

Chapter 13 Kinetics: Mechanisms and Rates of Reactions

Solutions to Problems in Chapter 13

13.1 In any sequence of steps, the slowest one will be rate-determining: (a) Pouring the coffee from the urn into the cup; (b) entering the items on the cash register (if the market has a good laser scanner, paying and receiving change may be rate-determining); and (c) preparing for the jump and passing through the door.

13.3 Every elementary reaction must depict actual molecular processes:

(a) $I_2 \rightarrow I + I$

(b) $H_2 + I_2 \rightarrow H_2I_2$

(c) $H_2 + I_2 \rightarrow H + HI_2$

13.5 A molecular picture of an elementary reaction shows the reactants, the products, and (if necessary) the intermediate collision complex:

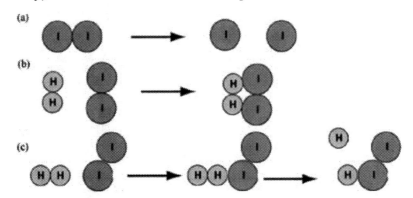

13.7 A satisfactory mechanism must consist entirely of reasonable elementary steps that sum to give the correct overall stoichiometry of the reaction:

(a) $I_2 \rightarrow I + I$

   $I + H_2 \rightarrow HI + H$

   $H + I \rightarrow HI$

(b) $H_2 + I_2 \rightarrow H_2I_2$

   $H_2I_2 \rightarrow HI + HI$

(c) $H_2 + I_2 \rightarrow H + HI_2$

   $H + HI_2 \rightarrow HI + HI$

13.9 The rate of a reaction can be expressed using the general expression:

$$\text{Rate} = \left(-\frac{1}{a}\right)\left(\frac{\Delta[A]}{\Delta t}\right)$$

(a) $\text{Rate} = -\dfrac{\Delta[Cl_2]}{\Delta t}$

(b) $-\dfrac{\Delta[Cl_2]}{\Delta t} = \left(\dfrac{1}{2}\right)\left(\dfrac{\Delta[NOCl]}{\Delta t}\right)$

(c) Use the result of (b) to calculate that NOCl appears at a rate of 94 M/s

13.11 In the reaction of NO and Cl_2, two NO molecules react for every Cl_2 that reacts, producing two NOCl molecules:

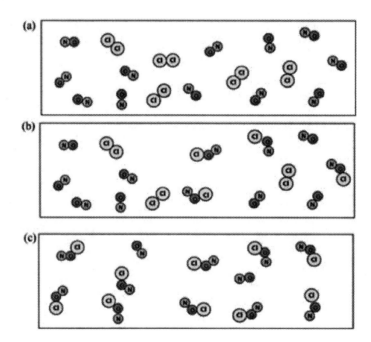

13.13 (a) To calculate the average rate of production, determine how many moles form during the time interval and divide by the time:

$$n = \frac{pV}{RT} = \frac{(0.15\ \text{bar})(5.0\ \text{L})}{(0.08314\ \text{L bar mol}^{-1}\ \text{K}^{-1})(550 + 273.15\ \text{K})} = 1.11 \times 10^{-2}\ \text{mol}$$

$$\frac{\Delta[CO_2]}{\Delta t} = \frac{1.11 \times 10^{-2}\ \text{mol}}{5.0\ \text{min}} = 2.2 \times 10^{-3}\ \text{mol/min}$$

(b) The reaction is $CaCO_3 \rightarrow CaO + CO_2$, so the amount of $CaCO_3$ decomposing is the same as the amount of CO_2 produced, 1.1×10^{-2} mol.

13.15 (a) Before the reaction begins:

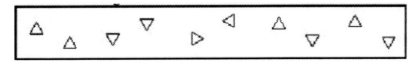

(b) After 20 minutes, the amount reacted is (20 min)(0.25 molecules/min) = 5 molecules:

13.17 (a) The number of decays is directly proportional to the starting amount of the isotope in first-order kinetics, and therefore there would be 6.0×10^6 decays in the 5.00 nmole case.

(b) The fraction decaying is the same in both cases:

$$\text{Fraction decaying} = \frac{\text{number of decays}}{\text{number of nuclei present}}$$

$$= \frac{1.2 \times 10^6 \text{ nuclei}}{(1.00 \text{ nmol})(10^{-9} \text{ mol/nmol})(6.022 \times 10^{23} \text{ nuclei/mol})}$$

$$= 1.0 \times 10^{-9}$$

(c) In first-order kinetics, the *number* reacting is proportional to the amount present, but the *fraction* reacting is independent of concentrations.

13.19 (a) The rate law for an elementary step contains the product of the reactant concentrations:

Rate = k[C][AB]

(b) The units of a rate constant have time in the denominator along with concentrations one power less than the number of reactants: units = (concentration)$^{-1}$ (time)$^{-1}$.

(c) The steps of the mechanism must sum to give the observed stoichiometry for the reaction. Intermediate A can react with AB, and then B and C can react:

$C + AB \rightarrow BC + A$

$A + AB \rightarrow B + A_2$

$B + C \rightarrow BC$

Net: $2 C + 2 AB \rightarrow 2 BC + A_2$

13.21 (a) The rate law for an elementary step contains the product of the reactant concentrations:

$$\text{Rate} = k[C][C][AB] = k[C]^2[AB]$$

(b) The units of a rate constant have time in the denominator along with concentrations one power less than the number of reactants: units = (concentration)$^{-2}$ (time)$^{-1}$

(c) The steps of the mechanism must sum to give the observed stoichiometry for the reaction. A bimolecular reaction between the intermediate (AC) and AB is the simplest possibility:

$$2\,C + AB \rightarrow BC + AC$$

$$AC + AB \rightarrow BC + A_2$$

$$\text{Net: } 2\,C + 2\,AB \rightarrow 2\,BC + A_2$$

13.23 According to the stated rate law, the rate of reaction is proportional to each concentration. Flask *a* contains six molecules of each type, whereas Flask *b* contains 10 NO and 2 O_3:

a: Rate = $k(6)(6) = 36\,k$

b: Rate = $k(10)(2) = 20\,k$

Flask *b* will react slower by a factor of 20/36 = 0.56.

13.25 (a) One way to treat experimental data is by plotting: for first-order behaviour, $\ln\left(\dfrac{[A]_0}{[A]}\right)$ vs. *t* is a straight line, and for second-order behaviour, $\dfrac{1}{[A]} - \dfrac{1}{[A]_0}$ vs. *t* is a straight line Here, the plot of $\ln\left(\dfrac{[A]_0}{[A]}\right)$ vs. *t* gives a straight line, so this reaction is first order.

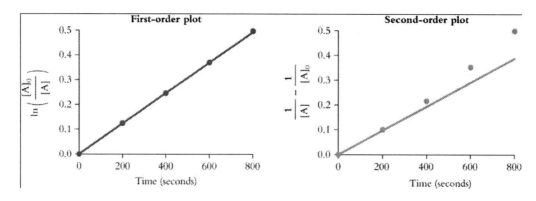

(b) For a first-order reaction, k = slope of the graph:

$$k = \text{slope} = \frac{\Delta y}{\Delta x} = \frac{0.491 - 0}{800 \text{ s} - 0 \text{ s}} = 6.14 \times 10^{-4} \text{ s}^{-1}$$

(c) To find concentration at any particular time, use $\ln[A] = \ln[A]_0 - kt$

$$\ln[A] = \ln[2.50] - (6.14 \times 10^{-4} \text{ s}^{-1})(1600 \text{ s}) = -0.0661$$

$$[A] = e^{-0.0661} = 0.936 \text{ bar}$$

(d) To find the time at which concentration reaches a particular value, use

$$kt = \ln\left(\frac{[A]_0}{[A]}\right) \qquad t = \frac{\ln\left(\frac{[A]_0}{[A]}\right)}{k} = \frac{\ln\left(\frac{2.50 \text{ bar}}{0.500 \text{ bar}}\right)}{6.14 \times 10^{-4} \text{ s}^{-1}} = 2.62 \times 10^3 \text{ s}$$

13.27 This is stated to be a second-order reaction, so rate = $k[\text{NOBr}]^2$ and Equation 13-5 applies:

$$\frac{1}{[A]} - \frac{1}{[A]_0} = kt$$

The problem states that k = 25 M^{-1} min^{-1}.

(a) $kt = \dfrac{1}{0.010 \text{ M}} - \dfrac{1}{0.025 \text{ M}} = 60 \text{ M}^{-1}$

$t = \dfrac{60 \text{ M}^{-1}}{25 \text{ M}^{-1} \text{ min}^{-1}} = 2.4 \text{ min}$

(b) $\dfrac{1}{[A]} = \left(\dfrac{1}{0.025 \text{ M}}\right) + (25 \text{ M}^{-1} \text{ min}^{-1})(125 \text{ min}) = 40 \text{ M}^{-1} + 3125 \text{ M}^{-1} = 3165 \text{ M}^{-1}$

$[A] = 3.2 \times 10^{-4} \text{ M}$

13.29 This is stated to be a first-order reaction, so rate = $k[\text{C}_5\text{H}_{11}\text{Br}]$ and Equation 13-3 applies:

$$kt = \ln\left(\frac{[A]_0}{[A]}\right)$$

(a) $kt = \ln\left(\dfrac{0.125}{1.25 \times 10^{-3}}\right) = 4.61 \qquad\qquad t = \dfrac{4.61}{0.385 \text{ h}^{-1}} = 12.0 \text{ h}$

b) $\ln[A] = \ln[0.125] - (3.5 \text{ h})(0.385 \text{ h}^{-1}) = -3.43$

$[A] = e^{-3.43} = 0.0324 \text{ M or } 3.24 \times 10^{-2} \text{ M}$

13.31 The isolation method requires that one concentration be substantially smaller than all the others, so the experimental reaction order follows the order with respect to that one reactant. We are told that it is possible to track the concentration of N_2O_5, so this reactant has to be the one with the low concentration (otherwise the concentration would not change enough to give good data). Do two experiments; in each, $[H_2O]_0 > 100 [N_2O_5]_0$ but with two different values for $[H_2O]_0$.

Plot $\ln\left(\dfrac{[N_2O_5]_0}{[N_2O_5]}\right)$ and $\dfrac{1}{[N_2O_5]} - \dfrac{1}{[N_2O_5]_0}$ vs. t to determine order with respect to N_2O_5, and use the ratio of slope values and H_2O concentrations to determine order with respect to H_2O.

After determining the reaction order, rate constant can be determined from the slope of the linear plot (as described in Example 13-7 in the text).

13.33 From the data provided, we recognize this as an initial rate problem. The essential feature of the initial rate method is that we can take ratios of initial rates under different conditions. First, apply this technique to Experiments 1 and 2, which have the same initial concentration of $S_2O_8^{2-}$:

$$\text{Initial rate}_1 = 4.4 \times 10^{-2} \text{ M/min} = k (0.125 \text{ M})^x (0.150 \text{ M})^y$$

$$\text{Initial rate}_2 = 1.3 \times 10^{-1} \text{ M/min} = k (0.375 \text{ M})^x (0.150 \text{ M})^y$$

When we take the ratio of the second initial rate to the first, the rate constant and the initial concentration term for $S_2O_8^{2-}$ cancel:

$$\frac{\text{Initial rate}_2}{\text{Initial rate}_1} = \frac{1.3 \times 10^{-1} \text{ M/min}}{4.4 \times 10^{-2} \text{ M/min}} = \frac{k(0.375 \text{ M})^x (0.150 \text{ M})^y}{k(0.125 \text{ M})^x (0.150 \text{ M})^y} = \frac{(0.375 \text{ M})^x}{(0.125 \text{ M})^x}$$

Simplifying, we find

$3.0 = (3.0)^x$, from which $x = 1$.

Now repeat this analysis for the third experiment and the first experiment, for which the initial concentrations of I^- are the same:

$$\frac{\text{Initial rate}_3}{\text{Initial rate}_1} = \frac{1.5 \times 10^{-2} \text{ M/min}}{4.4 \times 10^{-2} \text{ M/min}} = \frac{k(0.125 \text{ M})^x (0.050 \text{ M})^y}{k(0.125 \text{ M})^x (0.150 \text{ M})^y} = \frac{(0.050 \text{ M})^x}{(0.150 \text{ M})^x}$$

$0.34 = (0.33)^y$, from which $y = 1$.

The reaction is first order in each reactant, so the rate law is rate $= k[S_2O_8^{2-}][I^-]$ Use any of the experiments to evaluate the rate constant k:

$$4.4 \times 10^{-2} \text{ M/min} = k (0.125 \text{ M})(0.150 \text{ M})$$

$$k = \frac{4.4 \times 10^{-2} \text{ M/min}}{(0.125 \text{ M})(0.150 \text{ M})} = 2.3 \text{ M}^{-1} \text{ min}^{-1}$$

13.35 The rate law should relate the rate of reaction to the concentration of the reactants. Reactive intermediates should not be shown in the rate law:

    C + AB⇌ BC + A, followed by

    A + AB → B + A$_2$

    B + C → BC

    Net: 2 C + 2 AB → 2 BC + A$_2$

The rate law is determined by the rate-determining step: Rate = k_2[A][AB]. This is not satisfactory, however, because it contains the concentration of an intermediate (A). Set the rates equal for the forward and reverse first step:

$$k_1[\text{C}][\text{AB}] = k_{-1}[\text{BC}][\text{A}]$$

Solve this equality for [A]: $[\text{A}] = \left(\frac{k_1}{k_{-1}}\right)\frac{[\text{C}][\text{AB}]}{[\text{BC}]}$

Substitute into the rate expression: Rate $= k_2 \left(\frac{k_1}{k_{-1}}\right)\frac{[\text{C}][\text{AB}]^2}{[\text{BC}]}$

13.37 (a) The steps of a mechanism must sum to give the observed overall stoichiometry of the reaction. For ozone decomposition, this is 2 O$_3$ → 3 O$_2$. The two steps proposed by the student consume 1 O$_3$, produce 1 O$_2$, and generate an O atom, which must be consumed. Thus, the third step is O$_3$ + O → 2 O$_2$.

(b) The rate law is determined by the rate-determining step: Rate = k_2[O$_5$]. This is not satisfactory, however, because it contains the concentration of an intermediate. Set the rates equal for the forward and reverse first step:

$$k_1[\text{O}_3][\text{O}_2] = k_{-1}[\text{O}_5]$$

Solve this equality for [O$_5$]: $[\text{O}_5] = \left(\frac{k_1}{k_{-1}}\right)[\text{O}_3][\text{O}_2]$

Substitute into the rate expression:

$$\text{Rate} = k_2 \left(\frac{k_1}{k_{-1}}\right)[\text{O}_3][\text{O}_2]$$

(c) Atmospheric chemists would consider this mechanism to be molecularly unreasonable because fragmentation of O$_5$ in the second step (the breaking of two bonds simultaneously) is highly unlikely.

13.39 (a) The rate law is determined by the rate-determining step: Rate = $k_2[N_2O_2][O_2]$. This is not satisfactory, however, because it contains the concentration of an intermediate. Set the rates equal for the forward and reverse first step:

$$k_1[NO]^2 = k_{-1}[N_2O_2]$$

Solve this equality for $[N_2O_2]$: $[N_2O_2] = \left(\dfrac{k_1}{k_{-1}}\right)[NO]^2$

Substitute into the rate expression: $\text{Rate} = k_2\left(\dfrac{k_1}{k_{-1}}\right)[NO]^2[O_2]$

(b) This rate expression has an overall order of $(2 + 1) = 3$, so the mechanism is consistent with third-order behaviour.

(c) The intermediate species is N_2O_2. Two NO molecules could bind in several ways:

In the second step, O_2 collides with the intermediate and reacts to form two NO_2 molecules. The ONNO arrangement is the only intermediate for which the new set of bonds can easily occur:

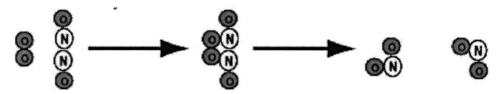

13.41 The rate constant for a reaction depends on temperature according to the Arrhenius equation (Equation 13-6): $k = Ae^{-E_a/RT}$. When $E_a = 0$, the exponent is $e^0 = 1$ and k is independent of temperature. A zero activation energy is the case when a reaction can occur without first breaking any chemical bonds. The most common example is the combination of two free radicals, such as $H_3C\bullet + \bullet CH_3 \rightarrow H_3C–CH_3$.

13.43 An exothermic reaction has products lower in energy than reactants. In the activated complex, A will be bonded to both B and C:

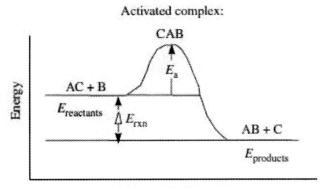

Reaction coordinate

13.45 An exothermic reaction is "downhill" from reactants to products, and the activation energy plot should show this:

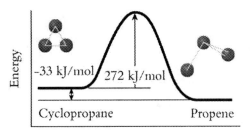

Reaction coordinate

13.47 Assume that the ratio of the rate constants is proportional to the ratio of the number of flashes, and then use the rearranged version of Equation 13-8 in your textbook:

$$E_a = R \ln\left(\frac{k_2}{k_1}\right)\left(\frac{1}{T_1} - \frac{1}{T_2}\right)^{-1} = \left(8.314 \text{ J mol}^{-1}\text{K}^{-1}\right)\ln\left(\frac{2.7}{3.3}\right)\left(\frac{1}{(29+279 \text{ K})} - \frac{1}{(23+273 \text{ K})}\right)^{-1}$$

$$E_a = \left(\frac{-1.67 \text{ J mol}^{-1} \text{ K}^{-1}}{-1.3 \times 10^{-4} \text{ K}^{-1}}\right)\left(\frac{10^{-3} \text{ kJ}}{1 \text{ J}}\right) = 13 \text{ kJ/mol}$$

13.49 Hydrogen gas adsorbs on the catalyst's surface as H atoms, which then react with CO molecules much more easily (and quickly) than direct reaction of CO molecules with H_2 molecules. Bonds that must be broken for this reaction to occur (depicted by the squiggly lines) are two H–H single bonds and both the π bonds in the CO molecule. Bonds that are formed (dashed lines) are three C–H bonds and one O–H bond.

Bond breakage (wavy lines):   Bond formation (dotted lines):

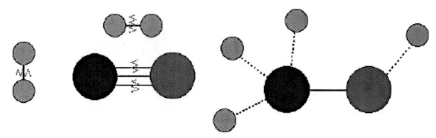

13.51 (a) The effect of a catalyst on a reaction is to reduce the activation energy without changing the energy of the reactants or the products. Thus, the energies of reactants and products are the same for both curves. Only the "bump" in the activation energy diagram changes, being lower in the presence of the catalyst than in its absence:

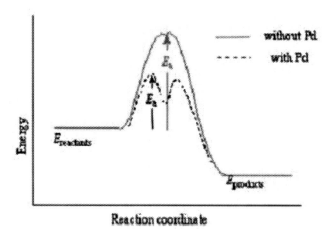

(b) A catalyst is present at the beginning and end of the reaction but does not appear in the net reaction. Here, Pd metal acts as a catalyst. An intermediate is anything that is produced in one step of the mechanism and then used up in another step. The intermediates for this reaction are the H atoms that form when H_2 gas absorbs onto the Pd metal surface. The valley on the activation energy diagram represents this intermediate stage.

(c)

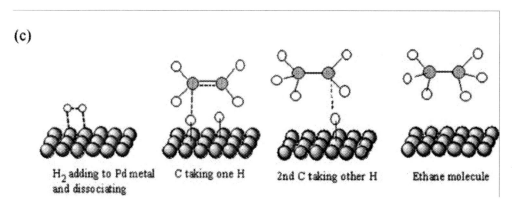

13.53 (a) The second step must use up the intermediate, O:

$$O + NO \rightarrow NO_2.$$

(b)

13.55 (a) A rate expression relates rate to changes: $\text{Rate} = \left(\dfrac{\Delta[C_6H_6]}{\Delta t}\right) = -\dfrac{1}{3}\left(\dfrac{\Delta[C_2H_2]}{\Delta t}\right)$

(b) Rate laws must always be determined experimentally. Thus, there is insufficient information to write the rate law. Experiments would have to be carried out measuring the rate as a function of $[C_2H_2]$ and the data analyzed using techniques described in your textbook.

13.57 The essential units of information needed to construct an activation energy diagram are the energy change and activation energy. Here, the reaction is a formation reaction and the change in moles of gas is zero during the reaction, so $\Delta E \cong \Delta H_f^0$:

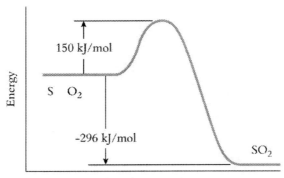

Reaction coordinate

13.59 When the concentration of a reactant increases by a factor of 3 (triples), the rate of reaction changes by 3^n, where n is the order with respect to that concentration. (a) Nine-fold increase; (b) no change; (c) rate increases by 5.2 times.

13.61 To determine the order of a reaction from a set of experimental data, prepare plots

of $\ln\left(\dfrac{[A]_0}{[A]}\right)$ vs. t and $\dfrac{1}{[A]} - \dfrac{1}{[A]_0}$ vs. t .

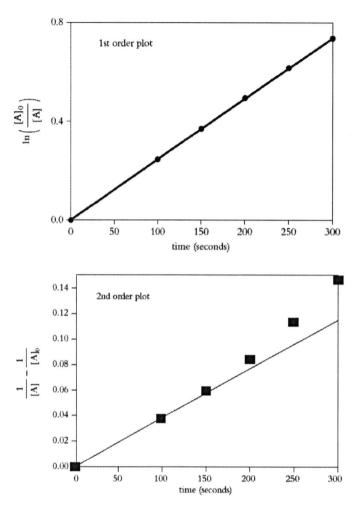

The first-order plot is linear, whereas the second-order plot is not. This reaction is first order. Determine the rate constant from the slope:

$$k = \text{slope} = \frac{\Delta y}{\Delta x} = \frac{0.766 - 0}{300 \text{ s} - 0 \text{ s}} = 2.55 \times 10^{-3} \text{ s}^{-1}$$

13.63 The mechanism has the rate law, rate = $k[X_2]$, first order in X_2 and independent of Y. Flask *a* contains 5 X_2 and 8 Y, whereas Flask *b* contains 10 X_2 and 8 Y. The rate for Flask *b* will be twice that for Flask *a*, because the concentration of X_2 is twice as great.

13.65 The initial concentration of N_2O_5 is much smaller than the initial concentration of H_2O in both experiments, so this is an example of the isolation technique. Assuming that the rate law has the form, rate = $k[N_2O_5]^x[H_2O]^y$, we have experimental rate = $k_{obs}[N_2O_5]^x$, where $k_{obs} = k[H_2O]_0^y$.

We need to plot the data: if $x = 1$, a log plot will be linear, whereas if $x = 2$, a $1/[N_2O_5]$ plot will be linear. Here is the first-order plot for Experiment A:

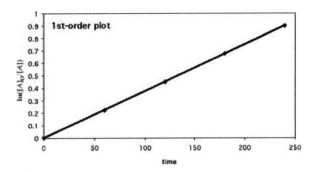

This plot is linear, with slope = k_{obs} = 0.0038 s^{-1}.

We do not need to plot the data for Experiment B, in which the initial concentration of H_2O is twice as large as in Experiment A. Instead, we note that the reaction is going twice as fast: in Experiment B, it takes only 60 seconds for the concentration to fall to the value that is reached in 120 seconds in Experiment A. Doubling [H_2O] leads to a doubling of the rate, so the reaction is first order in H_2O as well as first order in N_2O_5. The rate law is as follows:

$$\text{Rate} = k[N_2O_5][H_2O]$$

To evaluate k, we can use either set of conditions. Here we use Experiment A:

$$k_{obs} = k[H_2O]_0 = 0.0038 \text{ s}^{-1} \text{ and } [H_2O]_0 = 0.025 \text{ M}$$

$$k = \frac{k_{obs}}{[H_2O]_0} = \frac{0.0038 \text{ s}^{-1}}{0.025 \text{ M}} = = 0.15 \text{ M}^{-1} \text{ s}^{-1}$$

13.67   Reaction times for first-order reactions can be calculated using Equation 13-3, suitably rearranged:

$$t = \frac{\ln\left(\dfrac{[A]_0}{[A]}\right)}{k}$$

For 10.0% decomposition, [A] = 0.900[A]$_0$ and

$$\ln\left(\frac{[A]_0}{[A]}\right) = \ln\left(\frac{1.000}{0.900}\right) = 0.105.$$

$$t = \frac{0.105}{5.5 \times 10^{-4} \text{ s}^{-1}} = 1.9 \times 10^2 \text{ s}$$

For 50.0% decomposition, [A] = 0.500[A]$_0$ and $\ln\left(\dfrac{[A]_0}{[A]}\right) = \ln\left(\dfrac{1.000}{0.500}\right) = 0.693.$

$$t = \frac{0.693}{5.5 \times 10^{-4} \text{ s}^{-1}} = 1.3 \times 10^3 \text{ s}$$

For 99.9% decomposition, $[A] = 0.001[A]_0$ and $\ln\left(\dfrac{[A]_0}{[A]}\right) = \ln\left(\dfrac{1.000}{0.001}\right) = 6.91$.

$$t = \frac{6.91}{5.5 \times 10^{-4}\ \text{s}^{-1}} = 1.3 \times 10^4\ \text{s}$$

13.69 (a) False (overall reaction would give fourth-order kinetics)

(b) False (rate constants must be measured for at least two different temperatures to calculate E_a)

(c) True (rates of reaction increase with increasing temperature)

(d) True (a unimolecular step has first-order kinetics)

13.71 Reaction orders are given by the exponents on the concentrations that appear in the rate law. Overall order is the sum of those exponents: (a) First order in N_2O_5 and first order overall; (b) second order in NO, first order in H_2, and third order overall; and (c) first order in enzyme and first order overall.

13.73 The speed of a chemical reaction refers to how fast it proceeds. The spontaneity of a chemical reaction refers to whether or not the reaction can go in the direction written without outside intervention. A spontaneous reaction may nevertheless have a very slow speed.

13.75 Ultraviolet light causes chlorofluorocarbons to fragment, producing Cl atoms that catalyze the destruction of ozone:

$$CF_2Cl_2 \xrightarrow{hv} CF_2Cl + Cl$$
$$O_3 \xrightarrow{hv} O_2 + O$$
$$Cl + O_3 \rightarrow ClO + O_2$$
$$ClO + O \rightarrow Cl + O_2$$

Because the fourth reaction regenerates a Cl atom, the second through fourth reactions occur many hundreds of times for every CF_2Cl_2 molecule that fragments.

13.77 (a) Obtain the overall stoichiometry by adding the three steps, ignoring the reverse reaction of Step 1, which leads to no net change: $2\ NO + 2\ H_2 \rightarrow N_2 + 2\ H_2O$.

(b) The rate law is determined by the rate-determining step: Rate = $k_2[N_2O_2][H_2]$. This is not satisfactory, however, because it contains the concentration of an intermediate. Set the rates equal for the forward and reverse first step:

$$k_1[NO]^2 = k_{-1}[N_2O_2]$$

Solve this equality for $[N_2O_2]$: $[N_2O_2] = \left(\dfrac{k_1}{k_{-1}}\right)[NO]^2$

Substitute into the rate expression: $\text{Rate} = k_2 \left(\dfrac{k_1}{k_{-1}} \right) [NO]^2 [H_2]$

13.79 (a) A catalyst binds to one or more of the reactants in a way that weakens chemical bonds and makes it easier for bonds to rearrange to form the products.

(b) When temperature increases, the average energies of the molecules increase, with the result that enough energy is present for a larger fraction of the molecules to have sufficient energy to overcome the activation energy barrier.

(c) When concentration increases, the molecular density increases. There are more molecules to react, leading to a higher rate of molecular collisions. Both factors contribute to a greater rate of reaction.

13.81 Flask 1 contains five molecules, and Flask 2 contains 10 molecules, so the concentration is doubled. The problem states that Flask 2 reacts four times as fast. This is (2)(2), so the rate law is rate = $k[A]^2$. At the molecular level, the rate-determining step could be a reaction between two A molecules. For this process, the rate increases because the higher concentration results in a higher rate of collisions.

13.83 (a) The rate law is that for an elementary bimolecular reaction: Rate = $k[H_2][X_2]$.

(b) When a first step is rate-determining, it determines the rate law: Rate = $k[X_2]$.

(c) The rate law is determined by the rate-determining step: Rate = $k_2[X][H_2]$.

Set the rates equal for the forward and reverse first step: $k_1[X_2] = k_{-1}[X]^2$

Solve this equality for $[X]$: $[X] = \left(\dfrac{k_1}{k_{-1}} \right)^{1/2} [X_2]^{1/2}$

Substitute into the rate expression: $\text{Rate} = k_2 \left(\dfrac{k_1}{k_{-1}} \right)^{1/2} [H_2][X_2]^{1/2}$

13.85 (a) The net reaction can be obtained by summing the three steps, because when the reverse step occurs there is no net change: $Cl_2 + CHCl_3 \rightarrow HCl + CCl_4$.

(b) Intermediates are produced in early steps and consumed in later steps: Cl and CCl_3.

(c) The rate law is determined by the rate-determining step: Rate = $k_2[CHCl_3][Cl]$. This is not satisfactory, however, because it contains the concentration of an intermediate

Set the rates equal for the forward and reverse first step: $k_1[Cl_2] = k_{-1}[Cl]^2$

Solve this equality for [Cl]: $[Cl] = \left(\dfrac{k_1}{k_{-1}} \right)^{1/2} [Cl_2]^{1/2}$

Substitute into the rate expression:

$$\text{Rate} = k_2 \left(\dfrac{k_1}{k_{-1}} \right)^{1/2} [CHCl_3][Cl_2]^{1/2}$$

13.87 Activation energies are calculated from the Arrhenius equation using Equation 13-8:

$$E_a = R \ln \left(\dfrac{k_2}{k_1} \right) \left(\dfrac{1}{T_1} - \dfrac{1}{T_2} \right)^{-1}$$

Development time is inversely proportional to rate constant, so $\dfrac{k_2}{k_1} = \dfrac{t_1}{t_2}$

(a) $\dfrac{t_1}{t_2} = 2$ $T_1 = 20 + 273 = 293$ K $T_2 = 293 + 10 = 303$ K

$$\left(\dfrac{1}{T_1} - \dfrac{1}{T_2} \right)^{-1} = \left((3.413 \times 10^{-3} \text{ K}^{-1}) - (3.300 \times 10^{-3} \text{ K}^{-1}) \right)^{-1} = 8850. \text{ K}$$

$$E_a = \left(\dfrac{8.314 \text{ J}}{1 \text{ mol K}} \right) \left(\dfrac{10^{-3} \text{ kJ}}{1 \text{ J}} \right) (\ln 2)(8850. \text{K}) = 51 \text{ kJ/mol}$$

(b) To determine the time it takes at 25 °C, use $\ln \left(\dfrac{t_1}{t_2} \right) = \dfrac{E_a}{R} \left(\dfrac{1}{T_1} - \dfrac{1}{T_2} \right)$:

$E_a = 51 \text{ kJ/mol}$ $t_1 = 10 \text{ min}$

$T_1 = 20 + 273 = 293 \text{ K}$ $T_2 = 25 + 273 = 298 \text{ K}$

$$\ln \left(\dfrac{10 \text{ min}}{t_2} \right) = \left(\dfrac{1}{293 \text{ K}} - \dfrac{1}{298 \text{ K}} \right) \left(\dfrac{1 \text{ mol K}}{8.314 \times 10^{-3} \text{ kJ}} \right) \left(\dfrac{51 \text{ kJ}}{1 \text{ mol}} \right) = 0.351$$

$$\dfrac{10 \text{ min}}{t_2} = 1.42 \quad \text{from which} \quad t_2 = \dfrac{10 \text{ min}}{1.42} = 7.0 \text{ min}$$

13.89 (a) Prepare first-order and second-order plots and look for linear behaviour:

t	s	0	1000	2000	3000	4000

c	M	0.250	0.118	0.0770	0.0572	0.0455
$\ln\left(\dfrac{[A]_0}{[A]}\right)$		0.000	0.751	1.18	1.47	1.70
$\dfrac{1}{[A]} - \dfrac{1}{[A]_0}$	M^{-1}	0.00	4.47	8.99	13.5	18.0

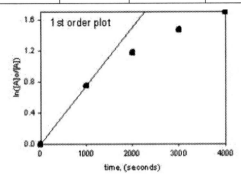 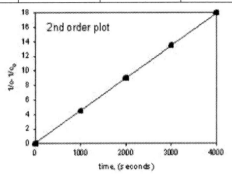

The second-order plot is linear, so rate $= k[CH_3CHO]^2$

(b) Determine the rate constant from the slope of the second-order plot:

$$k = \text{slope} = \frac{18.0\ M^{-1} - 0.00\ M^{-1}}{4000\ s - 0\ s} = 4.5 \times 10^{-3}\ M^{-1}\ s^{-1}$$

(c) Use Equation 13-5, suitably rearranged:

$$kt = \frac{1}{[A]} - \frac{1}{[A]_0}$$

$$[A]_0 = 0.250\ M \text{ and } [A] = 0.250\ M\left(\frac{100\% - 75\%}{100\%}\right) = 0.0625\ M$$

$$t = \frac{16.0\ M^{-1} - 4.00\ M^{-1}}{4.5 \times 10^{-3}\ M^{-1}\ s^{-1}} = 2.7 \times 10^3\ s$$

13.91 The hint for this problem suggests using the Arrhenius equation, $k = Ae^{-E_a/RT}$.

Evaluate $\dfrac{E_a}{RT}$ for the catalyzed and uncatalyzed situations, using

$T = 21 + 273 = 294$ K:

$$\left(\frac{E_a}{RT}\right)_{\text{uncat}} = \frac{125\ kJ\ mol^{-1}}{(8.314 \times 10^{-3}\ kJ\ mol^{-1}\ K^{-1})(294\ K)} = 51.1$$

$$\left(\frac{E_a}{RT}\right)_{\text{cat}} = \frac{46\ kJ\ mol^{-1}}{(8.314 \times 10^{-3}\ kJ\ mol^{-1}\ K^{-1})(294\ K)} = 18.8$$

$$k_{\text{uncat}} = Ae^{-51.1} \text{ and } k_{\text{cat}} = Ae^{-18.8}$$

Divide one of these by the other to eliminate A and find the ratio of the rate constants:

$$\frac{k_{cat}}{k_{uncat}} = e^{(51.1-18.8)} = e^{32.3} = 1.1 \times 10^{14}$$

13.93 Neither intermediates nor catalysts appear in the overall stoichiometry of the reaction, so any species that appears in the mechanism but not in the overall stoichiometry is either an intermediate or a catalyst. Catalysts are consumed in early steps and regenerated in later steps, whereas intermediates are produced in early steps and consumed in later steps.

13.95 The only true statements are (d) and (f). Statements (a) and (b) are false because $\Delta E_{reaction} = C - A$. Here is an energy diagram showing the quantities involved:

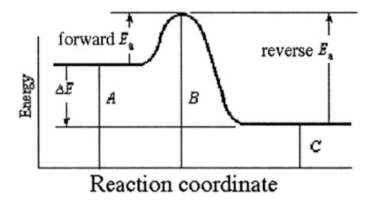

13.97 When an enzyme binds a reactant, bonds between the enzyme and reactant result in a reduction in the strength of the bonds that need to be broken in order for the catalyzed reaction to occur. This reduction in strength in turn reduces the amount of energy that must be supplied for the reaction to occur, accounting for the decrease in the activation energy for the enzyme-catalyzed reaction.

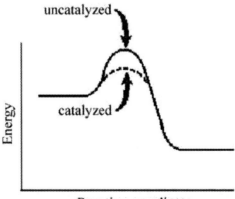

13.99 Your explanation should include the main features of the induced-fit model: a cavity into which the molecule fits that adjusts its shape to the target molecule (square) and then distorts to catalyze decomposition of the square:

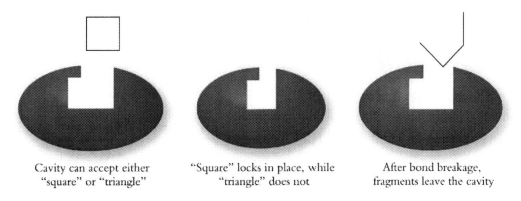

Cavity can accept either "square" or "triangle"   "Square" locks in place, while "triangle" does not   After bond breakage, fragments leave the cavity

13.101 (a) The catalyst, Br^-, is consumed in the first step, so it must be regenerated in a later step. Thus, A = Br^-. For the stoichiometry to be correct, B = O_2.

(b) For the overall reaction:

$$\Delta E \cong \Delta H = \left[2 \text{ mol}(-285.83 \text{ kJ/mol}) + 1 \text{ mol}(0 \text{ kJ/mol})\right] - 2 \text{ mol}(-187.8 \text{ kJ/mol}) = -196.1 \text{ kJ}$$

For the first step ($\Delta n_g = 0$):

$$\Delta E = \Delta H = \left[1 \text{ mol}(-285.83 \text{ kJ/mol}) + 1 \text{ mol}(-94.1 \text{ kJ/mol})\right]$$
$$-\left[1 \text{ mol}(-187.8 \text{ kJ/mol}) + 1 \text{ mol}(-121.4 \text{ kJ/mol}) = -70.73 \text{ kJ}\right]$$

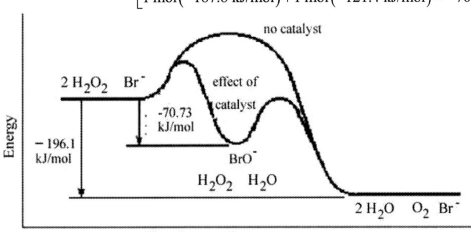

13.103 (a) The data are in the form of initial rate information, so use the ratios of initial rates to determine the order with respect to each reagent:

In Experiments 1 and 2, only $[I^-]$ changes: $\left(\dfrac{0.0060 \text{ M}}{0.0025 \text{ M}}\right)^x = \left(\dfrac{9.1 \times 10^{-2} \text{ M/s}}{3.8 \times 10^{-2} \text{ M/s}}\right)$

Simplifying, $(2.4)^x = 2.4$ $\qquad\qquad x = 1$

The reaction is first order in I^-.

In Experiments 1 and 3, only $[OCl^-]$ changes: $\left(\dfrac{0.037\ M}{0.025\ M}\right)^y = \left(\dfrac{5.6\ \times\ 10^{-2}\ M/s}{3.8\ \times\ 10^{-2}\ M/s}\right)$

Simplifying, $(1.5)^y = 1.5$ $\qquad\qquad\qquad\qquad y = 1$

The reaction is first order in OCl^-.

In Experiments 3 and 4, only $[OH^-]$ changes: $\left(\dfrac{0.10\ M}{0.20\ M}\right)^z = \left(\dfrac{5.6\ \times\ 10^{-2}\ M/s}{2.8\ \times\ 10^{-2}\ M/s}\right)$

Simplifying, $(0.50)^z = 2.0$; the rate *doubles* when the concentration *is cut in half.*

This means that the rate is inversely dependent on $\left[OH^-\right]$: $\quad z = -1$

The rate law is

$$\text{Rate} = k\frac{[I^-][OCl^-]}{[OH^-]}$$

(b) Use any of the initial rate data to evaluate k:

$$k = \frac{(\text{initial rate})[OH^-]}{[OCl^-][I^-]} = \frac{(3.8\ \times\ 10^{-2}\ M/s)(0.10\ M)}{(0.025\ M)(0.0025\ M)} = 61\ s^{-1}$$

(c) The inverse dependence indicates a complicated mechanism. The problem states that one reactant is in rapid equilibrium with water. Because HI is a strong acid, I^- is not a good choice:

$$OCl^- + H_2O \underset{k_{-1}}{\overset{k_1}{\rightleftharpoons}} HOCl + OH^- \qquad\qquad (\text{rapid equilibrium})$$

Now let HOCl react with I^- in a slow, rate-determining step:

$$I^- + HOCl \xrightarrow{\ k_2\ } HOI + Cl^- \ (\text{slow})$$

HOI can react with hydroxide to give the correct overall stoichiometry:

$$OH^- + HOI \rightarrow H_2O + OI^- \ \ (\text{fast})$$

$$\text{Rate} = k_2[HOCl][I^-]$$

Set the rates of the forward and reverse first reaction equal to each other and solve for [HOCl]:

$$k_1[OCl^-][H_2O] = k_{-1}[OH^-][HOCl] \quad \text{so} \quad [HOCl] = \frac{k_1[OCl^-][H_2O]}{k_{-1}[OH^-]}$$

Substitute this into the rate expression:

$$\text{Rate} = \frac{k_2 k_1[OCl^-][H_2O][I^-]}{k_{-1}[OH^-]}$$

This rate law has the correct first-order dependence on the two reactants and an inverse dependence on [OH⁻], in agreement with the experimental rate law.

(d) The predicted rate law is first order in water concentration. In order to test for this dependence, we could do kinetic experiments in mixed solvents. For example, we could use a 1:1 molar mixture of ethanol and water. In this solvent, the rate should be half as fast, if the proposed mechanism is the correct one.

13.105 The decomposition of di-t-butyl peroxide (DTBP) is a common kinetics experiment in undergraduate physical chemistry. The molecule decomposes into acetone and ethane via the first order reaction:

$$C_8H_{12}O_2 \text{ (g)} \rightarrow 2 \text{ (CH}_3)CO + C_2H_6$$

The reaction is started with pure di-t-butyl peroxide at 150°C and 1.04 bar pressure in a constant-volume apparatus. The half-life of the reaction is found to be 78.0 minutes. (a) Calculate the rate constant (in min⁻¹). (b) How long (in hours) will it take for the partial pressure of DTBP to decrease to 0.0700 bar? (c) How long (in hours) will it take for the total pressure to increase to 2.75 bar?

(a) The rate constant is found from the half-life:

$$k = \frac{0.693}{t_{1/2}} = \frac{0.693}{78.0 \text{ min}} = 8.88 \times 10^{-3} \text{ min}^{-1}$$

(b) The time required is calculated from the integrated rate equation for this first order reaction:

$$p_{DTBP} = p^0_{DTBP}e^{-kt}$$

$$\frac{p_{DTBP}}{p^0_{DTBP}} = e^{-kt}$$

$$\ln\left(\frac{p_{DTBP}}{p^0_{DTBP}}\right) = -kt$$

$$t = \frac{\ln\left(\frac{p_{DTBP}}{p^0_{DTBP}}\right)}{-k}$$

$$= \frac{\ln\left(\frac{0.0700 \text{ bar}}{1.04 \text{ bar}}\right)}{-8.88 \times 10^{-3} \text{ min}^{-1}}$$

$$= 304 \text{ min}$$

(c) The total pressure increases as the reaction progresses because three moles of gas (two moles of acetone plus one mole of ethane) are produced from one mole of DTBP. The total pressure is equal to the sum of the partial pressures of the three gases. If the partial pressure of the DTBP decreases by x (to $p^0_{DTBP} - x$), then the partial pressure of acetone increases to 2x and that of ethane to x. The total pressure is then:

$$p_{tot} = p_{DTBP} + p_{acetone} + p_{ethane}$$

$$= (p^o_{DTBP} - x) + 2x + x$$

$$= p^o_{DTBP} + 2x$$

Rearranging and solving for x:

$$2x = p_{tot} - p^o_{DTBP} = 2.75 \text{ bar} - 1.04 \text{ bar} = 1.71 \text{ bar}.$$
Thus, x = 1.71 bar / 2 = 0.855 bar.

The time required can be then calculated from the first order integrated rate equation:

$$t = \frac{\ln\left(\dfrac{p_{DTBP}}{p^0_{DTBP}}\right)}{-k}$$

$$= \frac{\ln\left(\dfrac{(p^o_{DTBP} - x)}{p^o_{DTBP}}\right)}{-k}$$

$$= \frac{\ln\left(\dfrac{(1.04 \text{ bar} - 0.855 \text{ bar})}{1.04 \text{ bar}}\right)}{-8.88 \times 10^{-3} \text{ min}^{-1}}$$

$$= 194 \text{ min}$$

Chapter 14 Principles of Chemical Equilibrium

Solutions to Problems in Chapter 14

14.1 The reaction proceeds, converting *cis*-butene into *trans*-butene, but as the partial pressure of *trans*-butene builds up, the reverse reaction becomes increasingly important until, when the ratio of partial pressures is 3, the rates are equal. Then there is no net change. A plot of pressure vs. time appears similar to Figure 13-2, with 0.75 bar for the final pressure of *trans*-butene and 0.25 bar for the final pressure of *cis*-butene:

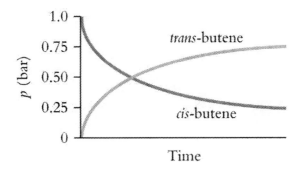

14.3 (a) According to the principle of reversibility, every elementary reaction can run in either direction, so as the system approaches equilibrium, the reverse reactions become important:

$$Cl^- (aq) + ClO_2^- (aq) \xrightarrow{k_{-1}} ClO^- (aq) + ClO^- (aq)$$

$$Cl^- (aq) + ClO_3^- (aq) \xrightarrow{k_{-2}} ClO^- (aq) + ClO_2^- (aq)$$

(b) The overall reaction is $3ClO^- (aq) \rightleftharpoons 2Cl^- (aq) + ClO_3^- (aq)$.

By inspection, the equilibrium constant expression is $K_{eq} = \dfrac{[ClO_3^-]_{eq}[Cl^-]_{eq}^2}{[ClO^-]_{eq}^3}$

(c) The overall reaction is the sum of the two elementary steps, so the equilibrium constant is the product of the rate ratios of the two steps: $K_{eq} = \dfrac{k_1 k_2}{k_{-1} k_{-2}}$

14.5 A molecular picture of an elementary reaction shows the reactants, the products, and (if necessary) the intermediate collision complex.

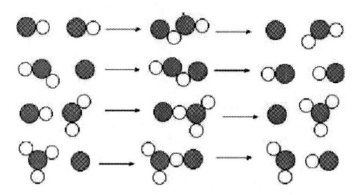

14.7 To test the reversibility of a reaction, set up a system containing the products and observe if reactants form. Here, a solution containing Cl^- and ClO_3^- ions should react to form some ClO^- ions. (If the reaction of ClO^- to form Cl^- and ClO_3^- goes virtually to completion, this experiment may not succeed.)

14.9 Equilibrium constant expressions can be written by inspection of the overall stoichiometry, remembering that pure liquids and solids do not appear in the expression. Equilibrium constant expressions contain product concentrations over reactant concentrations raised to their stoichiometric coefficient:

(a) $K_{eq} = \dfrac{(p_{IF_5})^2_{eq}}{(p_{F_2})^5_{eq}}$

(b) $K_{eq} = \dfrac{1}{(p_{O_2})^5_{eq}}$

(c) $K_{eq} = (p_{CO})^2_{eq}$

(d) $K_{eq} = \dfrac{1}{(p_{CO})_{eq}(p_{H_2})^2_{eq}}$

(e) $K_{eq} = \dfrac{[H_3O^+]^3_{eq}[PO_4^{3-}]_{eq}}{[H_3PO_4]_{eq}}$

14.11 Equilibrium constant expressions can be written by inspection of the overall stoichiometry, remembering that pure liquids and solids do not appear in the expression:

(a) $2IF_5\ (g) \rightleftharpoons I_2\ (s) + 5\ F_2\ (g)$ $K_{eq} = \dfrac{(p_{F_2})^5_{eq}}{(p_{IF_5})^2_{eq}}$

(b) $P_4O_{10}\ (s) \rightleftharpoons P_4\ (s) + 5\ O_2\ (g)$ $K_{eq} = (p_{O_2})^5_{eq}$

(c) $BaO\ (s) + 2\ CO\ (g) \rightleftharpoons BaCO_3\ (s) + C\ (s)$ $K_{eq} = \dfrac{1}{(p_{CO})_{eq}^2}$

(d) $CH_3OH\ (l) \rightleftharpoons CO\ (g) + 2\ H_2\ (g)$ $K_{eq} = (p_{CO})_{eq}(p_{H_2})_{eq}^2$

(e) $PO_4^{3-}\ (aq) + 3\ H_3O^+\ (aq) \rightleftharpoons H_3PO_4\ (aq) + 3\ H_2O\ (l)$ $K_{eq} = \dfrac{[H_3PO_4]_{eq}}{[H_3O^+]_{eq}^3[PO_4^{3-}]_{eq}}$

14.13 The standard states of gases are gases at 1 bar, the standard states of solutes are solutions at 1 M, and the standard states of solvents and pure liquids and solids are unit mole fractions, $X = 1$:

(a) $p = 1$ bar for F_2 and IF_5; $X = 1$ for I_2

(b) $p = 1$ bar for O_2; $X = 1$ for P_4 and P_4O_{10}

(c) $p = 1$ bar for CO; $X = 1$ for others

(d) $p = 1$ bar for CO and H_2; $X = 1$ for CH_3OH

(e) $c = 1$ M for H_3PO_4, H_3O^+, and PO_4^{3-}; $X = 1$ for H_2O

14.15 Your views should be like those in Figure 14-4, showing two different size samples with equal sized portions of each highlighted to show that equal volumes contain equal numbers of atoms:

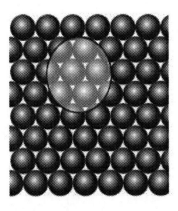

14.17 To determine an equilibrium constant at standard temperature from thermodynamic tables, calculate $\Delta G^{\circ}_{reaction}$ from tabulated values for ΔG°_{f} and then use Equation 14-3:

$$\Delta G^{\circ} = -RT \ln K_{eq}$$

The solubility reaction is $LiCl (s) \leftrightarrows Li^+ (aq) + Cl^- (aq)$ $\qquad K_{eq} = K_{sp}$

$$\Delta G^o_{reaction} = [1(-293.31 \text{ kJ/mol}) + 1(-131.0 \text{ kJ/mol})] - [1(-384.4 \text{ kJ/mol})]$$
$$= -39.91 \text{ kJ/mol}$$

$$\ln K_{sp} = -\frac{-3.991 \times 10^4 \text{ J/mol}}{(8.314 \text{ J mol}^{-1} \text{ K}^{-1})(298 \text{ K})} = 16.11 \qquad K_{sp} = e^{16.11} = 9.9 \times 10^6$$

14.19 To determine an equilibrium constant at standard temperature from thermodynamic tables, calculate $\Delta G^o_{reaction}$ from tabulated values for ΔG^o_f and then use Equation 13-3:

$$\Delta G^o = -RT \ln K_{eq}$$

(a) $\Delta G^o_{reaction} = [1(0 \text{ kJ/mol}) + 1(-394.4 \text{ kJ/mol})]$

$\qquad\qquad -[1(-137.2 \text{ kJ/mol}) + 1(-237.1 \text{ kJ/mol})] = -20.1 \text{ kJ/mol}$

$$\ln K_{eq} = -\frac{-2.01 \times 10^4 \text{ J/mol}}{(8.314 \text{ J mol}^{-1} \text{ K}^{-1})(298 \text{ K})} = 8.11 \qquad K_{eq} = e^{8.11} = 3.3 \times 10^3$$

(b) $\Delta G^o_{reaction} = [2(-394.4 \text{ kJ/mol})] - [1(0 \text{ kJ/mol}) + 2(-137.2 \text{ kJ/mol})]$

$\qquad\qquad = -514.4 \text{ kJ/mol}$

$$\ln K_{eq} = -\frac{-5.144 \times 10^5 \text{ J/mol}}{(8.314 \text{ J mol}^{-1} \text{ K}^{-1})(298 \text{ K})} = 207.6 \qquad K_{eq} = e^{207.6} = 1 \times 10^{90}$$

(c) $\Delta G^o_{reaction} = [1(-520.3 \text{ kJ/mol}) + 2(-137.2 \text{ kJ/mol})]$

$\qquad\qquad -[1(0 \text{ kJ/mol}) + 1(-1134.4 \text{ kJ/mol})] = 339.7 \text{ kJ/mol}$

$$\ln K_{eq} = -\frac{3.397 \times 10^5 \text{ J/mol}}{(8.314 \text{ J mol}^{-1} \text{ K}^{-1})(298 \text{ K})} = -137.1 \qquad K_{eq} = e^{-137.1} = 3 \times 10^{-60}$$

(d) $\Delta G^o_{reaction} = [3(-237.1 \text{ kJ/mol}) + 1(74.62 \text{ kJ/mol})]$

$\qquad\qquad -[6(0 \text{ kJ/mol}) + 3(-137.2 \text{ kJ/mol})] = -225.1 \text{ kJ/mol}$

$$\ln K_{eq} = -\frac{-2.251 \times 10^5 \text{ J/mol}}{(8.314 \text{ J mol}^{-1} \text{ K}^{-1})(298 \text{ K})} = 90.86 \qquad K_{eq} = e^{90.86} = 2.9 \times 10^{39}$$

14.21 To estimate the equilibrium constant at a temperature different from 298 K, calculate $\Delta H^\circ_{reaction}$ and $\Delta S^\circ_{reaction}$ at 298 K and then use Equations 12-10 and 14-3:

$$\Delta G^\circ = \Delta H^\circ - T\Delta S^\circ \qquad \Delta G^\circ = -RT \ln K_{eq}$$

14.19 (b) $\Delta H^\circ_{reaction} = [2(-393.5 \text{ kJ/mol})] - [1(0 \text{ kJ/mol}) + 2(-110.5 \text{ kJ/mol})] = -566.0 \text{ kJ/mol}$

$\Delta S^\circ_{reaction} = 2(213.8 \text{ J mol}^{-1} \text{ K}^{-1}) - [1(205.15 \text{ J mol}^{-1} \text{ K}^{-1}) + 2(197.7 \text{ J mol}^{-1}$

$$K^{-1})] = -173.0 \text{ J mol}^{-1} \text{ K}^{-1}$$

$$\Delta G^{\circ}_{\text{reaction, 250 K}} \cong (-566.0 \text{ kJ/mol}) - (250 \text{ K})(-0.1730 \text{ kJ mol}^{-1} \text{ K}^{-1}) = -522.8 \text{ kJ/mol}$$

$$\ln K_{\text{eq}} = -\frac{-5.228 \times 10^5 \text{ J/mol}}{(8.314 \text{ J mol}^{-1} \text{ K}^{-1})(250 \text{ K})} = 251.5 \qquad K_{\text{eq}} = e^{251.5} = 2 \times 10^{109}$$

14.19 (c) $\Delta H^{\circ}_{\text{reaction}} = [1(-548.0 \text{ kJ/mol}) + 2(-110.5 \text{ kJ/mol})] - [1 \ (0 \text{ kJ/mol}) + 1(-1213.0 \text{ kJ/mol})] = 444.0 \text{ kJ/mol}$

$$\Delta S^{\circ}_{\text{reaction}} = [1(72.1 \text{ J mol}^{-1} \text{ K}^{-1}) + 2(197.7 \text{ J mol}^{-1} \text{ K}^{-1})] - [1(5.7 \text{ J mol}^{-1} \text{ K}^{-1}) + 1(112.1 \text{ J mol}^{-1} \text{ K}^{-1})] = 349.7 \text{ J mol}^{-1} \text{ K}^{-1}$$

$$\Delta G^{\circ}_{\text{reaction, 250 K}} \cong (444.0 \text{ kJ/mol}) - (250 \text{ K})(0.3497 \text{ kJ mol}^{-1} \text{ K}^{-1}) = 356.6 \text{ kJ/mol}$$

$$\ln K_{\text{eq}} = -\frac{3.566 \times 10^5 \text{ J/mol}}{(8.314 \text{ J mol}^{-1} \text{ K}^{-1})(250 \text{ K})} = -171.6 \qquad K_{\text{eq}} = e^{-171.6} = 3 \times 10^{-75}$$

14.23 To estimate the equilibrium constant at a temperature different from 298 K, calculate $\Delta H^{\circ}_{\text{reaction}}$ and $\Delta S^{\circ}_{\text{reaction}}$ at 298 K and then use Equations 12-10 and 14-3:

$$\Delta G^{\circ} = \Delta H^{\circ} - T\Delta S^{\circ} \qquad \Delta G^{\circ} = -RT \ln K_{\text{eq}}$$

14.19 (a)

$$\Delta H^{\circ}_{\text{reaction}} = [1(-393.5 \text{ kJ/mol}) + 1(0 \text{ kJ/mol})] - [1(-110.5 \text{ kJ/mol}) + 1(-241.83 \text{ kJ/mol})] = -41.2 \text{ kJ/mol}$$

$$\Delta S^{\circ}_{\text{reaction}} = [1(213.8 \text{ J/mol K}) + 1(130.680 \text{ J mol}^{-1} \text{ K}^{-1})] - [1(197.7 \text{ J mol}^{-1} \text{ K}^{-1}) + 1(188.835 \text{ J mol}^{-1} \text{ K}^{-1})] = -42.1 \text{ J mol}^{-1} \text{ K}^{-1}$$

$$\Delta G^{\circ}_{\text{reaction, 395 K}} \cong (-41.2 \text{ kJ/mol}) - (395 \text{ K})(-0.0421 \text{ kJ mol}^{-1} \text{ K}^{-1}) = -24.6 \text{ kJ/mol}$$

$$\ln K_{\text{eq}} = -\frac{-2.46 \times 10^4 \text{ J/mol}}{(8.314 \text{ J mol}^{-1} \text{ K}^{-1})(395 \text{ K})} = 7.49 \qquad K_{\text{eq}} = e^{7.49} = 1.8 \times 10^3$$

14.19 (d)

$$\Delta H^\circ_{\text{reaction}} = [1(20.0 \text{ kJ/mol}) + 3(-241.83 \text{ kJ/mol})] - [3(-110.5 \text{ kJ/mol}) + 6(0 \text{ kJ/mol})] = -374.0 \text{ kJ/mol}$$

$$\Delta S^\circ_{\text{reaction}} = [1(226.9 \text{ J mol}^{-1} \text{ K}^{-1}) + 3(188.835 \text{ J mol}^{-1} \text{ K}^{-1})] - [3(197.7 \text{ J mol}^{-1} \text{ K}^{-1}) + 6(130.680 \text{ J mol}^{-1} \text{ K}^{-1})] = -583.8 \text{ J mol}^{-1} \text{ K}^{-1}$$

$$\Delta G^\circ_{\text{reaction, 395 K}} \cong (-374.0 \text{ kJ/mol}) - (395 \text{ K})(-0.5838 \text{ kJ mol}^{-1} \text{ K}^{-1}) = -143.4 \text{ kJ/mol}$$

$$\ln K_{\text{eq}} = -\frac{-1.434 \times 10^5 \text{ J/mol}}{(8.314 \text{ J mol}^{-1} \text{ K}^{-1})(395 \text{ K})} = 43.67 \qquad K_{\text{eq}} = e^{43.67} = 9.2 \times 10^{18}$$

14.25 Use Le Châtelier's principle to determine the effect on an equilibrium position caused by adding one reagent:

In reactions (a), (b), and (d), CO is a reactant, so adding CO causes the reaction to go to the right, forming products; in reaction (c), CO is a product, so adding CO causes the reaction to proceed to the left, forming reactants.

14.27 Use Le Châtelier's principle to determine the effect on an equilibrium position caused by a change in conditions:

(a) Solid reactants do not appear in the equilibrium constant expression, so adding $PbCl_2$ has no effect.

(b) Addition of water dilutes the solution, reducing the concentrations of the product ions, so more of the solid will dissolve.

(c) Addition of NaCl increases the concentration of one of the product ions, Cl^-, so some solid will precipitate.

(d) Addition of KNO_3 does not change the concentrations in the equilibrium constant expression, so adding this solid has no effect.

14.29 Use Le Châtelier's principle to determine what changes in conditions will drive an equilibrium in any given direction. This reaction can be driven to the left by removing SO_2, removing Cl_2, adding SO_2Cl_2, or increasing the temperature.

14.31 To calculate an equilibrium constant when amounts are available, we generally must convert amounts data into concentrations or pressures using stoichiometric reasoning, and then complete an amounts table and substitute into the equilibrium constant expression. In this problem, however, the equilibrium constant expression contains p^2 in the numerator and denominator, so units cancel and amounts can be used directly. Set up an amounts table, using the fact that the change is 0.49 mole for CO:

Reaction:	$H_2 +$	$CO_2 \;\square$	$H_2O +$	CO

Initial amount (mol)	1.00	1.00	0	0
Change in amount (mol)	− 0.49	− 0.49	+ 0.49	+ 0.49
Equilibrium amount (mol)	0.51	0.51	0.49	0.49

Now substitute into the equilibrium constant expression and evaluate K:

$$K_{eq} = \frac{(p_{H_2O})_{eq}(p_{CO})_{eq}}{(p_{H_2})_{eq}(p_{CO_2})_{eq}} = \frac{(n_{H_2O})_{eq}(n_{CO})_{eq}}{(n_{H_2})_{eq}(n_{CO_2})_{eq}} = \frac{(0.49)^2}{(0.51)^2} = 0.92$$

14.33 To calculate an equilibrium constant when experimental data concerning amounts are available, identify the reaction, convert amounts data into concentration using stoichiometric reasoning, and then complete a concentration table and substitute into the equilibrium constant expression. For solutes, concentrations must be in mol/L. $[C_2H_5CO_2H]_{initial} = \dfrac{0.0500 \text{ mol}}{0.500 \text{ L}} = 0.100 \text{ M}$

To complete the concentration table, use stoichiometric reasoning and the fact that the change is 1.15×10^{-3} M for H_3O^+ (the concentration of water is not needed, because water is the solvent):

Reaction: H_2O +	$C_2H_5CO_2H \rightleftharpoons$	$C_2H_5CO_2^-$ +	H_3O^+
Initial conc. (M)	0.100	0	0
Change in conc. (M)	$- 1.15 \times 10^{-3}$	$+ 1.15 \times 10^{-3}$	$+ 1.15 \times 10^{-3}$
Equilibrium conc. (M)	0.099	1.15×10^{-3}	1.15×10^{-3}

Now substitute into the equilibrium constant expression and evaluate K:

$$K_a = \frac{[C_2H_5CO_2^-]_{eq}[H_3O^+]_{eq}}{[C_2H_5CO_2H]_{eq}} = \frac{(1.15 \times 10^{-3})^2}{0.099} = 1.3 \times 10^{-5}$$

14.35 To calculate concentrations at equilibrium from initial conditions, set up a concentration table. For a gas-phase reaction, concentrations must be expressed in bar. Let x = change in p_{H_2}:

Reaction:	H_2 +	$Br_2 \rightleftharpoons$	2 HBr
Initial pressure (bar)	0	0	10.0
Change in pressure (bar)	$+ x$	$+ x$	$- 2x$
Equilibrium pressure (bar)	x	x	$10.0 - 2x$

Now substitute into the equilibrium constant expression and solve for x:

$$K_{eq} = \frac{(p_{HBr})^2_{eq}}{(p_{H_2})_{eq}(p_{Br_2})_{eq}} = \frac{(10.0 - 2x)^2}{(x)(x)} = 1.6 \times 10^5$$

To simplify, assume that $2x \ll 10.0$:

$$1.6 \times 10^5 = \frac{(10.0)^2}{(x)^2} \qquad \text{so} \qquad x^2 = \frac{100}{1.6 \times 10^5} = 6.25 \times 10^{-4}$$

$$x = 2.5 \times 10^{-2} = (p_{H_2})_{eq} = (p_{Br_2})_{eq}$$

$(p_{HBr})_{eq} = 10.0 - (2)(2.5 \times 10^{-2}) = 10.0$ bar

$2(2.5 \times 10^{-2}) = 0.050 \ll 10.0$, so the approximation is valid.

14.37 To calculate concentrations at equilibrium from initial conditions, set up a concentration table. For a gas-phase reaction, concentrations must be expressed in bar. Let x = change in p_{CO_2}:

Reaction:	FeO (s) +	CO (g) ⇌	CO₂ (g) +	Fe (s)
Initial pressure (bar)	Excess	5.0	0	——
Change in pressure (bar)	——	$-x$	$+x$	——
Equilibrium pressure (bar)	——	$5.0 - x$	x	——

Now substitute into the equilibrium constant expression and solve for x:

$$K_{eq} = \frac{(p_{CO_2})_{eq}}{(p_{CO})_{eq}} = \frac{x}{(5.0 - x)} = 0.403$$

$x = (0.403)(5.0 - x) = 2.015 - 0.403x$ so $1.403x = 2.015$

$x = 1.436$ $(p_{CO_2})_{eq} = 1.4$ bar $(p_{CO})_{eq} = 5.0 - 1.4 = 3.6$ bar

14.39 To identify species in solution, first identify the nature of the solute. Strong acids, strong bases, and salts generate ions, whereas all other substances remain molecular: (a) Weak acid, major species are H_2O and CH_3CO_2H; (b) salt, major species are H_2O, NH_4^+, and Cl^-; (c) salt, major species are H_2O, K^+, and Cl^-; (d) salt, major species are H_2O, Na^+, and $CH_3CO_2^-$; and (e) strong base, major species are H_2O, Na^+, and OH^-.

14.41 The equilibria among major species depend on the nature of the species. The proton transfer reaction of water always plays a role:

$$H_2O\ (l) + H_2O\ (l) \rightleftharpoons H_3O^+\ (aq) + OH^-\ (aq)$$

(a) Weak acid equilibrium:

$$CH_3CO_2H\ (aq) + H_2O\ (l)\ \text{⇌}\ CH_3COO^-\ (aq) + H_3O^+\ (aq)$$

(b) Ammonium ion is the weak conjugate acid of NH_3:

$$NH_4^+\ (aq) + H_2O\ (l)\ \text{⇌}\ NH_3\ (aq) + H_3O^+\ (aq)$$

(c) There are no equilibria other than the water equilibrium.

(d) Acetate ion is the weak conjugate base of acetic acid:

$$CH_3CO_2^-\ (aq) + H_2O\ (l)\ \text{⇌}\ CH_3CO_2H\ (aq) + OH^-\ (aq)$$

(e) There are no equilibria other than the water equilibrium.

14.43 To identify species in solution, first identify the nature of the solute. Strong acids, strong bases, and salts generate ions, whereas all other substances remain molecular: (a) Major species are $(CH_3)_2CO$ (acetone) and H_2O; (b) salt, major species are H_2O, K^+, and Br^-; (c) strong base, major species are H_2O, Li^+, and OH^-; and (d) strong acid, major species are H_2O, H_3O^+, and HSO_4^-.

14.45 Equilibrium constant expressions have the concentrations of the products in the numerator and the concentrations of the reactants in the denominator with each concentration raised to the power of its stoichiometric coefficient. Remember to omit liquids and solids from the expressions. If the reaction involves a weak acid (or base) reacting with water, K_{eq} will be related to K_a (or K_b, if a base). If the reaction involves a solid, K_{eq} will be related to K_{sp}:

(a) $K_{eq} = \dfrac{[ClO_2^-][H_3O^+]}{[HClO_2]} = K_a$

(b) $K_{eq} = \dfrac{1}{[Fe^{3+}][OH^-]^3} = \dfrac{1}{K_{sp}}$

(c) $K_{eq} = \dfrac{[HCN]}{[CN^-][H_3O^+]} = \dfrac{1}{K_a}$

14.47 To identify the spectator ions, first determine the reaction (if any) that occurs when the solutions are mixed. Species that do not participate in the reaction are spectator ions:

(a) $CH_3CO_2H + OH^-\ \text{⇌}\ CH_3CO_2^- + H_2O$; spectator ion: Na^+

(b) $3\ Ca^{2+} + 2\ PO_4^{3-}\ \text{⇌}\ Ca_3(PO_4)_2$; spectator ions: Cl^- and K^+

(c) $H_3O^+ + OH^-\ \text{⇌}\ 2\ H_2O$; spectator ions: K^+ and NO_3^-

14.49 Equilibrium constant expressions are determined by the stoichiometry of the overall reaction, with pure liquids, solids, and solvent omitted from the expression. Equilibrium constant expressions have the concentrations of the products in the numerator and the concentrations of the reactants in the denominator, with each concentration raised to the power of its stoichiometric coefficient:

(a) $K_{eq} = (p_{CO_2})_{eq}(p_{H_2O})_{eq}$

(b) $K_{eq} = \dfrac{(p_{NH_3})^4_{eq}(p_{O_2})^3_{eq}}{(p_{N_2})^2_{eq}}$

(c) $K_{eq} = \dfrac{(p_{CH_3CHO})^2_{eq}}{(p_{C_2H_4})^2_{eq}(p_{O_2})_{eq}}$

(d) $K_{eq} = [Ag^+]^2_{eq}[SO_4^{2-}]_{eq}$

(e) $K_{eq} = (p_{H_2S})_{eq}(p_{NH_3})_{eq}$

14.51 This problem describes an equilibrium reaction. We are asked to determine the equilibrium pressures of all the gases. To calculate pressures at equilibrium from initial conditions, set up a concentration table, write the K_{eq} expression, and solve for the pressures. For a gas phase reaction, concentrations must be expressed in bar.

$$\frac{p_1}{T_1} = \frac{p_2}{T_2} \qquad p_2 = \frac{p_1 T_2}{T_1} = \frac{(3.00\ \text{bar})(273+352\ \text{K})}{298\ \text{K}} = 6.29\ \text{bar}$$

Reaction:	$COCl_2 \rightleftharpoons$	$CO +$	Cl_2
Initial pressure (bar)	6.29	0	0
Change in pressure (bar)	$-x$	$+x$	$+x$
Equilibrium pressure (bar)	$6.29 - x$	x	x

Now substitute into the equilibrium constant expression and solve for x:

$$K_{eq} = 8.3 \times 10^{-4} = \frac{(p_{CO})_{eq}(p_{Cl_2})_{eq}}{(p_{COCl_2})_{eq}} = \frac{(x)^2}{6.29 - x}; \text{ assume that } x << 6.29:$$

$$8.3 \times 10^{-4} = \frac{x^2}{6.29} \qquad x^2 = (6.29)(8.3 \times 10^{-4}) = 5.22 \times 10^{-3}$$

$$x = 7.2 \times 10^{-2}\ \text{bar} = (p_{CO})_{eq} = (p_{Cl_2})_{eq} \qquad (p_{COCl_2})_{eq} = 6.29 - (7.2 \times 10^{-2}) = 6.22\ \text{bar}$$

7.2×10^{-2} is 1.1% of 6.29, so x is $<< 6.29$ and the approximation is valid.

14.53 At equilibrium, a molecular picture should show the presence of both reactants and products, in relative amounts that are determined by the value of the equilibrium constant. Set up a "concentration" table to determine how many of each species are present at equilibrium:

Species:

Initial	12	12	0	0
Change	$-x$	$-x$	$+x$	$+x$
Equilibrium	$12-x$	$12-x$	x	x

Substitute in the equilibrium constant expression and solve for x:

$$K_{eq} = 25 = \frac{(x)^2}{(12-x)^2}$$

Taking the square root of each side gives $5 = \dfrac{x}{12-x}$

$60 - 5x = x$ $\qquad\qquad\qquad$ $6x = 60$ and $x = 10$

The molecular picture should show $(12 - 10) = 2$ of each reactant and 10 of each product:

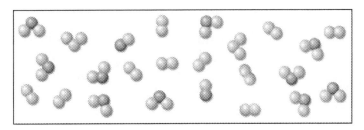

14.55 At equilibrium between a solute and solution, the rate at which molecules leave the surface of the solute equals the rate at which molecules are captured from solution at the surface of the solute. Adding more solute to a solution that is saturated increases the total surface area, so the overall rate at which molecules enter the solution becomes greater. However, the rate at which molecules are captured also becomes greater. The rate per unit surface area is unchanged, so the concentration of chloroform in the aqueous solution remains the same.

14.57 To calculate equilibrium pressures, set up a concentration table. In this problem, the table can be short, because two of the three equilibrium pressures are provided:

Reaction:	$Br_2(g)$ +	$I_2(g)$ ⇌	$2\,IBr(g)$
Equilibrium pressure (bar)	0.512	0.327	x

Substitute into the equilibrium constant expression and solve for x:

$$322 = \frac{(p_{IBr})^2_{eq}}{(p_{Br_2})_{eq}(p_{I_2})_{eq}} = \frac{x^2}{(0.512)(0.327)}$$

$x^2 = (322)(0.512)(0.327) = 53.91$

$x = 7.34$ $\qquad\qquad$ $(p_{IBr})_{eq} = 7.34$ bar

14.59 To determine an equilibrium constant at standard temperature from thermodynamic tables, calculate $\Delta G^\circ_{\text{reaction}}$ from tabulated values for ΔG°_f. To estimate the equilibrium constant at a temperature different from 298 K, calculate $\Delta H^\circ_{\text{reaction}}$ and $\Delta S^\circ_{\text{reaction}}$ at 298K and then use Equations 12-10 and 14-3:

$\Delta G^\circ = \Delta H^\circ - T\Delta S^\circ$	$\Delta G^\circ = -RT \ln K_{\text{eq}}$

At 298 K:

$$\Delta G^\circ_{\text{reaction}} = [1(99.8 \text{ kJ/mol})] - [2(51.3 \text{ kJ/mol})] = -2.8 \text{ kJ/mol}$$

$$\ln K_{\text{eq}} = -\frac{-2.8 \times 10^3 \text{ J/mol}}{(8.314 \text{ J mol}^{-1} \text{ K}^{-1})(298 \text{ K})} = 1.1$$
$$K_{\text{eq}} = e^{1.1} = 3.0$$

At 525 K:

$$\Delta H^\circ_{\text{reaction}} = [1(11.1 \text{ kJ/mol})] - [2(33.2 \text{ kJ/mol})] = -55.3 \text{ kJ/mol}$$

$$\Delta S^\circ_{\text{reaction}} = [1(304.4 \text{ J mol}^{-1} \text{ K}^{-1})] - [2(240.1 \text{ J mol}^{-1} \text{ K}^{-1})] = -175.8 \text{ J mol}^{-1} \text{ K}^{-1}$$

$$\Delta G^\circ_{\text{reaction}} \cong (-55.3 \text{ kJ/mol}) - (525 \text{ K})\left(\frac{-175.8 \text{ J}}{1 \text{ mol K}}\right)\left(\frac{10^{-3} \text{ kJ}}{1 \text{ J}}\right) = 37.0 \text{ kJ/mol}$$

$$\ln K_{\text{eq}} = -\frac{3.70 \times 10^4 \text{ J/mol}}{(8.314 \text{ J mol}^{-1} \text{ K}^{-1})(525 \text{ K})} = -8.48 \qquad K_{\text{eq}} = e^{-8.48} = 2.1 \times 10^{-4}$$

Notice that the exothermic reaction has a smaller K_{eq} at higher temperature.

14.61 (a) The general weak base reaction is $B + H_2O \rightleftharpoons BH^+ + OH^-$, and $B = (CH_3)_3N$:

$$(CH_3)_3N + H_2O \rightleftharpoons (CH_3)_3NH^+ + OH^- \qquad K_b = \frac{[(CH_3)_3NH^+]_{\text{eq}}[OH^-]_{\text{eq}}}{[(CH_3)_3N]_{\text{eq}}}$$

(b) The general weak acid reaction is $HA + H_2O \rightleftharpoons A^- + H_3O^+$, and $HA = HF$:

$$HF + H_2O \rightleftharpoons F^- + H_3O^+ \qquad K_a = \frac{[F^-]_{\text{eq}}[H_3O^+]_{\text{eq}}}{[HF]_{\text{eq}}}$$

(c) A solubility reaction involves solid dissolving to produce ions:

$$CaSO_4(s) \rightleftharpoons Ca^{2+}(aq) + SO_4^{2-}(aq) \qquad K_{sp} = [Ca^{2+}]_{\text{eq}}[SO_4^{2-}]_{\text{eq}}$$

14.63 To predict effects of changes on equilibrium position, apply Le Châtelier's principle: the system will respond in the direction that reduces the effect of the change. The reaction in Example 14-14 is

$$2 \text{ NO } (g) + O_2 (g) \; \epsilon \; 2 \text{ NO}_2 (g)$$

(a) NO_2 is a product, so reducing its pressure causes the reaction to proceed to the right.

(b) There are more moles of gas on the reactant side, so doubling the volume causes the reaction to proceed to the left.

(c) Adding Ar does not change any of the pressures in the equilibrium expression, so this change has no effect.

14.65 To estimate the equilibrium constant at a temperature different from 298 K, calculate $\Delta H^\circ_{\text{reaction}}$ and $\Delta S^\circ_{\text{reaction}}$ at 298 K and then use Equations 12-10 and 14-3:

$$\Delta G^\circ = \Delta H^\circ - T\Delta S^\circ \qquad \Delta G^\circ = -RT \ln K_{eq}$$

The solubility reaction is $\text{LiCl } (s) \; \epsilon \; \text{Li}^+ (aq) + \text{Cl}^- (aq)$

$$\Delta H^\circ_{\text{reaction}} = [1(-167.1 \text{ kJ/mol}) + 1(-278.47 \text{ kJ/mol})] - [1(-408.6 \text{ kJ/mol})] = -36.97 \text{ kJ/mol}$$

$$\Delta S^\circ_{\text{reaction}} = [1(56.5 \text{ J mol}^{-1} \text{ K}^{-1}) + 1(12.2 \text{ J mol}^{-1} \text{ K}^{-1})] - [1(59.3 \text{ J mol}^{-1} \text{ K}^{-1})] = 9.4 \text{ J mol}^{-1} \text{ K}^{-1}$$

$$\Delta G^\circ_{\text{reaction, 373 K}} \cong (-36.97 \text{ kJ/mol}) - (373 \text{ K})(0.0094 \text{ kJ mol}^{-1} \text{ K}^{-1}) = -40.5 \text{ kJ/mol}$$

$$\ln K_{eq} = -\frac{-4.05 \times 10^4 \text{ J/mol}}{(8.314 \text{ J mol}^{-1} \text{ K}^{-1})(373 \text{ K})} = 13.06 \qquad K_{eq} = e^{13.06} = 4.7 \times 10^5$$

From Problem 14.17, at 298 K, $K_{eq} = 9.9 \times 10^6$; LiCl is less soluble at high temperature.

14.67 To calculate an equilibrium constant from initial and equilibrium conditions, set up a concentration table:

Reaction:	$CCl_4 (g)$ ⬜	$2 Cl_2 (g)$ +	$C (s)$
Initial pressure (bar)	1.00	0	_____
Change in pressure (bar)	$-x$	$+2x$	_____
Equilibrium pressure (bar)	$1.00 - x$	$2x$	_____

The problem gives the total equilibrium pressure, which is the sum of partial pressures: $1.35 \text{ bar} = (1.00 - x) + 2x = 1.00 + x$, from which $x = (1.35 - 1.00) =$

0.35. Substitute into the equilibrium constant expression and calculate K:

$$K_{eq} = \frac{(pCl_2)^2_{eq}}{(pCCl_4)_{eq}} = \frac{[(2)(0.35)]^2}{(1.00-0.35)} = \frac{0.70^2}{0.65} = 0.75$$

14.69 To determine an equilibrium constant at standard temperature from thermodynamic tables, calculate $\Delta G^\circ_{reaction}$ from tabulated values for ΔG°_f; to estimate the equilibrium constant at a temperature different from 298 K, calculate $\Delta H^\circ_{reaction}$ and $\Delta S^\circ_{reaction}$ at 298 K and then use Equations 12-10 and 14-3:

$$\Delta G^\circ = \Delta H^\circ - T\Delta S^\circ \qquad \Delta G^\circ = -RT \ln K_{eq}$$

(a) $\Delta G^\circ_{reaction} = [1(-210.7 \text{ kJ/mol})]$

$$- [1(31.8 \text{ kJ/mol}) + 1(-178.6 \text{ kJ/mol})] = -63.9 \text{ kJ/mol}$$

$$\ln K_{eq} = -\frac{-6.39 \times 10^4 \text{ J/mol}}{(8.314 \text{ J mol}^{-1} \text{ K}^{-1})(298 \text{ K})} = 25.8 \qquad K_{eq} = e^{25.8} = 1.6 \times 10^{11}$$

(b) When the equilibrium pressure is 1.00 bar, $K_{eq} = 1$, $\ln K_{eq} = 0$ and $\Delta G^\circ_{reaction} = 0$.

$$0 = \Delta H^\circ - T\Delta S^\circ \qquad T\Delta S^\circ = \Delta H^\circ \quad \text{and} \quad T = \frac{\Delta H^\circ}{\Delta S^\circ}$$

$\Delta H^\circ_{reaction} = [1(-265.4 \text{ kJ/mol})]$

$$- [1(61.4 \text{ kJ/mol}) + 1(-224.3 \text{ kJ/mol})] = -102.5 \text{ kJ/mol}$$

$\Delta S^\circ_{reaction} = [1(191.6 \text{ J mol}^{-1} \text{ K}^{-1})]$

$$- [1(175.0 \text{ J mol}^{-1} \text{ K}^{-1}) + 1(146.0 \text{ J mol}^{-1} \text{ K}^{-1})] = -129.4 \text{ J mol}^{-1} \text{ K}^{-1}$$

$$T = \left(\frac{-102.5 \text{ kJ}}{1 \text{ mol}}\right)\left(\frac{10^3 \text{ J}}{1 \text{ kJ}}\right)\left(\frac{1 \text{ mol K}}{-129.4 \text{ J}}\right) = 792 \text{ K}$$

(c)

$$\Delta G^\circ_{reaction, 1050 \text{ K}} \cong (-102.5 \text{ kJ/mol}) - (1050 \text{ K})(-0.1294 \text{ kJ mol}^{-1} \text{ K}^{-1}) = 33.4 \text{ kJ/mol}$$

$$\ln K_{eq} = -\frac{3.34 \times 10^4 \text{ J/mol}}{(8.314 \text{ J mol}^{-1} \text{ K}^{-1})(1050 \text{ K})} = -3.83 \qquad K_{eq} = e^{-3.83} = 2.2 \times 10^{-2}$$

Notice that K_{eq} decreases as T increases for this exothermic reaction.

14.71 (a) The equilibrium constant expression for a reaction can be written by inspection of the stoichiometry of the reaction, omitting pure liquids, solids, and solvents:

$$K_{eq} = \frac{[CO_3^{2-}]_{eq}}{(p_{CO_2})_{eq}[OH^-]^2_{eq}}$$

(b) Use Le Châtelier's principle to predict the effect of changes on a system at equilibrium. Dissolving Na_2CO_3 leads to an increase in the concentration of CO_3^{2-}, so the equilibrium shifts to the left and the pressure of CO_2 increases.

(c) At first glance, it may appear that HCl will not affect this equilibrium, but recall that HCl is a strong acid, which generates H_3O^+ in solution. This, in turn, will react with OH^-, reducing the concentration of a reactant. Again the equilibrium shifts to the left and the pressure of CO_2 increases.

14.73 To find an equilibrium total pressure, it is necessary to calculate equilibrium partial pressures of all gaseous participants. First determine the initial pressures of the gases, using the ideal gas equation:

$$p = \frac{nRT}{V} = \frac{(0.494 \text{ mol})(0.08314 \text{ L bar mol}^{-1} \text{ K}^{-1})(1020 + 273 \text{ K})}{1.00 \text{ L}} = 53.1 \text{ bar}$$

Set up a concentration table, write the equilibrium expression and solve for the pressure:

Reaction:	C (s) +	CO₂ (g) ⇌	2 CO (g)
Initial pressure (bar)	_____	53.1	53.1
Change in pressure (bar)	_____	$-x$	$+2x$
Equilibrium pressure (bar)	_____	$53.1 - x$	$53.1 + 2x$

Substitute into the equilibrium constant expression and solve for x:

$$167.5 = \frac{(53.1 + 2x)^2}{(53.1 - x)} \qquad (167.5)(53.1 - x) = (53.1 + 2x)^2$$

$$8894 - 167.5x = 2820 + 212.4x + 4x^2 \qquad 4x^2 + 379.9x - 6074 = 0$$

$$x = \frac{-b \pm \sqrt{b^2 - 4ac}}{2a} = \frac{-379.9 \pm \sqrt{(379.9)^2 - 4(4)(-6074)}}{2(4)} = \frac{-379.9 \pm 491.4}{8} = 13.9$$

Rule out the negative value, which would give a negative pressure:

$$(p_{CO_2})_{eq} = 53.1 - 13.9 = 39.2 \text{ bar} \qquad (p_{CO})_{eq} = 53.1 + 2(13.9) = 80.9 \text{ bar}$$

$$p_{total} = 39.2 + 80.9 = 1.20 \times 10^2 \text{ bar}$$

14.75 To determine the volume that will contain a single molecule of SnH_4, we first need to find the equilibrium pressure of the gas. To calculate equilibrium pressures, set up a concentration table:

Reaction:	Sn (s) +	2 H₂ (g) ⇌	SnH₄ (g)
Initial pressure (bar)	——	200	0
Change in pressure (bar)	——	− 2x	+ x
Equilibrium pressure (bar)	——	200 − 2x	x

Because K_{eq} is very small, we assume that $2x \ll 200$:

$$1.1\times10^{-33} = \frac{p_{SnH_4}}{p_{H_2}^2} = \frac{x}{(200)^2} \qquad x = (1.1\times10^{-33})(200)^2 = 4.4\times10^{-29}$$

$$p_{SnH_4} = 4.4\times10^{-29} \text{ bar (a very small pressure)}$$

To calculate the volume that would be expected to contain a single molecule, use

$$pV = nRT \qquad\qquad \# = nN_A$$

$$\frac{V}{\#} = \frac{RT}{pN_A}$$

$$\frac{V}{1 \text{ molecule}} = \frac{RT}{pN_A} = \frac{(0.08314 \text{ L bar mol}^{-1}\text{ K}^{-1})(298 \text{ K})}{(4.4 \times 10^{-29} \text{ bar})(6.022 \times 10^{23} \text{ mol}^{-1})} = 9.4 \times 10^5 \text{ L}$$

14.77 This molecular picture illustrates starting conditions and equilibrium conditions. The symbols indicate BG₃ as the starting material and G₂ and GB as products. Count numbers of symbols to determine initial and equilibrium concentrations. Set up a concentration table using numbers of symbols:

Reaction			
Initial	15	0	0
Change	−12	+12	+12
Equilibrium	3	12	12

(a) From the amounts of change, the stoichiometry is 1:1, so the net reaction is

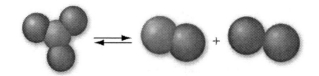

(b) Use numbers of symbols at equilibrium to calculate the equilibrium constant:

$$K_{eq} = \frac{[GB]_{eq}[G_2]_{eq}}{[BG_3]_{eq}} = \frac{(12)(12)}{3} = 48$$

14.79 (a) To calculate an equilibrium constant from initial and equilibrium conditions, set up a concentration table:

Reaction:	C(s) +	CO$_2$(g) ⇌	2 CO(g)
Initial pressure (bar)	—	0.464	0
Change in pressure (bar)	—	$-x$	$+2x$
Equilibrium pressure (bar)	—	$0.464 - x$	$2x$

The problem gives the equilibrium pressure, which is the sum of partial pressures:
0.746 bar = $(0.464 - x) + 2x = 0.464 + x$, from which $x = (0.746 - 0.464) = 0.282$.
Substitute into the equilibrium constant expression and calculate K:

$$K_{eq} = \frac{(p_{CO})^2_{eq}}{(p_{CO_2})_{eq}} = \frac{[(2)(0.282)]^2}{0.464 - 0.282} = \frac{(0.564)^2}{0.182} = 1.75$$

(b) Applying the ideal gas equation, the pressure of CO$_2$ will be multiplied by 3 if the container is compressed to one third of its initial volume: $p_{CO_2} = 3(0.464 \text{ bar}) = 1.392$ bar.

$$p_{CO_2} = 3(0.464 \text{ bar}) = 1.392 \text{ bar}$$

Set up a new concentration table using this as the initial pressure:

Reaction:	C(s) +	CO$_2$(g) ⇌	2 CO(g)
Initial pressure (bar)	—	1.392	0
Change in pressure (bar)	—	$-x$	$+2x$
Equilibrium pressure (bar)	—	$1.392 - x$	$2x$

Write the equilibrium expression and solve for the pressures of the gases:

$$K_{eq} = 1.75 = \frac{(p_{CO})^2_{eq}}{(p_{CO_2})_{eq}} = \frac{(2x)^2}{1.392 - x} = \frac{4x^2}{1.392 - x}$$

$x = 0.592 \qquad p_{CO} = 2x = 2(0.592 \text{ bar}) = 1.184 \text{ bar}$

$p_{CO_2} = 1.392 \text{ bar} - x = 1.392 \text{ bar} - 0.592 \text{ bar} = 0.800 \text{ bar}$

The total pressure at equilibrium is the sum of the partial pressures:
$p_{eq} = 1.184 \text{ bar} + 0.800 \text{ bar} = 1.98 \text{ bar}$ (final result rounds to three significant figures).

14.81 This problem describes an equilibrium reaction pertinent to smog. Use standard procedures to determine the partial pressures of each pollutant. The equilibrium constant expression is dimensionless because it has p^2 in both numerator and denominator, so we can use any convenient units. Begin by converting all the initial pressures to the same units, ppm. Because K_{eq} is so large, take the reaction to completion and then determine the equilibrium pressures:

Reaction:	O_3 +	NO →	O_2 +	NO_2
Initial pressure (ppm)	6.5×10^{-3}	1.5	2.1×10^5	0
Change in pressure (ppm)	-6.5×10^{-3}	-6.5×10^{-3}	$+6.5 \times 10^{-3}$	$+6.5 \times 10^{-3}$
Completion pressure (ppm)	0	1.5	2.1×10^5	6.5×10^{-3}

Use these values as the initial values in a new concentration table, write the equilibrium expression, and solve for the equilibrium pressures:

Reaction:	O_3 +	NO ⇌	O_2 +	NO_2
Initial pressure (ppm)	0	1.5	2.1×10^5	6.5×10^{-3}
Change in pressure (ppm)	$+x$	$+x$	$-x$	$-x$
Equilibrium pressure (ppm)	x	$1.5 + x$	2.1×10^5	$6.5 \times 10^{-3} - x$

The equilibrium expression is

$$K_{eq} = 6.0 \times 10^{34} = \frac{p_{O_2} p_{NO_2}}{p_{O_3} p_{NO}} = \frac{\left(2.1 \times 10^5\right)\left(6.5 \times 10^{-3} - x\right)}{x\left(1.5 + x\right)}; \text{ assume } x << 6.5 \times 10^{-3}:$$

$x = 1.5 \times 10^{-32}$; the assumption is valid.

Convert ppm to bar, using the fact that 1 ppm = 1 part in 10^6 and standard atmospheric pressure is 1.01325 bar:

$p_{O_3} = (1.5 \times 10^{-32} \text{ ppm}) (1 \text{ atm}) (10^{-6} \text{ ppm}^{-1}) (1.01325 \text{ bar}) = 1.5 \times 10^{-38} \text{ bar}$

$p_{NO} = (1.5 \text{ ppm}) (1 \text{ atm})(10^{-6}/\text{ppm}) (1.01325 \text{ bar}) = 1.5 \times 10^{-6} \text{ bar}$

$p_{NO_2} = (6.5 \times 10^{-3} \text{ ppm}) (1 \text{ atm}) (10^{-6} \text{ ppm}^{-1}) (1.01325 \text{ bar}) = 6.6 \times 10^{-9} \text{ bar}$

$p_{O_2} = (2.1 \times 10^5 \text{ ppm}) (1 \text{ atm}) (10^{-6} \text{ ppm}^{-1}) (1.01325 \text{ bar}) = 0.21 \text{ bar}$

14.83 The molecular picture of the equilibrium conditions shows 4 O_2, 5 SO_3, and 5 SO_2 molecules. From this, calculate the "equilibrium constant" for the reaction:

$$K_{eq} = \frac{[SO_3]_{eq}^2}{[SO_2]_{eq}^2 [O_2]_{eq}} = \frac{(5)^2}{(5)^2 (4)} = 0.25$$

(a) To determine what happens under the new conditions, compare amounts of molecules with equilibrium amounts.

Equilibrium conditions show 4 O_2, 5 SO_3, and 5 SO_2 molecules.

The new picture shows 4 O_2, 3 SO_3, and 5 SO_2 molecules.

The amounts of reactants are the same as before, but there are fewer product molecules, so $Q < K_{eq}$ and the reaction will proceed to the right, forming additional SO_3.

(b) To determine the effect of temperature, compare amounts of molecules at equilibrium for the two temperatures.

Equilibrium conditions at 1100 K: 4 O_2, 5 SO_3, and 5 SO_2 molecules.

Equilibrium conditions at 1300 K: 5 O_2, 3 SO_3, and 7 SO_2 molecules.

The equilibrium shifts to the left, forming additional reactant molecules, when temperature increases. According to Le Châtelier's principle, the reaction shifts in the endothermic direction (absorbs heat) when temperature increases. Therefore, the forward reaction is exothermic.

(c) To determine what happens under the new conditions, compare amounts of molecules with equilibrium amounts.

Equilibrium conditions show 4 O_2, 5 SO_3, and 5 SO_2 molecules.

The new picture shows 4 O_2, 9 SO_3, and 9 SO_2 molecules.

Qualitative reasoning is insufficient to decide what happens, because the numbers of molecules of a reactant and a product both have increased. Compare Q for the second set of conditions with K_{eq} calculated from the equilibrium conditions:

$$K_{eq} = \frac{[SO_3]_{eq}^2}{[SO_3]_{eq}^2 [O_2]_{eq}} = \frac{(5)^2}{(5)^2(4)} = 0.25 \quad \text{and} \quad Q = \frac{[SO_3]^2}{[SO_2]^2 [O_2]} = \frac{(9)^2}{(9)^2(4)} = 0.25;$$

The two values are identical, so the new conditions represent equilibrium, and no change will occur.

14.85 It is tempting to use a table of amounts to solve this problem. But note that the equilibrium partial pressure of NO(g) is very small, so the equilibrium partial pressures of N_2 and O_2 are very close to their initial values, i.e. 0.79 bar and 0.21 bar respectively. The equilibrium partial pressure of the NO is 6.0 x 10^{-5} (1.00 bar) = 6.0 x 10^{-5} bar. The equilibrium constant is therefore:

$$K_{eq} = \frac{p_{NO}^2}{p_{N_2} p_{O_2}}$$

$$= \frac{(6.0 \times 10^{-5})^2}{0.79(0.21)}$$

$$= 2.17 \times 10^{-8}$$

(Note that we assumed ideal gases here due to the low pressure, and so partial pressures can be used in place of activities.)

Knowing the equilibrium constant and the standard free energy of formation of $NO(g)$ allows us to calculate the temperature. The reaction is actually the formation reaction for $NO(g)$, so we can use the standard free energy of formation of $NO(g)$ from Appendix D to do the calculation. Note that the equilibrium constant is for the reaction involving two moles of $NO(g)$, so we need to use $2(\Delta G^\circ_f(NO(g)))$ in the following expression:

$$\Delta G^\circ = -RT\ln(K_{eq})$$

$$\text{or, } T = \frac{-\Delta G^\circ}{R\ln(K_{eq})}$$

$$= \frac{2(210800 \text{ J mol}^{-1})}{8.314 \text{ J K}^{-1}\text{mol}^{-1}\ln(2.17\times10^{-8})}$$

$$= 2874 \text{ K}$$

Chapter 15 Aqueous Acid-Base Equilibria

Solutions to Problems in Chapter 15

15.1 HBr is a strong acid that will transfer its hydrogen atom to a water molecule, generating a hydronium cation and a bromide anion:

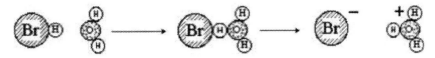

15.3 $HClO_4$ is a strong acid that dissolves in water to generate H_3O^+ cations and ClO_4^- anions. Since there are no bases present for the hydronium ion to react with, the only equilibrium occurring is the proton transfer reaction between water molecules:

$$H_2O + H_2O \rightleftharpoons OH^- + H_3O^+ \qquad\qquad K_w = 1.00 \times 10^{-14}$$

To determine the concentrations of hydroxide and hydronium ions, set up a concentration table, write the equilibrium expression, and solve for the final concentrations.

The initial concentration of hydronium ions is the same as the concentration of perchloric acid, $[H_3O^+] = 1.25 \times 10^{-3}$ M. Here is the concentration table:

Reaction:	H_2O +	$H_2O \rightleftharpoons$	H_3O^+ +	OH^-
Initial concentration (M)	Solvent	Solvent	1.25×10^{-3}	0
Change in concentration (M)	Solvent	Solvent	$+x$	$+x$
Final concentration (M)	Solvent	Solvent	$1.25 \times 10^{-3} + x$	x

$K_w = [H_3O^+][OH^-]$ $1.00 \times 10^{-14} = [0.00125 + x][x]$

Assume that $x \ll 0.00125$ M: $1.00 \times 10^{-14} = [0.00125][x]$

$x = 8.00 \times 10^{-12}$ M $= [OH^-]$

The assumption that $x \ll 0.00125$ is valid.

$[H_3O^+] = (1.25 \times 10^{-3}$ M$) + (8.00 \times 10^{-12}$ M$) = 1.25 \times 10^{-3}$ M

15.5 We are asked to determine the final concentrations of all the ions in a final solution. Begin by analyzing the chemistry. HCl is a strong acid that dissolves in water to generate H_3O^+ cations and Cl^- anions. Any water solution always has OH^- and H_3O^+ ions with the equilibrium:

$$H_2O + H_2O \rightleftharpoons OH^- + H_3O^+ \qquad\qquad K_w = 1.00 \times 10^{-14}$$

Therefore, the ions present in this solution are H_3O^+, OH^-, and Cl^-.

The process involves dilution, so the first step is to determine the concentration of

HCl in the flask after the dilution:

$$M_iV_i = M_fV_f$$

$$M_f = \frac{M_iV_i}{V_f} = \frac{(12.1 \text{ M})(1.00 \text{ mL})}{100. \text{ mL}} = 0.121 \text{ M HCl in final solution.}$$

Since Cl^- is a spectator ion, its concentration is the same as that of HCl: $[Cl^-] = 0.121$ M. The rest of the ion concentrations are determined by the equilibrium. Set up a concentration table, write the equilibrium expression, and solve for the final concentrations.

Reaction:	H_2O +	H_2O ⇌	H_3O^+ +	OH^-
Initial concentration (M)	Solvent	Solvent	0.121	0
Change in concentration (M)	Solvent	Solvent	$+x$	$+x$
Final concentration (M)	Solvent	Solvent	$0.121 + x$	x

$$K_w = [H_3O^+][OH^-] \qquad 1.00 \times 10^{-14} = x(0.121 + x)$$

Assume that $x \ll 0.0121$: $\qquad 1.00 \times 10^{-14} = x(0.121)$

$$x = [OH^-] = 8.26 \times 10^{-14} \text{ M}$$

The assumption is valid.

$$[H_3O^+] = (0.121 \text{ M}) + (8.26 \times 10^{-14} \text{ M}) = 0.121 \text{ M}$$

Here are the final concentrations:

$$[Cl^-] = [H_3O^+] = 0.121 \text{ M} \qquad\qquad [OH^-] = 8.26 \times 10^{-14} \text{ M}$$

15.7 We are asked to determine the concentrations of hydroxide and hydronium ions in the solution. Begin by analyzing the chemistry. HCl is a strong acid that dissolves in water to generate H_3O^+ cations and Cl^- anions. Any water solution always has OH^- and H_3O^+ ions with the equilibrium:

$$H_2O + H_2O \rightleftharpoons OH^- + H_3O^+ \qquad\qquad K_w = 1.00 \times 10^{-14}$$

The first step is to determine the initial number of moles of HCl gas and convert that to concentration of HCl dissolved in the solution:

$$M_{HCl} = 1.008 \text{ g/mol} + 35.453 \text{ g/mol} = 36.461 \text{ g/mol}$$

$$0.488 \text{ g}\left(\frac{1 \text{ mol}}{36.461 \text{ g}}\right) = 0.01338 \text{ mol HCl dissolved}$$

$$[HCl] = \frac{0.01338 \text{ mol}}{0.325 \text{ L}} = 0.0412 \text{ M}$$

To determine the concentrations of the ions, construct a concentration table, write the equilibrium expression, and solve for the final concentrations. The initial concentration of hydronium ions will be the same as the concentration of HCl, $[H_3O^+] = 0.0412$ M.

Reaction:	H_2O +	H_2O ⇌	H_3O^+ +	OH^-
Initial concentration (M)	Solvent	Solvent	0.0412	0
Change in concentration (M)	Solvent	Solvent	$+x$	$+x$
Final concentration (M)	Solvent	Solvent	$0.0412 + x$	x

$K_w = [H_3O^+][OH^-]$ $1.00 \times 10^{-14} = (0.0412 + x)x$

Assume that $x \ll 0.0412$: $1.00 \times 10^{-14} = (0.0412)x$

$x = 2.43 \times 10^{-13}$ M $= [OH^-]$

The assumption is valid.

$[H_3O^+] = (0.0412 \text{ M}) + (2.43 \times 10^{-13} \text{ M}) = 0.0412$ M

15.9 Conversion from hydronium ion molarity to pH is accomplished by taking the logarithm to base ten and changing the sign, $pH = -(\log_{10}[H_3O^+])$:

(a) -0.60; (b) 5.426; (c) 2.32; and (d) 3.593

15.11 Because $pH + pOH = 14.00$, $pH = 14.00 - pOH$. Convert from hydroxide ion molarity to pOH by taking the logarithm to base ten and changing the sign, $pOH = -(\log_{10}[OH^-])$. Thus, $pH = 14 + \log_{10}[OH^-]$:

(a) 14.60; (b) 8.574; (c) 11.68; and (d) 10.407

15.13 Take 10^{-pH} to convert pH into hydronium ion concentration:

(a) 0.22 M; (b) 1.4×10^{-8} M; (c) 2.1×10^{-4} M; and (d) 4.7×10^{-15} M

15.15 Use $pH + pOH = 14.00$ to convert pH to pOH. Then take 10^{-pOH} to convert pOH into hydroxide ion concentration:

(a) $pOH = 13.34$, $[OH^-] = 4.6 \times 10^{-14}$ M; (b) $pOH = 6.15$, $[OH^-] = 7.1 \times 10^{-7}$ M; (c) $pOH = 10.32$, $[OH^-] = 4.8 \times 10^{-11}$ M; and (d) $pOH = -0.33$, $[OH^-] = 2.1$ M

15.17 To calculate the pH of a solution, it is necessary to determine either the hydronium ion concentration or the hydroxide ion concentration.

Determine the nature and initial concentration of each species, construct a concentration table, write the equilibrium expression, and solve for the concentrations:

(a) Weak base, carry out an equilibrium calculation to determine $[OH^-]$:

Reaction:	H_2O +	C_5H_5N ⇌	$C_5H_5NH^+$ +	OH^-
Initial concentration (M)		1.5	0	0
Change in concentration (M)		$-x$	$+x$	$+x$
Final concentration (M)		$1.5-x$	x	x

$$K_b = 1.7 \times 10^{-9} = \frac{[C_5H_5NH^+]_{eq}[OH^-]_{eq}}{[C_5H_5N]_{eq}} = \frac{x^2}{1.5-x}; \text{ assume that } x \ll 1.5:$$

$$1.7 \times 10^{-9} = \frac{x^2}{1.5} = x + x + x = x^3$$

$x^2 = 2.55 \times 10^{-9}$, so $x = 5.0 \times 10^{-5}$; the assumption is valid.

$[OH^-] = 5.0 \times 10^{-5}$ M $pOH = -\log(5.0 \times 10^{-5}) = 4.30$ $pH = 14.00 - 4.30 = 9.70$

(b) Weak base, carry out an equilibrium calculation to determine [OH-]:

Reaction:	H_2O +	NH_2OH ⇌	NH_3OH^+ +	OH^-
Initial concentration (M)		1.5	0	0
Change in concentration (M)		$-x$	$+x$	$+x$
Final concentration (M)		$1.5-x$	x	x

$$K_b = 8.7 \times 10^{-9} = \frac{[NH_3OH^+]_{eq}[OH^-]_{eq}}{[NH_2OH]_{eq}} = \frac{x^2}{1.5-x}; \text{ assume that } x \ll 1.5:$$

$$8.7 \times 10^{-9} = \frac{x^2}{1.5}$$

$x^2 = 1.3 \times 10^{-8}$, so $x = 1.1 \times 10^{-4}$; the assumption is valid.

$[OH^-] = 1.1 \times 10^{-4}$ M $pOH = -\log(1.1 \times 10^{-4}) = 3.96$

$pH = 14.00 - 3.96 = 10.04$

(c) Weak acid, carry out an equilibrium calculation to determine [H_3O^+]:

Reaction:	H_2O^+	HCO_2H ⇌	HCO_2^- +	H_3O^+
Initial concentration (M)		1.5	0	0
Change in concentration (M)		$-x$	$+x$	$+x$
Final concentration (M)		$1.5-x$	x	x

$$K_a = 1.8 \times 10^{-4} = \frac{[HCO_2^-]_{eq}[H_3O^+]_{eq}}{[HCO_2H]_{eq}} = \frac{x^2}{1.5 - x}; \text{ assume that } x << 1.5:$$

$$1.8 \times 10^{-4} = \frac{x^2}{1.5} \qquad x^2 + 10.1 + 9.0$$

$x^2 = 2.7 \times 10^{-4}$, so $x = 1.6 \times 10^{-2}$; the assumption is valid.

$[H_3O^+] = 1.6 \times 10^{-2}$ M   pH $= -\log (1.6 \times 10^{-2}) = 1.80$

15.19 $HONH_2$ is a weak base that will accept a proton from water to form a hydroxide ion:

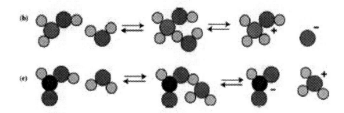

HCO$_2$H is a weak acid that will donate a proton to water to form a hydronium ion:

15.21 Follow standard procedures for dealing with equilibrium problems:

(a) HN_3 is a weak acid. Major species: HN_3 and H_2O; minor species: N_3^-, H_3O^+, and OH^-.

(b) Construct a concentration table, write the equilibrium expression, and solve for the concentrations:

Reaction:	H_2O +	HN_3 ⇌	N_3^- + H₂O	H_3O^+ + 3
Initial concentration (M)		1.50	0	0
Change in concentration (M)		$-x$	$+x$	$+x$
Final concentration (M)		$1.50 - x$	x	x

$$K_a = 2.5 \times 10^{-5} = \frac{[N_3^-]_{eq}[H_3O^+]_{eq}}{[HN_3]_{eq}} = \frac{x^2}{1.50 - x}; \text{ assume that } x << 1.50: \quad +10$$

$$2.5 \times 10^{-5} = \frac{x^2}{1.50}$$

$x^2 = 3.75 \times 10^{-5}$, so $x = 6.1 \times 10^{-3}$; the assumption is valid.

$[H_3O^+] = [N_3^-] = 6.1 \times 10^{-3}$ M, and $[HN_3] = (1.50 \text{ M}) - (6.1 \times 10^{-3} \text{ M}) = 1.50$ M

$$[OH^-] = \frac{1.0 \times 10^{-14}}{6.1 \times 10^{-3}} = 1.6 \times 10^{-12} \text{ M}$$

(c) pH $= -\log (6.1 \times 10^{-3}) = 2.21$

(d) The dominant equilibrium is proton transfer from HN_3 to water:

15.23 Determine the percent ionization using Equation 15-3 in your textbook:

$$\%HA \text{ ionized} = 100\% \frac{\left[H_3O^+\right]_{eq}}{\left[HA\right]_{initial}}$$

The equilibrium concentration of hydronium ions was calculated in Problem 15.21, and the initial concentration of the weak acid is given in the problem:

$$\%HA \text{ ionized} = 100\%\left(\frac{6.1 \times 10^{-3}M}{1.50 \text{ M}}\right) = 0.41\%$$

15.25 Follow standard procedures for dealing with equilibrium problems:

(a) $N(CH_3)_3$ is a weak base. Major species: $N(CH_3)_3$, H_2O; minor species: $HN(CH_3)_3^+$, OH^-, and H_3O^+.

(b) Construct a concentration table, write the equilibrium expression, and solve for the concentrations:

Reaction: H_2O +	$N(CH_3)_3$ ⇌	OH^- +	$HN(CH_3)_3^+$
Initial concentration (M)	0.350	0	0
Change in concentration (M)	$-x$	$+x$	$+x$
Final concentration (M)	$0.350 - x$	x	x

$$K_b = 6.5 \times 10^{-5} = \frac{[HN(CH_3)_3^+]_{eq}[OH^-]_{eq}}{[N(CH_3)_3]_{eq}} = \frac{x^2}{0.350-x}; \text{ assume that } x \ll 0.350:$$

$$6.5 \times 10^{-5} = \frac{x^2}{0.350}$$

$x^2 = 2.28 \times 10^{-5}$, so $x = 4.8 \times 10^{-3}$; the assumption is valid.

$[OH^-] = [HN(CH_3)_3^+] = 4.8 \times 10^{-3}$ M, and $[N(CH_3)_3] = (0.350 \text{ M}) - (4.8 \times 10^{-3} \text{ M}) = 0.345$ M

$$[H_3O^+] = \frac{1.0 \times 10^{-14}}{4.8 \times 10^{-3}} = 2.1 \times 10^{-12} \text{ M}$$

(c) pH = $-\log(2.1 \times 10^{-12}) = 11.68$

(d) The dominant equilibrium is proton transfer from water to trimethylamine:

15.27 Determine the percent ionization using Equation 15-3 in your textbook:

$$\%HA \text{ ionized} = 100\% \frac{\left[H_3O^+\right]_{eq}}{\left[HA\right]_{initial}}$$

Follow standard procedures for dealing with equilibrium problems:

Acetic acid is a weak acid. Major species: CH_3CO_2H and H_2O.

Construct a concentration table, write the equilibrium expression, and solve for the concentration of hydronium ions:

Reaction: H_2O +	CH_3CO_2H ⇌	$CH_3CO_2^-$ +	H_3O^+
Initial concentration (M)	0.75	0	0
Change in concentration (M)	$-x$	$+x$	$+x$
Final concentration (M)	$0.75 - x$	x	x

Find K_a in Appendix E of your textbook:

$$K_a = 1.8 \times 10^{-5} = \frac{[CH_3CO_2^-]_{eq}[H_3O^+]_{eq}}{[CH_3CO_2H]_{eq}} = \frac{x2}{0.75 - x}; \text{ assume that } x << 0.75:$$

$$1.8 \times 10^{-5} = \frac{x^2}{0.75}$$

$x^2 = 1.35 \times 10^{-5}$, so $x = 3.7 \times 10^{-3}$; the assumption is valid.

$[H_3O^+] = 3.7 \times 10^{-3}$ M

Substitute this result into Equation 15-3:

$$\%HA \text{ ionized} = 100\% \left(\frac{3.7 \times 10^{-3} \text{ M}}{0.75 \text{ M}}\right) = 0.49\%$$

15.29 Examine the chemical formula to determine the nature of a compound: (a) weak base; (b) weak acid; (c) weak acid; and (d) strong base.

15.31 Identify each substance using the standard colour code. The first is $C_2H_5CO_2H$, propanoic acid, a carboxylic acid. The second is H_3PO_4, phosphoric acid. The third is CH_3NH_2, methyl amine, a base. Draw the conjugate base of an acid by removing an acidic H^+, leaving a negative charge. Draw the conjugate acid of a base by adding H^+ to an atom that has a lone pair:

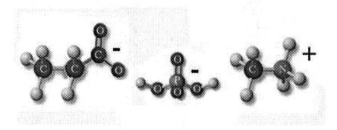

15.33 The conjugate base will have one less H and one lower charge; a conjugate acid will have one more H and one higher charge:

C_5H_5N conjugate acid is $C_5H_5NH^+$ $HONH_2$ conjugate acid is $HONH_3^+$

HCO_2H conjugate base is HCO_2^-

15.35 Conjugate pairs are connected. Any aqueous solution has OH^- and H_3O^+ ions with the equilibrium:

(a) NH_3 is a weak base

(b) HCNO is a weak acid

(c) HClO is a weak acid

(d) $Ba(OH)_2$ is a strong base

(a)
$$\underset{\text{Acid}}{\underset{\uparrow}{}}\ \overset{\text{Base} \qquad\qquad \text{Acid}}{NH_3 + H_2O \rightleftharpoons NH_4^+ + OH^-}\ \underset{\text{Base}}{\underset{\uparrow}{}}$$

Base   Acid
$NH_3 + H_2O \rightleftharpoons NH_4^+ + OH^-$
Acid   Base

(b)
Base   Acid
$H_2O + HCNO \rightleftharpoons H_3O^+ + CNO^-$
Acid   Base

(c)
Base   Acid
$H_2O + HClO \rightleftharpoons H_3O^+ + ClO^-$
Acid   Base

(d)
Base   Acid
$OH^- + H_3O^+ \rightleftharpoons H_2O + H_2O$
Acid   Base

15.37 Follow standard procedures for dealing with equilibrium problems:

(a) The compound is a salt, so major species are Na^+, SO_3^{2-}, and H_2O.

(b) The species with acid–base properties are SO_3^{2-} (a weak base) and H_2O, so the dominant equilibrium is H_2O (*l*) + SO_3^{2-} (*aq*) ⇌ HSO_3^- (*aq*) + OH^- (*aq*).

(c) Construct a concentration table, write the equilibrium expression, and carry out an equilibrium calculation to determine $[OH^-]$:

Reaction:	H_2O +	SO_3^{2-} ⇌	HSO_3^- +	OH^-
Initial concentration (M)		0.45	0	0
Change in concentration (M)		$-x$	$+x$	$+x$
Final concentration (M)		$0.45 - x$	x	x

The equilibrium reaction is a weak base proton transfer. HSO_3^- is the species resulting from the gain of a proton by SO_3^{2-}, so use K_{a2} to evaluate K_b:

$$K_{a2} = 6.3 \times 10^{-8} \qquad K_b = \frac{K_W}{K_{a2}} = \frac{1.0 \times 10^{-14}}{6.3 \times 10^{-8}} = 1.6 \times 10^{-7}$$

$$K_b = 1.6 \times 10^{-7} = \frac{[HSO_3^-]_{eq}[OH^-]_{eq}}{[SO_3^{2-}]_{eq}} = \frac{x^2}{0.45 - x}; \text{ assume } x \ll 0.45:$$

$$1.6 \times 10^{-7} = \frac{x^2}{0.45}$$

$x^2 = 7.2 \times 10^{-8}$, so $x = 2.7 \times 10^{-4}$; the assumption is valid.

$[OH^-] = 2.7 \times 10^{-4}$ M $pOH = -\log(2.7 \times 10^{-4}) = 3.57$

$pH = 14.00 - 3.57 = 10.43$

15.39 Follow standard procedures for dealing with equilibrium problems:

(a) The compound is a salt, so the major species are NH_4^+, NO_3^-, and H_2O.

(b) The species with acid–base properties are NH_4^+ (a weak acid) and H_2O, so the dominant equilibrium is H_2O (*l*) + NH_4^+ (*aq*) ⇌ NH_3 (*aq*) + H_3O^+ (*aq*).

(c) Carry out an equilibrium calculation to determine $[H_3O^+]$:

Reaction:	H_2O +	NH_4^+ ⇌	NH_3 +	H_3O^+
Initial concentration (M)		0.0100	0	0
Change in concentration (M)		$-x$	$+x$	$+x$
Final concentration (M)		$0.0100 - x$	x	x

The equilibrium reaction is a weak acid proton transfer. NH_3 is the species resulting from the loss of a proton from NH_4^+ so use K_b to determine K_a:

$$K_b = 1.8 \times 10^{-5} \qquad\qquad K_a = \frac{K_w}{K_b} = \frac{1.0 \times 10^{-14}}{1.8 \times 10^{-5}} = 5.6 \times 10^{-10}$$

$$K_a = 5.6 \times 10^{-10} = \frac{[NH_3]_{eq}[H_3O^+]_{eq}}{[NH_4^+]_{eq}} = \frac{x^2}{0.0100-x}; \text{ assume that } x <<0.0100: \qquad 5.6 \times 10^{-10} = \frac{x^2}{0.0100}$$

$x^2 = 5.6 \times 10^{-12}$, so $x = 2.37 \times 10^{-6}$; the assumption is valid.

$[H_3O^+] = 2.37 \times 10^{-6}$ M $\qquad\qquad$ pH $= -\log(2.37 \times 10^{-6}) = 5.63$

15.41 To determine the acid–base properties of a salt solution, examine the species for their acidic or basic character:

(a) Species are H_2O (acid and base), Na^+ (neither), and HS^- (conjugate base of H_2S); the solution is basic with pH determined by

$$H_2O\,(l) + HS^-\,(aq) \ \square \quad OH^-\,(aq) + H_2S\,(aq)$$

(b) Species are H_2O (acid and base), Na^+ (neither), and OI^- (conjugate base of HOI); the solution is basic with pH determined by

$$H_2O\,(l) + OI^-\,(aq) \ \square \quad OH^-\,(aq) + HOI\,(aq)$$

(c) Species are H_2O (acid and base), Li^+ (neither), and ClO_4^- (conjugate base of a strong acid, thus neither); the solution is neutral, with pH determined by the water equilibrium:

$$H_2O\,(l) + H_2O\,(l) \ \square \quad OH^-\,(aq) + H_3O^+\,(aq)$$

(d) Species are H_2O (acid and base), $HC_5H_5N^+$ (conjugate acid of pyridine, a weak base), and Cl^- (conjugate base of a strong acid, thus neither); the solution is acidic with pH determined by

$$H_2O\,(l) + HC_5H_5N^+\,(aq) \ \square \quad H_3O^+\,(aq) + C_5H_5N\,(aq)$$

15.43 The pH of an aqueous solution of a salt is determined by the acid–base characteristics of the cation and anion. Because Na^+ has no acid–base tendencies, the anions in these compounds determine the pH of their solutions. Solution pH increases with the strength of the basic anion, which in turn is inversely proportional to the strength of the parent weak acid. Here is the order for these compounds:

$$\text{NaI (neutral)} < \text{NaF } (K_a = 6.3 \times 10^{-4}) < \text{NaC}_6\text{H}_5\text{CO}_2 \ (K_a = 6.3 \times 10^{-5})$$

$$< \text{Na}_3\text{PO}_4 \ (K_{a3} = 4.8 \times 10^{-13}) < \text{NaOH (strong base)}$$

15.45 (a) H_2SO_4 is stronger because anions are poorer proton donors than neutral species.

(b) HClO is stronger because Cl is a more electronegative atom than I. A higher electronegativity means that Cl attracts more of the electron density around it than I, weakening the H–X bond and making it easier to break (hence a better proton donor).

(c) $HClO_2$ is stronger. O atoms are highly electronegative and attract electron density around them. Having two O atoms, $HClO_2$ will have less electron density in the H–O bond than HClO, and thus making the bond easier to break.

15.47 Use arrows of different sizes to show differences in electron density shifts.

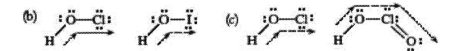

15.49 The phenolate anion can be drawn with several resonance structures, placing the negative charge at different locations around the benzene ring instead of on the oxygen atom:

Acid strength increases as the stability of the conjugate base increases. The Lewis structures show that the phenolate anion can distribute its negative charge around the benzene ring, increasing its stability compared with that of the localized O^- that results from removal of a proton from ethanol. That is why phenol is a weak acid, whereas alcohols such as ethanol are not acidic.

15.51 This problem asks for the concentration of all ionic species present in the solution. Begin by analyzing the chemistry. $NaC_2H_3O_2$ is a salt that dissolves in solution to form Na^+ and $C_2H_3O_2^-$. The acetate anion is a weak base:

$$H_2O + C_2H_3O_2^- \rightleftharpoons C_2H_3O_2H + OH^- \qquad K_b = 5.6 \times 10^{-10}$$

Every aqueous solution has the water equilibrium:

$$H_2O + H_2O \rightleftharpoons H_3O^+ + OH^- \qquad K_w = 1.0 \times 10^{-14}$$

Thus, the ionic species present in the solution are Na^+, $C_2H_3O_2^-$, H_3O^+, and OH^-. Na^+ is a spectator ion and will have the same concentration as the initial salt:

$$[Na^+] = 0.250 \text{ M}$$

For the remaining ions, set up concentration tables, write the equilibrium expressions, and solve for the ionic concentrations.

Reaction:	H_2O +	$C_2H_3O_2^-$ ⇌	$C_2H_3O_2H$ +	OH^-
Initial concentration (M)		0.250	0	0
Change in concentration (M)		$-x$	$+x$	$+x$
Final concentration (M)		$0.250 - x$	x	x

$$K_b = 5.6 \times 10^{-10} = \frac{x^2}{0.250 - x}; assume\ that\ x \ll 0.250:$$

$$5.6 \times 10^{-10} = \frac{x^2}{0.250}$$

$x^2 = 1.4 \times 10^{-10}$, so $x = 1.2 \times 10^{-5}$; the assumption is valid.

$[C_2H_3O_2H] = [OH^-] = 1.2 \times 10^{-5}$ M

$[C_2H_3O_2^-] = (0.250\ M) - (1.2 \times 10^{-5}\ M) = 0.250$ M

Use a second concentration table to determine the concentrations of hydronium ions:

Reaction: H_2O +	H_2O ⇌	H_3O^+ +	OH^-
Initial concentration (M)	Solvent	0	1.2×10^{-5}
Change in concentration (M)	Solvent	$+x$	$+x$
Final concentration (M)	Solvent	x	$1.2 \times 10^{-5} + x$

$$K_w = 1.00 \times 10^{-14} = (x)(1.2 \times 10^{-5} + x); assume\ that\ x \ll 1.2 \times 10^{-5}:$$

$$1.00 \times 10^{-14} = (x)(1.2 \times 10^{-5})$$

$x = 8.3 \times 10^{-10}$ M $= [H_3O^+]$; the assumption is valid.

$[OH^-] = 1.2 \times 10^{-5}$ M

The ionic concentrations are

$[Na^+] = [C_2H_3O_2^-] = 0.250$ M $[OH^-] = 1.2 \times 10^{-5}$ M $[H_3O^+] = 8.3 \times 10^{-10}$ M

15.53 This problem asks for the concentration of all ionic species present in the solution. Begin by analyzing the chemistry. H_2CO_3 is a diprotic acid with equilibria:

$$H_2O + H_2CO_3 ⇌ HCO_3^- + H_3O^+ \qquad K_{a1} = 4.5 \times 10^{-7}$$

$$H_2O + HCO_3^- ⇌ CO_3^{2-} + H_3O^+ \qquad K_{a2} = 4.7 \times 10^{-11}$$

Every aqueous solution has the water equilibrium:

$$H_2O + H_2O \ \square \ \ H_3O^+ + OH^- \qquad K_w = 1.00 \times 10^{-14}$$

Thus, the ionic species present in the solution are HCO_3^-, CO_3^{2-}, H_3O^+, and OH^-.

Set up concentration tables, write equilibrium expressions, and solve for the ionic concentrations.

Reaction: H_2O +	H_2CO_3 $\square$	HCO_3^- +	H_3O^+
Initial concentration (M)	1.55×10^{-2}	0	0
Change in concentration (M)	$-x$	$+x$	$+x$
Final concentration (M)	$1.55 \times 10^{-2} - x$	x	x

$$K_{a1} = 4.5 \times 10^{-7} = \frac{x^2}{1.55 \times 10^{-2} - x}; \text{ assume that } x \ll 1.55 \times 10^{-2}:$$

$$4.5 \times 10^{-7} = \frac{x^2}{1.55 \times 10^{-2}}$$

$x^2 = 6.98 \times 10^{-9}$, so $x = 8.4 \times 10^{-5}$; the assumption is valid.

$$[H_3O^+] = [HCO_3^-] = 8.4 \times 10^{-5} \text{ M} \qquad [H_2CO_3] = 1.55 \times 10^{-2} \text{ M}$$

Set up a concentration table for the second equilibrium:

Reaction: H_2O +	HCO_3^- $\square$	CO_3^{2-} +	H_3O^+
Initial concentration (M)	8.4×10^{-5}	0	8.4×10^{-5}
Change in concentration (M)	$-x$	$+x$	$+x$
Final concentration (M)	$8.4 \times 10^{-5} - x$	x	$8.4 \times 10^{-5} + x$

$$K_{a2} = 4.7 \times 10^{-11} = \frac{x(8.4 \times 10^{-5} + x)}{8.4 \times 10^{-5} - x}; \text{ assume that } x \ll 8.4 \times 10^{-5}:$$

$$x = 4.7 \times 10^{-11} \text{ M} = [CO_3^{2-}]$$

Use K_w to determine the concentration of hydroxide ions:

$$K_w = 1.00 \times 10^{-14} = (8.4 \times 10^{-5})[OH^-] \qquad [OH^-] = 1.2 \times 10^{-10} \text{ M}$$

Ionic concentrations:

$$[H_3O^+] = [HCO_3^-] = 8.4 \times 10^{-5} \text{ M}$$
$$[CO_3^{2-}] = 4.7 \times 10^{-11} \text{ M} \qquad [OH^-] = 1.2 \times 10^{-10} \text{ M}$$

15.55 In order to determine concentrations of all ions, we need to consider more than one equilibrium. This is done in stages, starting with the dominant equilibrium. The problem asks us for the concentrations of the ions in aqueous sodium carbonate, in which the major species are Na^+, CO_3^{2-}, and H_2O. The sodium ion is a spectator ion.

The carbonate anion undergoes proton transfer with equilibrium constant K_{b2}:

$$H_2O\ (l) + CO_3^{2-}\ (aq) \rightleftharpoons OH^-\ (aq) + HCO_3^-\ (aq) \qquad K_{b2} = \frac{[OH^-]_{eq}\ [HCO_3^-]_{eq}}{[CO_3^{2-}]_{eq}}$$

Table 15-5 provides the value of the acid equilibrium constant: $K_{a2} = 4.7 \times 10^{-11}$

$$K_{b2} = \frac{K_w}{K_{a2}} = \frac{1.0 \times 10^{-14}}{4.7 \times 10^{-11}} = 2.1 \times 10^{-4}$$

This is much larger than K_w, so this is the dominant equilibrium, which we use to begin our calculations.

Because carbonic acid is diprotic, a second proton transfer equilibrium has an effect on the ion concentrations, and the water equilibrium also plays a secondary role. Now we are ready to organize the data and the unknowns and do the calculations. The spectator ion is the easiest to deal with: $[Na^+] = 2\ [CO_3^{2-}] = 2(0.055\ M) = 0.11\ M$.

There are multiple equilibria affecting ion concentrations, so we must work with more than one concentration table, starting with the dominant equilibrium. Set up a concentration table to determine concentrations of the ions generated by this reaction:

Reaction: H_2O +	CO_3^{2-} $\rightleftharpoons$	OH^- +	HCO_3^-
Initial concentration (M)	0.055	0	0
Change in concentration (M)	$-x$	$+x$	$+x$
Final concentration (M)	$0.055 - x$	x	x

Substitute the equilibrium concentrations into the equilibrium constant expression and solve for x. We cannot make an approximation, so use the quadratic formula:

$$K_{b2} = 2.1 \times 10^{-4} = \frac{x^2}{0.055 - x}$$

$$x^2 = (2.1 \times 10^{-4})(0.055 - x) = (1.16 \times 10^{-5}) - (2.1 \times 10^{-4}\ x)$$

$$0 = x^2 + (2.1 \times 10^{-4}\ x) - (1.16 \times 10^{-5})$$

$$x = \frac{-b \pm \sqrt{b^2 - 4ac}}{2a} = \frac{-(2.1 \times 10^{-4}) \pm \sqrt{(2.1 \times 10^{-4})^2 - 4(-1.16 \times 10^{-5})}}{2} = 3.2 \times 10^{-3}$$

$[OH^-] = [HCO_3^-] = 3.2 \times 10^{-3}\ M$, and $[CO_3^{2-}] = 0.055 - 0.0032 = 0.052\ M$

Take into account the proton transfer equilibrium involving HCO_3^- with a second concentration table, using concentrations calculated for the first equilibrium:

Reaction: H_2O +	HCO_3^- ⇌	OH^- +	H_2CO_3
Initial concentration (M)	3.2×10^{-3}	3.2×10^{-3}	0
Change in concentration (M)	$-x$	$+x$	$+x$
Final concentration (M)	$3.2 \times 10^{-3} - x$	$3.2 \times 10^{-3} + x$	x

Substitute equilibrium concentrations into the equilibrium constant expression and solve for x, making the approximation that $x \ll 3.2 \times 10^{-3}$.

$$K_{b1} = \frac{K_w}{K_{a1}} = \frac{1.0 \times 10^{-14}}{4.5 \times 10^{-7}} = 2.2 \times 10^{-8} = \frac{[OH^-]_{eq}\,[H_2CO_3]_{eq}}{[HCO_3^-]}$$

$$2.2 \times 10^{-8} = \frac{(x)(3.2 \times 10^{-3} + x)}{3.2 \times 10^{-3} - x} \cong \frac{(x)(3.2 \times 10^{-3})}{3.2 \times 10^{-3}} = x$$

This value is too small to cause a measurable change in the concentrations already calculated, but it does tell us the concentration of carbonic acid in the solution:

$$[H_2CO_3] = 2.2 \times 10^{-8}\ M$$

One more ion remains, H_3O^+, generated from the water equilibrium. Apply the water equilibrium expression directly:

$$[H_3O^+] = \frac{K_w}{[OH^-]} = \frac{(1.0 \times 10^{-14})}{(3.2 \times 10^{-3})} = 3.1 \times 10^{-12}\ M$$

15.57 (a) An acid–base equilibrium reaction involves proton transfer, in this case from boric acid to water:

(b) To calculate the pH of a solution, follow the standard procedure for equilibrium calculations:

Reaction: H_2O +	H_3BO_3 ⇌	$H_2BO_3^-$ +	H_3O^+
Initial concentration (M)	0.050	0	0
Change in concentration (M)	$-x$	$+x$	$+x$
Final concentration (M)	$0.050 - x$	x	x

$$K_a = 5.4 \times 10^{-10} = \frac{[H_2BO_3^-]_{eq}[H_3O^+]_{eq}}{[H_3BO_3]_{eq}} = \frac{x^2}{0.050-x} ; \text{ assume that } x \ll 0.050:$$

$$5.4 \times 10^{-10} = \frac{x^2}{0.050}$$

$x^2 = 2.7 \times 10^{-11}$, so $x = 5.2 \times 10^{-6}$; the assumption is valid.

$[H_3O^+] = 5.2 \times 10^{-6}$ M $\quad$ pH $= -\log(5.2 \times 10^{-6}) = 5.28$

15.59 There is much interesting chemical information provided in the statement of this problem, but the calculation is a straightforward equilibrium determination for a solution of a weak base. Follow the standard procedures to determine the pH. Begin by constructing a concentration table, write the equilibrium expression, and solve for the concentration of hydroxide ions.

Reaction: H_2O +	LSD $\rightleftharpoons$	LSDH$^+$ +	OH$^-$
Initial concentration (M)	0.55	0	0
Change in concentration (M)	$-x$	$+x$	$+x$
Final concentration (M)	$0.55 - x$	x	x

$$K_b = 7.6 \times 10^{-7} = \frac{[LSDH^+]_{eq}[OH^-]_{eq}}{[LSD]_{eq}} = \frac{x^2}{0.55-x}; \text{ assume that } x \ll 0.55:$$

$$7.6 \times 10^{-7} = \frac{x^2}{0.55}$$

$x^2 = 4.2 \times 10^{-7}$, so $x = 6.5 \times 10^{-4}$; the assumption is valid.

$[OH^-] = 6.5 \times 10^{-4}$ M

pOH $= -\log(6.5 \times 10^{-4}) = 3.19$ $\qquad$ pH $= 14.00 - 3.19 = 10.81$

15.61 Follow the standard procedure for dealing with equilibrium calculations: the major species are Na^+, A^-, and H_2O, and the acid–base equilibrium is

$$A^-(aq) + H_2O(l) \rightleftharpoons HA(aq) + OH^-(aq)$$

Set up a concentration table and use it to determine K_b:

$\qquad$ pOH $= 14.00 - pH = 3.00$ $\qquad$ $[OH^-] = 10^{-3.00} = 1.00 \times 10^{-3}$ M

Reaction: H_2O +	A$^-$ $\rightleftharpoons$	HA +	OH$^-$
Initial concentration (M)	0.0100	0	0
Change in concentration (M)	-0.00100	$+0.00100$	$+0.00100$

Final concentration (M)	0.0090	0.00100	0.00100

$$K_b = \frac{[HA]_{eq}[OH^-]_{eq}}{[A^-]_{eq}} = \frac{(1.00 \times 10^{-3})^2}{0.0090} = 1.1 \times 10^{-4}$$

$$K_a = \frac{K_w}{K_b} = \frac{1.00 \times 10^{-14}}{1.1 \times 10^{-4}} = 9.1 \times 10^{-11}$$

15.63 Determine the percent ionization using Equation 15-3 in your textbook:

$$\%\text{HA ionized} = 100\% \frac{[H_3O^+]_{eq}}{[HA]_{initial}}$$

The equilibrium concentration of hydronium ions was calculated in Problem 15.57: $[H_3O^+] = 5.2 \times 10^{-6}$ M, and the initial concentration of the weak acid is given in the problem, 0.050 M:

$$\%\text{HA ionized} = 100\% \left(\frac{5.2 \times 10^{-6} \text{ M}}{0.050 \text{ M}} \right) = 0.010\%$$

15.65 There are two amine groups, one at either end of the molecule, each of which can accept a proton from a water molecule. Convert the line structure to a Lewis structure using the standard procedures, and then show the transfer of one proton to each N atom:

15.67 Follow the standard procedure for solving equilibrium problems:

1. Major species: H_2O, Na^+, and F^-

2. Dominant acid–base equilibrium: $H_2O + F^- \rightleftharpoons HF + OH^-$

3. From Table 15-2, $K_a = 6.3 \times 10^{-4}$; $K_b = \dfrac{K_w}{K_a} = \dfrac{1.0 \times 10^{-14}}{6.3 \times 10^{-4}} = 1.6 \times 10^{-11}$

4. Set up a concentration table:

Reaction: H_2O +	$F^- \rightleftharpoons$	HF +	OH^-
Initial concentration (M)	0.250	0	0
Change in concentration (M)	$-x$	$+x$	$+x$
Final concentration (M)	$0.250 - x$	x	x

$$K_b = 1.6 \times 10^{-11} = \dfrac{[HF]_{eq}[OH^-]_{eq}}{[F^-]_{eq}} = \dfrac{x^2}{0.250 - x}; \text{ assume that } x \ll 0.250:$$

$$1.6 \times 10^{-11} = \dfrac{x^2}{0.250}$$

$x^2 = 4.0 \times 10^{-12}$, so $x = 2.0 \times 10^{-6}$; the assumption is valid.

$[OH^-] = 2.0 \times 10^{-6}$ M $pOH = -\log(2.0 \times 10^{-6}) = 5.70$ $pH = 14.00 - 5.70 = 8.30$

15.69 To determine concentrations of species in a solution, follow the standard procedure:

1. This is a strong acid. Major species are H_2O, H_3O^+, and HSO_4^-.

2. The dominant acid–base equilibrium is $H_2O + HSO_4^- \rightleftharpoons SO_4^{2-} + H_3O^+$.

3. From Table 15-5, $K_{a2} = 1.0 \times 10^{-2}$.

4. In this solution, there are hydronium ions from the strong acid present initially:

Reaction: H_2O +	$HSO_4^- \rightleftharpoons$	SO_4^{2-} +	H_3O^+
Initial concentration (M)	2.00	0	2.00
Change in concentration (M)	$-x$	$+x$	$+x$
Final concentration (M)	$2.00 - x$	x	$2.00 + x$

$$K_{a2} = 1.0 \times 10^{-2} = \frac{[SO_4{}^{2-}]_{eq}[H_3O^+]_{eq}}{[HSO_4{}^-]_{eq}} = \frac{x(2.00 + x)}{2.00 - x}; \text{ assume that } x \ll 2.00:$$

$$1.0 \times 10^{-2} = \frac{(2.00)(x)}{(2.00)}, \text{ so } x = 1.0 \times 10^{-2}; \text{ the assumption is valid.}$$

$[H_3O^+] = 2.00 + 0.010 = 2.01 \text{ M} \qquad [SO_4{}^{2-}] = 1.0 \times 10^{-2} \text{ M}$

$[HSO_4{}^-] = 2.00 - x = 1.99 \text{ M}$

15.71 (a) H_2SO_4 is a strong acid, so the major species in solution are H_2O, $HSO_4{}^-$, and H_3O^+. The hydrogen sulphate ion is a weak acid, so the equilibrium reaction that determines the pH is $H_2O + HSO_4{}^- \rightleftharpoons SO_4{}^{2-} + H_3O^+$.

(b) Na_2SO_4 is a salt, so the major species in solution are H_2O, $SO_4{}^{2-}$, and Na^+. The sulphate ion is the conjugate base of a weak acid, so the equilibrium reaction that determines pH is $SO_4{}^{2-} + H_2O \rightleftharpoons HSO_4{}^- + OH^-$.

(c) $NaHSO_4$ is a salt, so the major species in solution are H_2O, $HSO_4{}^-$, and Na^+. The hydrogen sulphate ion is a weak acid, so the equilibrium reaction that determines the pH is $H_2O + HSO_4{}^- \rightleftharpoons SO_4{}^{2-} + H_3O^+$.

(d) NH_4Cl is a salt, so the major species in solution are H_2O, Cl^-, and $NH_4{}^+$. The ammonium ion is the conjugate acid of a weak base, so the equilibrium reaction that determines pH is $H_2O + NH_4{}^+ \rightleftharpoons NH_3 + H_3O^+$.

15.73 Tabulated equilibrium constants for acid–base reactions always refer to reactions in which H_2O is one of the reactants. The reaction in this problem is the reverse of a base reaction:

$$HPO_4{}^{2-} \ (aq) + OH^- \ (aq) \rightleftharpoons PO_4{}^{3-} \ (aq) + H_2O \ (l)$$

Table 15-5 lists K_a values for phosphoric acid:

$$HPO_4{}^{2-} \ (aq) + H_2O \ (l) \rightleftharpoons PO_4{}^{3-} \ (aq) + H_3O^+ \ (aq) \qquad K_{a3} = 4.8 \times 10^{-13}$$

K_a and K_b for a conjugate acid–base pair are related through $K_a K_b = K_w$:

$$K_b = \frac{1.0 \times 10^{-14}}{4.8 \times 10^{-13}} = 2.1 \times 10^{-2}$$

$$K_{eq} = \frac{1}{K_b} = 48$$

15.75 Molecular pictures must show the correct relative numbers of the various species in the solution. From the starting condition (six molecules of oxalic acid), make appropriate changes and then draw new pictures:

(a) Hydroxide ions react with oxalic acid to form water and hydrogen oxalate ions: $H_2C_2O_4 + OH^- \rightleftharpoons H_2O + HC_2O_4^-$. The picture shows two molecules of oxalic acid and four each of water and hydrogen oxalate:

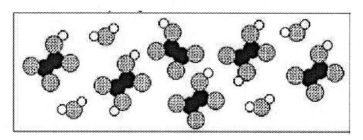

(b) When all oxalic acid has reacted, hydroxide ions react with hydrogen oxalate ions to form water and oxalate ions: $HC_2O_4^- + OH^- \rightleftharpoons H_2O + C_2O_4^{2-}$. The picture shows four hydrogen oxalate ions, eight water molecules, and two oxalate ions:

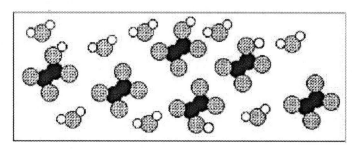

(c) NH$_3$, a weak base, accepts a proton from oxalic acid, a weak acid: $H_2C_2O_4 + NH_3 \rightleftharpoons NH_4^+ + HC_2O_4^-$

The picture shows two oxalic acid molecules and four each of ammonium and hydrogen oxalate ions:

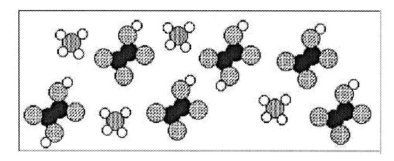

15.77 The chemical reaction that occurs is

$P_4O_{10} + 6\ H_2O \rightarrow 4\ H_3PO_4$

(a) The major species present are H_2O and H_3PO_4.

(b) To determine the ranking of the minor species, consider what reactions generate them. H_3PO_4 undergoes proton transfer with water to form $H_2PO_4^-$ and H_3O^+ in equal concentrations, but $H_2PO_4^-$ undergoes further proton transfer with water to form HPO_4^{2-} and H_3O^+. HPO_4^{2-}, in turn, generates a tiny amount of PO_4^{3-}, and the water equilibrium generates a tiny amount of OH^-. The minor species are (in order of highest concentration to lowest concentration): H_3O^+, $H_2PO_4^-$, HPO_4^{2-}, OH^-, and PO_4^{3-}.

(c) The dominant equilibrium that determines the pH is

$$H_3PO_4 + H_2O \rightleftharpoons H_2PO_4^- + H_3O^+$$

Set up a concentration table, solve for hydronium ion concentration, and then calculate the pH. Determine the initial concentration using standard stoichiometric procedures:

$$n_{H_3PO_4} = 3.5 \text{ g } P_4O_{10} \left(\frac{1 \text{ mol}}{283.88 \text{ g}}\right)\left(\frac{4 \text{ mol } H_3PO_4}{1 \text{ mol } P_4O_{10}}\right) = 0.0493 \text{ mol}$$

$$[H_3PO_4] = \frac{0.0493 \text{ mol}}{1.50 \text{ L}} = 0.033 \text{ M}$$

Reaction: H_2O +	H_3PO_4 $\rightleftharpoons$	$H_2PO_4^-$ +	H_3O^+
Initial concentration (M)	0.033	0	0
Change in concentration (M)	$-x$	$+x$	$+x$
Final concentration (M)	$0.033 - x$	x	x

Now substitute into the equilibrium constant expression and solve for x:

$$K_{a1} = 0.0069 = \frac{[H_2PO_4^-]_{eq}[H_3O^+]_{eq}}{[H_3PO_4]_{eq}} = \frac{x^2}{0.033 - x}; \text{ solve by the quadratic equation.}$$

$$x = [H_3O^+] = 0.012 \text{ M} \qquad pH = -\log(0.012) = 1.92$$

15.79 Proton transfer occurs from the carboxylic acid O–H (shown screened below) and the amino nitrogen atom:

15.81 Net ionic equations show only the reacting species. Remember that strong acids generate H_3O^+ in solution and react to completion with weak bases, and strong bases generate OH⁻ in solution and react to completion with weak acids:

(a) Strong base reacting with weak acid: $OH^- + C_6H_5CO_2H \rightleftharpoons H_2O + C_6H_5CO_2^-$

(b) Strong acid reacting with weak base: $H_3O^+ + (CH_3)_3N \rightleftharpoons H_2O + (CH_3)_3NH^+$

(c) Weak base reacting with weak acid: $SO_4^{2-} + CH_3CO_2H \rightleftharpoons HSO_4^- + CH_3CO_2^-$

HSO_4^-, $pK_a = 1.99$; CH_3CO_2H, $pK_a = 4.75$; HSO_4^- is stronger, so this reaction proceeds to a small extent.

(d) Strong base reacting with weak acid: $OH^- + NH_4^+ \rightleftharpoons H_2O + NH_3$

(e) Weak base reacting with weak acid: $HPO_4^{2-} + NH_3 \rightleftharpoons PO_4^{3-} + NH_4^+$

HPO_4^{2-}, $pK_a = 12.32$; NH_4^+, $pK_a = 9.25$; NH_4^+ is stronger, so this reaction proceeds to a small extent.

15.83 Follow the standard procedure for dealing with equilibrium calculations:
(a) The major species are Na^+, HCO_3^-, and H_2O.
There are two equilibria involving major species:

$$HCO_3^- \ (aq) + H_2O \ (l) \rightleftharpoons CO_3^{2-} \ (aq) + H_3O^+ \ (aq) \quad pK_{eq} = pK_{a2} = 10.33$$

$$HCO_3^- \ (aq) + H_2O \ (l) \rightleftharpoons H_2CO_3 \ (aq) + OH^- \ (aq)$$

$$pK_{b1} = pK_w - pK_{a1} = 14.00 - 6.35 = 7.65$$

The equilibrium with the larger K_{eq} (smaller pK_{eq}) dominates, making this solution basic.
(b) Set up a concentration table, solve for hydroxide ion concentration, and then calculate the pH. Determine the initial concentration using standard stoichiometric procedures:

$$\left[HCO_3^- \right] = \frac{n}{V} = \frac{0.0228 \text{ mol}}{0.150 \text{ L}} = 0.152 \text{ M}$$

Reaction: H_2O +	$HCO_3^- \rightleftharpoons$	H_2CO_3 +	OH^-
Initial concentration (M)	0.152	0	0
Change in concentration (M)	$-x$	$+x$	$+x$
Final concentration (M)	$0.152 - x$	x	x

  Now substitute into the equilibrium constant expression and solve for x:
$K_{eq} = 10^{-7.65} = 2.2 \times 10^{-8}$

$$2.2 \times 10^{-8} = \frac{\left[H_2CO_3 \right]_{eq} \left[OH^- \right]_{eq}}{\left[HCO_3^- \right]_{eq}} = \frac{x^2}{0.152 - x}; \text{ assume that } x \ll 0.152:$$

$x^2 = 3.3 \times 10^{-9}$, from which $x = 5.8 \times 10^{-5}$; the assumption is valid.
$[OH^-] = 5.8 \times 10^{-5}$ M
$pOH = -\log(5.8 \times 10^{-5}) = 4.24$ \qquad $pH = 14.00 - 4.24 = 9.76$

15.85 (a) HBr is a strong acid, so when this gas bubbles through water, it generates hydronium ions. The major species are H_2O, H_3O^+, Br^-, Ca^{2+}, and OH^-, and the reaction that goes to completion is $H_3O^+ + OH^- \rightarrow 2\,H_2O$.
(b) The major species are H_2O, Na^+, HSO_4^-, and OH^-, and the reaction that goes to completion is $HSO_4^- + OH^- \rightarrow H_2O + SO_4^{2-}$.
(c) The major species are H_2O, NH_4^+, I^-, Pb^{2+}, and NO_3^-, and the reaction that goes to completion is formation of PbI_2 precipitate: $Pb^{2+} + 2\,I^- \rightarrow PbI_2\,(s)$.

15.87 (a) $H_2SO_4\,(l) + H_2SO_4\,(l) \ \square \ H_3SO_4^+$ (solvated) $+ HSO_4^-$ (solvated)
(b) Lewis structure:

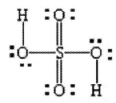

Either doubly-bonded oxygen atom can accept a proton.

(c) $H_2SO_4\,(l) + HClO_4\,(l) \ \square \ H_3SO_4^+$ (solvated) $+ ClO_4^-$ (solvated)

15.89 (a) A weak acid (HA) hydrolyzes in water according to:
$$HA(aq) + H_2O(l) \ ⇌ \ H_3O^+(aq) + A^-(aq)$$
For this reaction,
$$K_a = \frac{[H_3O^+][A^-]}{[HA]}$$
but $pK_a = -\log_{10}K_a$ and $[HA] = c$, thus
$$= -\log_{10}\left(\frac{[H_3O^+][A^-]}{c}\right)$$
$$= -\log_{10}[H_3O^+] - \log_{10}[A^-]) + \log_{10}c$$
But $[H_3O^+] = [A^-]$, and so:
$$pK_a = -2\log_{10}[H_3O^+] + \log_{10}c$$
$$= 2pH + \log_{10}c$$
rearranging, $pH = \tfrac{1}{2}pK_a - \tfrac{1}{2}\log_{10}c$
(b) For formic acid, the hydrolysis reaction is:
$$HCOOH(aq) + H_2O(l) \ ⇌ \ H_3O^+(aq) + HCOO^-(aq)$$
For which the concentration table is:

	HCOOH(*aq*)	H$_3$O$^+$(*aq*)	HCOO$^-$(*aq*)
Initial (M)	0.50	0	0
Change (M)	-x	+x	+x
Equilibrium (M)	0.50-x	x	x

At equilibrium,

$$K_a = \frac{[H_3O^+][HCOO^-]}{[HCOOH]} = 1.8 \times 10^{-4}$$

$$\frac{x(x)}{0.50-x} = 1.8 \times 10^{-4}$$

$$x^2 = 1.8 \times 10^{-4}(0.50-x)$$

$$x^2 + 1.8 \times 10^{-4}x - 0.9 \times 10^{-4} = 0$$

$$x = 0.00940 \; or \; -0.0096$$

Taking the positive root, $x = 0.00940$. Thus pH = -log$_{10}$(0.0094) = 2.03
Using the expression developed in part (a),
pH = ½ pK$_a$ – ½ log c
= ½ (-log$_{10}$(1.8 x 10^{-4})) – ½ log$_{10}$(0.50)
= 1.87 – (-0.15)
= 2.02
Thus the two methods give the same pH.

Chapter 16 Applications of Aqueous Equilibria

Solutions to Problems in Chapter 16

16.1 A buffer solution must contain both a weak acid and its conjugate weak base. Begin each analysis by calculating the initial amounts of acid and base.

(a) The strong base reacts with NH_4^+, a weak acid, to generate the conjugate weak base.

$$n_{NH_4^+} = (0.25 \text{ M})(0.100 \text{ L}) = 0.025 \text{ mol}$$

$$n_{OH^-} = (0.25 \text{ M})(0.150 \text{ L}) = 0.0375 \text{ mol}$$

There is excess strong base (0.0375 mole compared with 0.025 mole), so no weak acid remains after mixing and this solution does not have buffer properties.

(b) Initial amounts:

$$n_{NH_4^+} = (0.25 \text{ M})(0.100 \text{ L}) = 0.025 \text{ mol}$$

$$n_{OH^-} = (0.25 \text{ M})(0.050 \text{ L}) = 0.0125 \text{ mol}$$

The strong base reacts completely with the weak acid to generate the conjugate weak base:

Reaction:	NH_4^+ +	OH^- ⇌	NH_3 +	H_2O
Start amount (mol)	0.025	0.0125	0	---
Change in amount (mol)	− 0.0125	− 0.0125	+ 0.0125	---
Final amount (mol)	0.0125	0.0000	0.0125	Solvent

Since both acid and conjugate base exist, this solution is an NH_4^+/NH_3 buffer. Use the buffer equation, Equation 16-1, with amounts to determine the pH:

$$pH = pK_a + \log\left(\frac{n_{A^-}}{n_{HA}}\right) \qquad \text{For } NH_4^+, pK_a = 9.25.$$

$$pH = 9.25 + \log\left(\frac{0.0125 \text{ mol}}{0.0125 \text{ mol}}\right) = 9.25$$

(c) NH_4^+ is a weak acid and HCl is a strong acid. There is no conjugate base present in high concentration, so this solution does not have buffer properties.

(d) NH_3 is a weak base and HCl is a strong acid. The strong acid reacts completely with the weak base to generate the conjugate weak acid.

Initial amounts:

$$n_{NH_3} = (0.25 \text{ M})(0.100 \text{ L}) = 0.025 \text{ mol}$$

$$n_{H_3O^+} = (0.25 \text{ M})(0.050 \text{ L}) = 0.0125 \text{ mol}$$

Reaction:	$NH_3 +$	H_3O^+ ⮂	$NH_4^+ +$	H_2O
Start amount (mol)	0.025	0.0125	0	---
Change in amount (mol)	-0.0125	-0.0125	$+0.0125$	---
Final amount (mol)	0.0125	0.0000	0.0125	Solvent

This is an NH_4^+/NH_3 buffer. Use the buffer equation, Equation 16-1:

$$pH = pK_a + \log\left(\frac{n_{A^-}}{n_{HA}}\right) \qquad pK_a = 9.25$$

$$pH = 9.25 + \log\left(\frac{0.0125 \text{ mol}}{0.0125 \text{ mol}}\right) = 9.25$$

16.3 The two solutions that are buffered are (b) and (d). The concentration tables for these solutions show that they are identical in acid–base composition, despite being prepared differently. Thus only one calculation is required. The added strong acid reacts completely with the weak base of the buffer to generate the conjugate weak acid.

Reaction:	NH_3	H_3O^+ ⮂	$NH_4^+ +$	H_2O
Start amount (mol)	0.0125	0.0050	0.0125	---
Change in amount (mol)	-0.0050	-0.0050	$+0.0050$	---
Final amount (mol)	0.0075	0.0000	0.0175	Solvent

Use the buffer equation, Equation 16-1, with the new amounts:

$$pH = pK_a + \log\left(\frac{n_{A^-}}{n_{HA}}\right) = 9.25 + \log\left(\frac{0.0075 \text{ mol}}{0.0175 \text{ mol}}\right) = 9.25 - 0.37 = 8.88$$

16.5 The two solutions that are buffered are (b) and (d). The concentration tables for these solutions show that they are identical in acid–base composition, despite being prepared differently. Thus, only one calculation is required. Addition of a strong base consumes the weak acid, produces a weak base, and increases the pH of the buffer.

First use the buffer equation to calculate the acid:base ratio in a solution whose pH is greater by 0.10 unit than the initial buffer solution. Then set up a concentration ratio, using x = the amount of added base, and solve for x: new pH = 9.25 + 0.10 = 9.35.

$$[NH_3] = 0.0125 + x \qquad\qquad [NH_4^+] = 0.0125 - x$$

$$9.35 = 9.25 + \log\left(\frac{0.0125 + x}{0.0125 - x}\right) \qquad \log\left(\frac{0.0125 + x}{0.0125 - x}\right) = 0.10$$

$$\frac{0.0125 + x}{0.0125 - x} = 1.26$$

$$(0.0125 + x) = 1.26(0.0125 - x) = 0.0157 - 1.26x$$

$$0.0032 = 2.26x, \text{ and so } x = 1.4 \times 10^{-3} \text{ mol = amount of base that must be added}$$

16.7 The acid is HCO_3^-, and its conjugate base is CO_3^{2-}.

Consumption of hydronium ions: $CO_3^{2-} (aq) + H_3O^+ (aq) \rightarrow HCO_3^- (aq) + H_2O$ (l)

Consumption of hydroxide ions: $HCO_3^- (aq) + OH^- (aq) \rightarrow CO_3^{2-} (aq) + H_2O$ (l)

16.9 Buffer solutions must contain a weak acid and its conjugate base. A weak acid and its salt, a strong acid and a weak base, or a strong base and a weak acid can meet these requirements. To determine if a pair of reactants can generate a buffer solution, identify the acid–base properties of the pair:

(a) $NaHCO_3$ is both a weak base and a weak acid, and NaOH is a strong base. This pair generates a buffer solution containing HCO_3^- and CO_3^{2-}.

(b) NaOH is a strong base, and NH_3 is a weak base; no buffer possibilities.

(c) H_3PO_4 is a weak acid, and HCl is a strong acid; no buffer possibilities.

(d) HCl is a strong acid, and Na_2CO_3 is a weak base. This pair generates a buffer solution containing HCO_3^- and CO_3^{2-}.

16.11 A buffer solution requires a conjugate acid–base pair whose pK_a is within 1 pH unit of the desired pH. For pH = 3.50, the HCO_2H/HCO_2^- system (pK_a = 3.75) would be most suitable. Sodium formate, $NaHCO_2$, could be used along with HCl solution, which would protonate some formate anions to form the conjugate weak acid. For pH = 12.60, the HPO_4^{2-}/PO_4^{3-} system (pK_a = 12.32) would be most suitable. Potassium phosphate, K_3PO_4, could be used along with HCl solution, which would protonate some phosphate anions to form the conjugate weak acid.

16.13 This is a NH_3/NH_4^+ buffer. Use the buffer equation, Equation 16-1, to determine the amount of ammonium chloride needed:

$$M_{NH_4Cl} = 14.01 \text{ g/mol} + 4(1.01 \text{ g/mol}) + 35.45 \text{ g/mol} = 53.50 \text{ g/mol}$$

$$\log\left(\frac{n_{NH_3}}{n_{NH_4^+}}\right) = pH - pK_a = 8.90 - 9.25 = -0.35$$

$$\frac{n_{NH_3}}{n_{NH_4^+}} = 10^{-0.35} = 0.447$$

$$n_{NH_4^+} = \frac{n_{NH_3}}{0.447} = \frac{(1.25\ L)(0.25\ M)}{0.447} = 0.699\ mol = moles\ of\ NH_4Cl$$

$$m_{NH_4Cl} = (0.699\ mol)(53.50\ g/mol) = 37\ g$$

16.15 The question asks how much strong acid is required to lower the pH from 8.90 to 8.65. Use the buffer equation to determine the number of moles of acid required to cause this change, with x = moles of acid added:

$$\log\left(\frac{0.3125 - x}{0.699 + x}\right) = pH - pK_a = 8.65 - 9.25 = -0.60$$

$$\frac{0.3125 - x}{0.699 + x} = 10^{-0.60} = 0.251 \qquad x = 0.110\ moles\ HCl$$

$$V = \frac{0.110\ mol}{2.0\ mol/L} = 0.055\ L\ or\ 55\ mL$$

16.17 To prepare the buffer solution, dissolve the appropriate amount of Na_2CO_3 in water, add sufficient 0.500 M HCl solution, and make up to 1.5 L with additional water. Calculate the carbonate:hydrogen carbonate ratio using the buffer equation:

$$\log\left(\frac{[CO_3^{2-}]}{[HCO_3^-]}\right) = pH - pK_a = 9.85 - 10.33 = -0.48$$

$$\frac{[CO_3^{2-}]}{[HCO_3^-]} = 10^{-0.48} = 0.33$$

Thus, $[CO_3^{2-}] = 0.33\ [HCO_3^-]$ $[CO_3^{2-}] + [HCO_3^-] = 0.35\ M$

$0.33\ [HCO_3^-] + [HCO_3^-] = 0.35\ M$

$1.33\ [HCO_3^-] = 0.35\ M$ and $[HCO_3^-] = 0.263\ M$

$[CO_3^{2-}] = 0.33[HCO_3^-] = 0.087\ M$

$$n_{HCO_3^-} = (0.263\ M)(1.5\ L) = 0.3945\ mol$$

$$n_{CO_3^{2-}} = (0.087\ M)(1.5\ L) = 0.1305\ mol$$

All the carbonate and hydrogen carbonate must come from sodium carbonate:

$$n_{Na_2CO_3} = 0.3945 + 0.1305 = 0.525\ mol$$

$$m_{Na_2CO_3} = nM = (0.525 \text{ mol})(106 \text{ g/mol}) = 55.7 \text{ g}$$

The 0.3945 mole of hydrogen carbonate in the buffer solution is generated by adding strong acid:

$$V_{HCl} = \frac{n}{c} = \frac{0.3945 \text{ mol}}{0.500 \text{ mol/L}} = 0.789 \text{ L}$$

Dissolve 55.7 g of sodium carbonate in about 0.5 L of water, add 0.789 L of 0.500 M HCl and enough water to make 1.5 L (two significant figures accuracy would be sufficient for the final volume, but the reagents should be measured to three significant figures).

16.19 The question asks how much strong base is required to raise the pH by 0.15 units, from 9.85 to 10.00. Let x be the moles of added base, and use the buffer equation with a ratio in moles rather than concentration to determine the number of moles required to cause this change.

From the solution to Problem 16.17, the original buffer solution contains

$$n_{HCO_3^-} = (0.263 \text{ M})(1.5 \text{ L}) = 0.3945 \text{ mol}$$

$$n_{CO_3^{2-}} = (0.087 \text{ M})(1.5 \text{ L}) = 0.1305 \text{ mol}$$

$$\log\left(\frac{0.1305 + x}{0.3945 - x}\right) = pH - pK_a = 10.00 - 10.33 = -0.33$$

$$\frac{0.1305 + x}{0.3945 - x} = 10^{-0.33} = 0.468$$

$$(0.468)(0.3945 - x) = 0.1305 + x$$

$$0.184 - 0.468\,x = 0.1305 + x$$

$$0.0535 = 1.468\,x$$

$$x = 0.0364 \text{ moles of NaOH}$$

$$m_{NaOH} = nM = (0.0364 \text{ moles})(40.\ \text{g/mol}) = 1.5 \text{ g}$$

We round to two significant figures because that is the typical accuracy of the buffer equation. The resulting solution is still buffered, because the ratio of acid to conjugate base remains in the buffer region, between 0.1 and 10.

16.21 To determine the approximate pH at the stoichiometric point of an acid–base titration, examine the equilibrium that determines the pH in the vicinity of the stoichiometric point.

(a) This is a weak base–strong acid titration. At the stoichiometric point, the major acid–base species are NH_4^+ and H_2O, and the dominant equilibrium is

$$NH_4^+ \ (aq) + H_2O \ (l) \ \square \quad NH_3 \ (aq) + H_3O^+ \ (aq)$$

Hydronium ions are produced in this reaction, so pH < 7 at the stoichiometric point.

(b) This is a strong acid–strong base titration. The only acid–base species present at the stoichiometric point is H_2O, and the dominant equilibrium is

$$H_2O \ (l) + H_2O \ (l) \ \square \quad OH^- \ (aq) + H_3O^+ \ (aq)$$

Thus, pH = 7 at the stoichiometric point.

(c) This is a weak base–strong acid titration. At the stoichiometric point, the major acid–base species are CH_3CO_2H and H_2O, and the dominant equilibrium is

$$CH_3CO_2H \ (aq) + H_2O \ (l) \ \square \quad CH_3CO_2^- \ (aq) + H_3O^+ \ (aq)$$

Hydronium ions are produced in this reaction, so pH < 7 at the stoichiometric point.

16.23 This problem describes an acid–base titration. The major species differ at various points during a titration, so there are different dominant equilibria that must be identified before doing an equilibrium calculation to determine pH.

(a) Before titration begins, the major species are a weak acid, aspirin (HA) and H_2O, and the dominant equilibrium is $HA + H_2O \ \square \quad A^- + H_3O^+$, for which $K_a = 3.0 \times 10^{-4}$.

Reaction: H_2O +	$HA \ \square$	A^- +	H_3O^+
Initial concentration (M)	10^{-2}	0	0
Change in concentration (M)	$-x$	$+x$	$+x$
Final concentration (M)	$10^{-2} - x$	x	x

Now substitute into the equilibrium constant expression and solve for x:

$$K_a = 3.0 \times 10^{-4} = \frac{[A^-]_{eq}[H_3O^+]_{eq}}{[HA]_{eq}} = \frac{x^2}{10^{-2} - x}$$

$$x^2 = (3.0 \times 10^{-4})(10^{-2} - x) = (3.0 \times 10^{-6}) - (3.0 \times 10^{-4})x$$

$$x^2 + (3.0 \times 10^{-4})x - (3.0 \times 10^{-6}) = 0$$

$$x = \frac{-b \pm \sqrt{b^2 - 4ac}}{2a} = \frac{-(3.0 \times 10^{-4}) \pm \sqrt{(3.0 \times 10^{-4})^2 - 4(-3.0 \times 10^{-6})}}{2} = 1.6 \times 10^{-3}$$

$$[H_3O^+] = 1.6 \times 10^{-3} \ M \qquad pH = -\log(1.6 \times 10^{-3}) = 2.8$$

(b) At the stoichiometric point, the major acid–base species present are A^- and H_2O and the dominant equilibrium is

$$A^- + H_2O \rightleftharpoons HA + OH^- \qquad K_b = \frac{K_w}{K_a} = \frac{1.00 \times 10^{-14}}{3.0 \times 10^{-4}} = 3.3 \times 10^{-11}$$

Reaction: H_2O +	A^- ⇌	HA +	OH^-
Initial concentration (M)	10^{-2}	0	0
Change in concentration (M)	$-x$	$+x$	$+x$
Final concentration (M)	$10^{-2} - x$	x	x

Now substitute into the equilibrium constant expression and solve for x:

$$K_b = 3.3 \times 10^{-11} = \frac{[HA]_{eq}[OH^-]_{eq}}{[A^-]_{eq}} = \frac{x^2}{10^{-2} - x}; \text{ assume } x \ll 10^{-2}:$$

$x^2 = (10^{-2})(3.3 \times 10^{-11}) = 3.3 \times 10^{-13}$, so $x = 5.7 \times 10^{-7}$; the assumption is valid.

$$[OH^-]_{eq} = 5.7 \times 10^{-7} \text{ M}$$

$$pOH = -\log(5.7 \times 10^{-7}) = 6.24 \qquad pH = 14.00 - pOH = 14.00 - 6.24 = 7.8$$

(c) At the midpoint of the titration, half of the weak acid has been converted into its conjugate base, and $pH = pK_a = -\log(3.0 \times 10^{-4}) = 3.5$.

16.25 The best indicator for a titration is one for which $pK_{In} = pH_{stoichiometric point}$; the pH at the stoichiometric point was calculated in Problem 16.23(b) to be 7.8. The best indicator is phenol red, $pK_{In} = 7.9$.

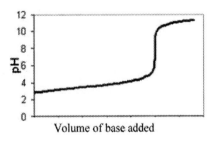

16.27 This is a titration type problem for a diprotic acid asking for information at the second stoichiometric point. At the second stoichiometric point, all the weak acid (H_2A) has been converted into A^{2-}, so the major acid–base species present are A^{2-} and H_2O and the dominant equilibrium is

$$A^{2-} (aq) + H_2O (l) \rightleftharpoons HA^- (aq) + OH^- (aq)$$

$$K_{b1} = \frac{K_w}{K_{a2}} = \frac{1.00 \times 10^{-14}}{3.9 \times 10^{-6}} = 2.6 \times 10^{-9}$$

Begin by determining the initial amount of A^{2-} (we can assume that dilution effects are negligible). Then construct an equilibrium table, write the equilibrium expression, and solve for the hydroxide concentration to determine the pH:

At the second stoichiometric point, $[A^{2-}]_{initial} = [H_2A]_{before\ titration} = 1.50 \times 10^{-2}$ M.

Here is the completed equilibrium table:

Reaction: H_2O +	A^{2-} ☐	HA^- +	OH^-
Initial concentration (M)	0.0150	0	0
Change in concentration (M)	$-x$	$+x$	$+x$
Final concentration (M)	$0.0150 - x$	x	x

Now substitute into the equilibrium constant expression and solve for x:

$$K_{b1} = 2.6 \times 10^{-9} = \frac{[HA^-]_{eq}[OH^-]_{eq}}{[A^{2-}]_{eq}} = \frac{x^2}{0.0150 - x}; \text{ assume that } x \ll 0.0150:$$

$x^2 = (2.6 \times 10^{-9})(0.0150) = 3.90 \times 10^{-11}$, so $x = 6.24 \times 10^{-6}$; the assumption is valid.

$[OH^-] = 6.24 \times 10^{-6}$ M $pOH = -\log(6.24 \times 10^{-6}) = 5.20$

pH = 14.00 − 5.20 = 8.80; a suitable indicator for this titration is thymol blue, $pK_{in} = 8.9$. (If dilution is taken into account, $[OH^-] = 6.14 \times 10^{-6}$ M, pOH = 5.21.)

16.29 The major species present are different at various points during a titration, so there are different dominant equilibria that must be identified before doing a calculation to determine pH. We also need to know the initial concentration of the solution:

$$n = \frac{m}{M} = \frac{1.51\ g}{120.\ g/mol} = 0.0126\ mol \qquad c = \frac{n}{V} = \frac{0.0126\ mol}{0.100\ L} = 0.126\ M$$

(a) Before titration begins, the major species are a weak acid, dihydrogen phosphate (HA) and H_2O, and the dominant equilibrium is $HA + H_2O \,\square\ A^- + H_3O^+$, for which $K_a = 6.2 \times 10^{-8}$ and $pK_a = 7.21$:

Reaction: H_2O +	HA ☐	A^- +	H_3O^+
Initial concentration (M)	0.126	0	0
Change in concentration (M)	$-x$	$+x$	$+x$
Final concentration (M)	$0.126 - x$	x	x

Now substitute into the equilibrium constant expression and solve for x:

$$K_a = 6.2 \times 10^{-8} = \frac{[A^-]_{eq}[H_3O^+]_{eq}}{[HA]_{eq}} = \frac{x^2}{0.126 - x}; \text{ assume that } x \ll 0.126:$$

$x^2 = (6.2 \times 10^{-8})(0.126) = 7.8 \times 10^{-9}$, so $x = 8.8 \times 10^{-5}$; the assumption is valid.

$[H_3O^+] = 8.8 \times 10^{-5}$ M   pH $= -\log(8.8 \times 10^{-5}) = 4.05$

(b) At the first midpoint, both the acid and conjugate base of dihydrogen phosphate are present in equal concentrations, the solution is buffered, and the pH can be calculated using the buffer equation:

$$pH = pK_{a1} + \log 1 = 7.21$$

(c) At the first stoichiometric point, Equation 16-2 from your textbook applies:

$[H_3O^+]_{eq,\ 1st\ stoichiometric\ point} = \sqrt{K_{a1}K_{a2}}$

$[H_3O^+] = \sqrt{(6.2 \times 10^{-8})(4.8 \times 10^{-13})} = \sqrt{2.976 \times 10^{-20}} = 1.7 \times 10^{-10}$ M

pH $= -\log(1.7 \times 10^{-10}) = 9.76$

(d) At the second midpoint, both the acid and conjugate base of hydrogen phosphate are present in equal concentrations, the solution is buffered, and the pH can be calculated using the buffer equation: pH $= pK_{a2} + \log 1 = 12.32$.

(e) At the second stoichiometric point, all the hydrogen phosphate has been converted into its conjugate base, so the major acid–base species present are phosphate (A^-) and H_2O and the dominant equilibrium is

$A^- + H_2O \rightleftharpoons HA + OH^-$

$K_b = \dfrac{K_w}{K_{a2}} = \dfrac{1.0 \times 10^{-14}}{4.8 \times 10^{-13}} = 2.1 \times 10^{-2}$

Reaction: H_2O +	$A^- \rightleftharpoons$	HA +	OH^-
Initial concentration (M)	0.126	0	0
Change in concentration (M)	$-x$	$+x$	$+x$
Final concentration (M)	$0.126 - x$	x	x

Now substitute into the equilibrium constant expression and solve for x:

$K_a = 2.1 \times 10^{-2} = \dfrac{[HA]_{eq}[OH^-]_{eq}}{[A^-]_{eq}} = \dfrac{x^2}{0.126 - x}$

$x^2 = (2.1 \times 10^{-2})(0.126 - x) = (2.6 \times 10^{-3}) - (2.1 \times 10^{-2})x$

$x^2 + (2.1 \times 10^{-2})x - (2.6 \times 10^{-3}) = 0$

$x = \dfrac{-b \pm \sqrt{b^2 - 4ac}}{2a} = \dfrac{-(2.1 \times 10^{-2}) \pm \sqrt{(2.1 \times 10^{-2})^2 - 4(-2.6 \times 10^{-3})}}{2} = 4.2 \times 10^{-2}$

$[OH^-]_{eq} = 4.2 \times 10^{-2}$ M

$$pOH = -\log(4.2 \times 10^{-2}) = 1.38 \qquad pH = 14.00 - pOH = 14.00 - 1.38 = 12.62$$

16.31 To write a solubility product equilibrium expression, first determine the chemical formula and stoichiometry of the salt:

(a) $AgCl\ (s) \rightleftharpoons Ag^+\ (aq) + Cl^-\ (aq)$ $\qquad K_{sp} = [Ag^+]_{eq}[Cl^-]_{eq}$

(b) $BaSO_4\ (s) \rightleftharpoons Ba^{2+}\ (aq) + SO_4^{2-}\ (aq)$ $\qquad K_{sp} = [Ba^{2+}]_{eq}[SO_4^{2-}]_{eq}$

(c) $Fe(OH)_2\ (s) \rightleftharpoons Fe^{2+}\ (aq) + 2\ OH^-\ (aq)$ $\qquad K_{sp} = [Fe^{2+}]_{eq}[OH^-]^2_{eq}$

(d) $Ca_3(PO_4)_2\ (s) \rightleftharpoons 3\ Ca^{2+}\ (aq) + 2\ PO_4^{3-}\ (aq$ $\qquad K_{sp} = [Ca^{2+}]^3_{eq}[PO_4^{3-}]^2_{eq}$

16.33 "Determine the mass that dissolves" means "calculate the amount present in solution at equilibrium." Initial amounts are zero, so it is easy to complete a concentration table. Calculate molarity from the solubility product expression, and then convert to mass using standard stoichiometric methods. Let x be the number of moles of salt dissolving in 1 L of solution.

(a)

Reaction:	AgCl (s) ⇌	Ag⁺ (aq) +	Cl⁻ (aq)
Initial concentration (M)	Solid	0	0
Change in concentration (M)	Solid	$+x$	$+x$
Final concentration (M)	Solid	x	x

$$K_{sp} = 1.8 \times 10^{-10} = [Ag^+]_{eq}[Cl^-]_{eq} = x^2 \qquad x = 1.34 \times 10^{-5}M$$

$$m_{AgCl} = 0.475\ L\left(\frac{1.34 \times 10^{-5}\ mol}{1\ L}\right)\left(\frac{143.32\ g}{1\ mol}\right) = 9.1 \times 10^{-4}\ g$$

(b)

Reaction:	BaSO₄ (s) ⇌	Ba²⁺ (aq) +	SO₄²⁻ (aq)
Initial concentration (M)	Solid	0	0
Change in concentration (M)	Solid	$+x$	$+x$
Final concentration (M)	Solid	x	x

$$K_{sp} = 1.1 \times 10^{-10} = [Ba^{2+}]_{eq}[SO_4^{2-}]_{eq} = x^2 \qquad x = 1.05 \times 10^{-5}M$$

$$m_{BaSO_4} = 0.475\ L\left(\frac{1.05 \times 10^{-5}\ mol}{1\ L}\right)\left(\frac{233.39\ g}{1\ mol}\right) = 1.2 \times 10^{-3}\ g$$

(c)

Reaction:	Fe(OH)₂ (s) ⇌	Fe²⁺ (aq) +	2 OH⁻ (aq)
Initial concentration (M)	Solid	0	0

Change in concentration (M)	Solid	$+x$	$+2x$
Final concentration (M)	Solid	x	$2x$

$$K_{sp} = 4.8 \times 10^{-17} = [Fe^{2+}]_{eq}[OH^-]^2_{eq} = 4x^3$$

$$x = 2.3 \times 10^{-6} \text{ M}$$

$$m_{Fe(OH)_2} = 0.475 \text{ L} \left(\frac{2.3 \times 10^{-6} \text{ mol}}{1 \text{ L}} \right) \left(\frac{89.87 \text{ g}}{1 \text{ mol}} \right) = 9.8 \times 10^{-5} \text{ g}$$

(d)

Reaction:	$Ca_3(PO_4)_2$ (s) ⇌	$3 Ca^{2+}$ (aq) +	$2 PO_4^{3-}$ (aq)
Initial concentration (M)	Solid	0	0
Change (M)	Solid	$+3x$	$+2x$
Final concentration (M)	Solid	$3x$	$2x$

$$K_{sp} = 2.0 \times 10^{-33} = [Ca^{2+}]^3_{eq}[PO_4^{3-}]^2_{eq} = 108x^5$$

$$x = 1.1 \times 10^{-7} \text{ M}$$

$$m_{Ca_3(PO_4)_2} = 0.475 \text{ L} \left(\frac{1.1 \times 10^{-7} \text{ mol}}{1 \text{ L}} \right) \left(\frac{310.18 \text{ g}}{1 \text{ mol}} \right) = 1.6 \times 10^{-5} \text{ g}$$

16.35 To calculate a solubility product from the mass that dissolves in solution, convert to molarity of ions and then evaluate the equilibrium constant expression. Begin by writing the chemical reaction and the equilibrium expression:

$$CaC_2O_4 \text{ (s)} ⇌ Ca^{2+}(aq) + C_2O_4^{2-} \text{ (aq)} \qquad K_{sp} = [Ca^{2+}]_{eq}[C_2O_4^{2-}]_{eq}$$

$$n = \frac{m}{M} = 6.1 \text{ mg} \left(\frac{10^{-3} \text{ g}}{1 \text{ mg}} \right) \left(\frac{1 \text{ mol}}{128.1 \text{ g}} \right) = 4.76 \times 10^{-5} \text{ mol}$$

Since the stoichiometry is 1:1, the moles of calcium ions and oxalate ions are the same as the moles of calcium oxalate:

$$[Ca^{2+}]_{eq} = [C_2O_4^{2-}]_{eq} = \frac{n}{V} = \frac{4.76 \times 10^{-5} \text{ mol}}{1.0 \text{ L}} = 4.76 \times 10^{-5} \text{ mol/L}$$

$$K_{sp} = [Ca^{2+}]_{eq}[C_2O_4^{2-}]_{eq} = (4.76 \times 10^{-5})^2 = 2.3 \times 10^{-9}$$

16.37 This problem describes a solution containing a common ion. To determine concentrations of ions at equilibrium, follow the five-step procedure for working equilibrium problems:

1. Species present initially are Pb^{2+}, NO_3^-, and $PbCl_2$

2. Reaction is a solubility process: $PbCl_2$ (s) ⇌ Pb^{2+} $(aq) + 2\, Cl^-$ (aq)

3. $K_{sp} = [Pb^{2+}]_{eq}[Cl^-]^2_{eq} = 1.7 \times 10^{-5}$

4. Set up and complete a concentration table, letting x be the increase in $[Pb^{2+}]$:

Reaction:	$PbCl_2$ (s) ⇌	Pb^{2+} $(aq) +$	$2\, Cl^-$ (aq)
Initial concentration (M)	Solid	0.650	0
Change in concentration (M)	Solid	$+x$	$+2x$
Final concentration (M)	Solid	$0.650 + x$	$2x$

5. Substitute into the equilibrium constant expression and solve for x:

$1.7 \times 10^{-5} = (0.650 + x)(2x)^2$; assume $x \ll 0.650$:

$$4x^2 = \frac{1.7 \times 10^{-5}}{0.650}$$

$x^2 = 6.54 \times 10^{-6}$, so $x = 2.56 \times 10^{-3}$

$2.56 \times 10^{-3} < 5\%$ of 0.650, so the assumption is valid.

The increase in lead ions is $\Delta[Pb^{2+}] = 2.56 \times 10^{-3}$ M.

To convert to moles Pb^{2+} that dissolve, multiply by the solution volume:

$n = cV = (2.56 \times 10^{-3}\ M)(0.750\ L) = 1.92 \times 10^{-3}$ mol

The amount of $PbCl_2$ that dissolves is determined by the amount of increase in Pb^{2+}; convert to mass by multiplying by the molar mass of $PbCl_2$:

$m = nM = (1.92 \times 10^{-3}\ mol)(278.1\ g/mol) = 0.53$ g

16.39 The chemical reactions are the following:

$$\begin{array}{ll}
CaSO_3(s) \rightleftharpoons Ca^{2+}\ (aq) + SO_3^{2-}\ (aq) & K_{sp} \\
SO_3^{2-}\ (aq) + H_3O^+\ (l) \rightleftharpoons HSO_3^-\ (aq) + H_2O\ (l) & 1/K_{a2} \\
\hline
CaSO_3\ (s) + H_3O^+\ (aq) \rightleftharpoons Ca^{2+}\ (aq) + HSO_3^-\ (aq) + H_2O\ (l) &
\end{array}$$

$$K_{eq} = \frac{[Ca^{2+}][HSO_3^-]}{[H_3O^+]} = [Ca^{2+}][SO_3^{2-}] \times \frac{[HSO_3^-]}{[H_3O^+][SO_3^{2-}]} = \frac{K_{sp}}{K_{a2}} = \frac{1.0 \times 10^{-4}}{6.3 \times 10^{-8}} = 1.6 \times 10^3$$

16.41 This problem describes a common-ion effect on solubility equilibrium. We are asked to determine the final concentration of calcium ions in solution. Begin by analyzing the chemistry. The starting materials are HCl (aq), a strong acid, and $CaSO_3$ (s), an insoluble salt. In addition to water, the major species in solution are H_3O^+, Cl^-, and

H_2O.

Hydronium ions will react with calcium sulphite (see Problem 16.39):

$$CaSO_3\ (s) + H_3O^+\ (aq) \rightleftharpoons Ca^{2+}\ (aq) + HSO_3^-\ (aq) + H_2O\ (l)$$

$$K_{eq} = \frac{[Ca^{2+}][HSO_3^-]}{[H_3O^+]} = \frac{K_{sp}}{K_{a2}} = 1.6 \times 10^3$$

Because the K_{eq} for the reaction is large, assume that the reaction goes to completion and then do the equilibrium calculations.

At completion, all of the hydronium has reacted to form Ca^{2+} and HSO_3^-:

$[Ca^{2+}] = [HSO_3^-] = 0.125$ M

Here is the concentration table for the return from completion to equilibrium:

Reaction	$CaSO_3\ (s)\ +$	$H_3O^+\ (aq)\ \rightleftharpoons$	$Ca^{2+}\ (aq)\ +$	$HSO_3^-\ (aq)\ +$	H_2O
Start concentration (M)	----	0	0.125	0.125	----
Change in concentration (M)	----	$+x$	$-x$	$-x$	----
Final concentration (M)	----	x	$0.125 - x$	$0.125 - x$	----

Now write the equilibrium expression and solve for the concentration of Ca^{2+} ions:

$$K_{eq} = 1.6 \times 10^3 = \frac{(0.125-x)^2}{x};\ \text{assume that } x << 0.125:$$

$1.6 \times 10^3 = \dfrac{0.125^2}{x}$, so $x = 9.8 \times 10^{-6}$; the assumption is valid.

$[Ca^{2+}] = (0.125) - (9.8 \times 10^{-6}) = 0.125$ M

16.43 Write the chemical formulas according to the procedures in your textbook. The coordination number is the number of bonds between ligands and the metal species. Both CN^- and NH_3 are monodentate (form one bond with the metal centre), and ethylenediamine is a bidentate ligand (forms two bonds to the metal centre).

(a) $[Fe(CN)_6]^{3-}$; (b) $[Zn(NH_3)_4]^{2+}$; (c) $[V(en)_3]^{3+}$

16.45 This is a complexation equilibrium problem. The problem gives information about the amounts of both starting materials, so this is a limiting reactant situation. Because K for complexation generally is large, we do a concentration table that takes the reaction to completion and then brings it back to equilibrium. We must calculate the initial concentration of each species, construct a table of amounts, and use the results to determine the final solution concentrations.

The reaction is Zn^{2+} $(aq) + 4\,NH_3$ $(aq) \rightleftharpoons$ $[Zn(NH_3)_4]^{2+}$ (aq) $K_f = 4.1 \times 10^8$

Calculate the initial concentration of Zn^{2+}:

$$n_{Zn^{2+}} = \frac{m}{M} = \frac{0.275\ g}{136.29\ g/mol} = 2.02 \times 10^{-3}\ mol$$

$$c = \frac{n}{V} = \frac{2.02 \times 10^{-3}\ mol}{(375\ mL)(10^{-3}\ L/mL)} = 5.4 \times 10^{-3}\ mol/L$$

Complete a concentration table after taking the reaction to completion:

Reaction:	Zn^{2+} $(aq) +$	$4\,NH_3$ $(aq) \rightleftharpoons$	$[Zn(NH_3)_4]^{2+}$ (aq)
Start concentration (M)	5.4×10^{-3}	0.250	0
Change in concentration (M)	-5.4×10^{-3}	$-4(5.4 \times 10^{-3})$	$+5.4 \times 10^{-3}$
Completion concentration (M)	0	0.228	5.4×10^{-3}
Change in concentration (M)	$+x$	$+4x$	$-x$
Equilibrium concentration (M)	x	$0.228 + 4x$	$5.4 \times 10^{-3} - x$

$$K_f = 4.1 \times 10^8 = \frac{0.0054 - x}{x(0.228 + 4x)^4}; \text{ assume that } x << 0.0054:$$

$$x(0.228)^4 (4.1 \times 10^8) = 0.0054$$

$$x = \frac{0.0054}{(0.228)^4 (4.1 \times 10^8)} = 4.9 \times 10^{-9}; \text{ the assumption is valid.}$$

Here are the concentrations at equilibrium:

$[NH_3]_{eq} = 0.228$ M; $[Zn^{2+}]_{eq} = 4.9 \times 10^{-9}$ M; $[[Zn(NH_3)_4]^{2+}]_{eq} = 5.4 \times 10^{-3}$ M.

16.47 This problem describes a common-ion effect solubility reaction. We are asked to determine the final ion concentrations in solution. Begin by analyzing the chemistry. The starting materials are NaCl, a salt, and Pb^{2+}. In addition to water, the major species in solution are Na^+, Cl^-, and Pb^{2+}.

The presence of lead(II) and chloride ions together as major species in solution will result in the following reaction:

$$Pb^{2+}\ (aq) + 4\,Cl^-\ (aq) \rightleftharpoons PbCl_4^{2-}\ (aq) \qquad K_f = \frac{[PbCl_4^{2-}]}{[Pb^{2+}][Cl^-]^4} = 2.5 \times 10^{15}$$

Because K_f for the reaction is so large, assume that the reaction goes to completion and then do the equilibrium calculations. The problem gives information about the amounts of both starting materials, so this is a limiting reactant situation. We must calculate the number of moles of each species, construct a table of amounts, and use the results to determine the final solution concentrations.

Calculations of initial amounts:

$$n_{Na^+} = n_{Cl^-} = 0.25 \text{ mol}$$

$$n_{Pb^{2+}} = (7.5 \times 10^{-3} \text{ M})(1.50 \text{ L}) = 0.0113 \text{ mol}$$

Na^+ is a spectator ion:

$$[Na^+] = \frac{0.25 \text{ mol}}{1.50 \text{ L}} = 0.167 \text{ M}$$

Complete a concentration table that takes the reaction to completion:

Reaction:	Pb^{2+} +	$4\ Cl^-$ ⇌	$PbCl_4^{2-}$
Start amount (mol)	0.0113	0.25	0
Change in amount (mol)	-0.0113	$-4(0.0113)$	$+0.0113$
Final amount (mol)	0	0.205	0.0113

Compute the concentrations by dividing the amount of moles by the volume, 1.50 L:

$$[Cl^-] = \frac{0.205 \text{ mol}}{1.50 \text{ L}} = 0.137 \text{ M}$$

$$[PbCl_4^{2-}] = \frac{0.0113 \text{ mol}}{1.50 \text{ L}} = 0.0075 \text{ M}$$

Use these amounts to complete the concentration table at equilibrium:

Reaction:	Pb^{2+} +	$4\ Cl^-$ ⇌	$PbCl_4^{2-}$
Start concentration (M)	0	0.137	0.0075
Change in concentration (M)	$+x$	$+4x$	$-x$
Equilibrium concentration (M)	x	$0.137 + 4x$	$0.0075 - x$

$$K_f = 2.5 \times 10^{15} = \frac{0.0075 - x}{x(0.137 + 4x)^4}; \text{ assume that } x << 0.0075:$$

$$x = \frac{0.0075}{(0.137)^4(2.5 \times 10^{15})} = 8.52 \times 10^{-15}; \text{ the assumption is valid.}$$

Here are the final concentrations:

$[Cl^-] = 0.14 \text{ M}$ $[Pb^{2+}] = 8.5 \times 10^{-15} \text{ M}$

$[Na^+] = 0.17 \text{ M}$ $[PbCl_4^{2-}] = 7.5 \times 10^{-3} \text{ M}$

16.49

$$Mn^{2+} + en \ \square \ [Mn(en)]^{2+}$$

$$[Mn(en)]^{2+} + en \ \square \ [Mn(en)_2]^{2+}$$

$$[Mn(en)_2]^{2+} + en \ \square \ [Mn(en)_3]^{2+}$$

16.51 This problem describes a multiple equilibrium situation. We are asked to determine the amount of $CaSO_3$ that will dissolve in the solution. Begin by analyzing the chemistry. The starting materials are $CaSO_3$ (*s*), an insoluble salt, and nitrilotriacetate. In addition to water, the major species in solution is NTA^{3-}.

The solid salt has a solubility equilibrium:

$$CaSO_3 \ (s) \ \square \ Ca^{2+} \ (aq) + SO_3^{2-} \ (aq) \qquad K_{sp} = 1.0 \times 10^{-4} \text{ (see Problem 15.39)}$$

The resulting calcium ions can coordinate with the NTA^{3-} to form a complex:

$$Ca^{2+} \ (aq) + 2 \, NTA^{3-} \ (aq) \ \square \ [Ca(NTA)_2]^{4-} \ (aq) \qquad K_f = 3.2 \times 10^{11}$$

Determine the overall chemical reaction and the equilibrium expression:

$$CaSO_3 \ (s) \ \square \ Ca^{2+} \ (aq) + SO_3^{2-} \ (aq)$$

$$\dfrac{Ca^{2+} \ (aq) + 2 \, NTA^{3-} \ (aq) \ \square \ [Ca(NTA)_2]^{4-} \ (aq)}{CaSO_3 \ (s) + 2 \, NTA^{3-} \ (aq) \ \square \ [Ca(NTA)_2]^{4-} \ (aq) + SO_3^{2-} \ (aq)}$$

$$K_{eq} = \dfrac{[[Ca(NTA)_2]^{4-}][SO_3^{2-}]}{[NTA^{3-}]^2} = [Ca^{2+}][SO_3^{2-}]\dfrac{[[Ca(NTA)_2]^{4-}]}{[NTA^{3-}]^2[Ca^{2+}]} = K_{sp}K_f$$

$$K_{eq} = (1.0 \times 10^{-4})(3.2 \times 10^{11}) = 3.2 \times 10^7$$

Because the equilibrium constant is large, take the reaction to completion and then return to equilibrium. At completion, all the ligands have reacted to form the two products in 1:2 stoichiometry.

Here is the concentration table for the return from completion to equilibrium:

Reaction:	$CaSO_3$ (*s*) +	$2NTA^{3-}$ $\square$	$[Ca(NTA)_2]^{4-}$ +	SO_3^{2-} (*aq*)
Start concentration (M)	----	0	0.125	0.125
Change in concentration (M)	----	$+ 2x$	$- x$	$- x$
Final concentration (M)	----	$2x$	$0.125 - x$	$0.125 - x$

$$K_{eq} = 3.2 \times 10^7 = \dfrac{(0.125 - x)^2}{(2x)^2}$$

Take the square root of each side: $5.65 \times 10^3 = \dfrac{0.125 - x}{2x}$

$2x(5.65 \times 10^3) = 0.125 - x$, so $x = 1.1 \times 10^{-5}$

$[SO_3^{2-}] = (0.125) - (1.1 \times 10^{-5}) = 0.125$ M

The amount of sulphite ions formed will equal the amount of solid dissolved:

$$n_{CaSO_3} = n_{SO_3^{2-}} = (0.125 \text{ M})(0.50 \text{ L}) = 6.25 \times 10^{-2} \text{ mol}$$

$$M_{CaSO_3} = 40.08 \text{ g/mol} + 32.07 \text{ g/mol} + 3(16.00 \text{ g/mol}) = 120.1 \text{ g/mol}$$

$$m_{CaSO_3} = (6.25 \times 10^{-2} \text{ mol})(120.1 \text{ g/mol}) = 7.5 \text{ g}$$

16.53 A molecular picture of a buffer solution should show molecules/ions of the conjugate acid–base pair in the correct proportions, as determined by the pH of the buffer solution. For formic acid, $pK_a = 3.75$. Use the buffer equation to calculate the base:acid ratio of a formic acid buffer with pH = 4.04:

$$\log\left(\frac{n_{A^-}}{n_{HA}}\right) = pH - pK_a = 4.04 - 3.75 = 0.29$$

$$\frac{n_{A^-}}{n_{HA}} = 2$$

The picture should show twice as many formate ions as formic acid molecules. The solution is acidic, so there should also be some hydronium ions, but the exact proportion depends on the total concentration of the buffer species.

Here is a view showing six formate ions, three formic acid molecules, and one hydronium ion:

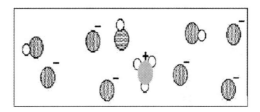

16.55 To determine if all the solid dissolves, calculate the value that Q would have if all the solid dissolves, and compare its value with the K_{sp} value. If $Q < K_{sp}$, the spontaneous direction of reaction is to the right even after all of the solid dissolves, so all of the solid dissolves, whereas if $Q > K_{sp}$, not all of the solid dissolves.

The reaction is $PbCl_2(s) \;\rightleftharpoons\; Pb^{2+}(aq) + 2\,Cl^-(aq)$

$K_{sp} = 2 \times 10^{-5}$ and $Q = [Pb^{2+}][Cl^-]^2$

Use mole–mass conversions to determine concentrations:

$$n_{Pb^{2+}} = \frac{m}{M} = \frac{0.50 \text{ g}}{278.1 \text{ g/mol}} = 1.80 \times 10^{-3} \text{ mol}$$

$$[Pb^{2+}] = \frac{1.80 \times 10^{-3} \text{ mol}}{0.300 \text{ L}} = 6.0 \times 10^{-3} \text{ M}$$

$$[Cl^-] = 2[Pb^{2+}] = 1.2 \times 10^{-2} \text{ M}$$

$$Q = (6.0 \times 10^{-3})(1.2 \times 10^{-2})^2 = 8.6 \times 10^{-7}; \; Q < K_{sp}, \text{ so all of the solid dissolves.}$$

16.57 This problem describes the $H_2PO_4^-/HPO_4^{2-}$ buffer solution.

(a) Use the buffer equation to calculate the pH of a buffer solution:

The conjugate acid–base pair is $H_2PO_4^-/HPO_4^{2-}$, $pK_a = pK_{a2} = 7.21$.

$$pH = pKa + \log\left(\frac{[A^-]}{[HA]}\right) = 7.21 + \log\left(\frac{0.20 \text{ M}}{0.50 \text{ M}}\right) = 7.21 - 0.40 = 6.81$$

(b) Calculate the change in molarity due to the added base, and then use the buffer equation to calculate the new pH. The added hydroxide reacts completely with the weak acid and generates the conjugate weak base:

$$n_{NaOH} = \frac{m}{M} = \frac{0.120 \text{ g}}{40.0 \text{ g/mol}} = 3.00 \times 10^{-3} \text{ mol}$$

$$\Delta c = \frac{n_{NaOH}}{V} = \frac{3.00 \times 10^{-3} \text{ mol}}{0.15 \text{ L}} = 0.020 \text{ M}$$

$$[HPO_4^{2-}] = 0.20 + 0.020 = 0.22 \text{ M}$$

$$[H_2PO_4^-] = 0.50 - 0.020 = 0.48 \text{ M}$$

$$pH = pK_a + \log\left(\frac{[A^-]}{[HA]}\right) = 7.21 + \log\left(\frac{0.22 \text{ M}}{0.48 \text{ M}}\right) = 7.21 - 0.34 = 6.87$$

$$\Delta pH = 6.87 - 6.81 = 0.06$$

(c) Adding acid reduces the concentration of weak base, increases the concentration of weak acid, and reduces the pH. Use the buffer equation to calculate the conjugate base:acid ratio at the limit of the pH range. Then calculate the change in molarity that this ratio represents, and convert to moles:

$$pH_{min} = 6.81 - 0.10 = 6.71 = 7.21 + \log\left(\frac{[A^-]}{[HA]}\right)$$

$$\log\left(\frac{[A^-]}{[HA]}\right) = 6.71 - 7.21 = -0.50$$

$$\frac{[A^-]}{[HA]} = 0.32 = \frac{0.20 - \Delta c}{0.50 + \Delta c}$$

$$(0.20 - \Delta c) = 0.32(0.50 + \Delta c) = 0.16 + 0.32\Delta c, \text{ so } 0.04 = 1.32\Delta c$$

$$\Delta c = 3.0 \times 10^{-2} \text{ M}$$

$$n_{\text{acid neutralized}} = \Delta cV = (3.0 \times 10^{-2} \text{ M})(0.25 \text{ L}) = 7.5 \times 10^{-3} \text{ mol}$$

16.59 The bromocresol purple indicator is purple when pH < 8.5 and yellow when pH > 8.5:

(a) HCl is a strong acid (pH << 5); therefore, the indicator colour is yellow.

(b) NaOH is a strong base (pH >> 7); therefore. the indicator colour is purple.

(c) KCl is a neutral ionic solution (pH =7); therefore, the indicator colour is purple.

(d) NH_3 is a weak base (pH >7); therefore; the indicator colour is purple.

16.61 In a titration of a weak acid by a strong base, pH = pK_a at the midpoint of the titration. Thus, pK_a = 2.36 for leucine and $K_a = 4.4 \times 10^{-3}$.

16.63 To determine if a precipitate forms, evaluate Q for the solution after mixing and compare its value for the value for K_{eq}. If $Q < K_{eq}$, the spontaneous direction of reaction is to the right after the solutions are mixed and a precipitate forms, whereas if $Q > K_{eq}$, no precipitate forms.

The reaction is Ca^{2+} (aq) + SO_4^{2-} (aq) ⇌ $CaSO_4$ (s)

$$K_{eq} = \frac{1}{K_{sp}} = \frac{1}{4.9 \times 10^{-5}} = 2.0 \times 10^4 \text{ and } Q = \frac{1}{[Ca^{2+}][SO_4^{2-}]}$$

The total volume of the solution is 350 mL + 150 mL = 500 mL

$$[Ca^{2+}] = 2.00 \times 10^{-2} \text{ M}\left(\frac{350 \text{ mL}}{500 \text{ mL}}\right) = 0.0140 \text{ M}$$

$$[SO_4^{2-}] = 1.50 \times 10^{-2} \text{ M}\left(\frac{150 \text{ mL}}{500 \text{ mL}}\right) = 0.00450 \text{ M}$$

$$Q = \frac{1}{(0.0140)(0.00450)} = 1.59 \times 10^4; Q < K_{eq}, \text{ so the precipitate forms.}$$

16.65 To calculate an equilibrium constant when experimental data concerning concentrations are available, identify the reaction, and then complete an amounts table and substitute into the equilibrium constant expression.

To complete the amounts table, use stoichiometric reasoning and the fact that the change is 2.0×10^{-3} M for Pb^{2+}:

Reaction:	PbF_2 (s) $\rightleftharpoons$	Pb^{2+} (aq) +	2 F⁻ (aq)
Start concentration (M)	Solid	0	0
Change in concentration (M)	Solid	2.0×10^{-3}	4.0×10^{-3}
Final concentration (M)	Solid	2.0×10^{-3}	4.0×10^{-3}

Now substitute into the equilibrium constant expression and evaluate K_{sp}:

$$K_{sp} = [Pb^{2+}]_{eq}[F^-]^2_{eq} = (2.0 \times 10^{-3})(4.0 \times 10^{-3})^2 = 3.2 \times 10^{-8}$$

16.67 The buffer equation is used for calculations involving buffer solutions. Concentrations of acid and conjugate base are needed. The weak acid in the ammonia buffer system is the ammonium ion, $pK_a = 9.25$;

(a) $[NH_3] = 1.00$ M

$$n_{NH_4^+} = \frac{35.0 \text{ g}}{53.49 \text{ g/mol}} = 0.654 \text{ mol} \qquad [NH_4^+] = \frac{n}{V} = \frac{0.654 \text{ mol}}{1.00 \text{ L}} = 0.654 \text{ M}$$

$$pH = pKa + \log\left(\frac{[A^-]}{[HA]}\right) = 9.25 + \log\left(\frac{1.00 \text{ M}}{0.654 \text{ M}}\right) = 9.25 + 0.18 = 9.43$$

(b) First calculate the ratio of base to acid in the new solution, and then use the new ratio to determine the amount of change from the original ratio.

The new pH is $9.43 - 0.05 = 9.38$.

$$\log\left(\frac{[A^-]}{[HA]}\right) = pH - pK_a = 9.38 - 9.25 = 0.13 \qquad \frac{[A^-]}{[HA]} = 10^{0.13} = 1.35$$

Thus, $[NH_3] = 1.35[NH_4^+]$; from before, the total concentration is

$[NH_4^+] + [NH_3] = 1.654$ M

$[NH_4^+] + 1.35[NH_4^+] = 1.654$ M, so $2.35[NH_4^+] = 1.654$ M and $[NH_4^+] = 0.704$ M

The moles of acid required for this change are

$\Delta n = \Delta cV = (0.704 \text{ M} - 0.654 \text{ M})(1.00 \text{ L}) = 0.050$ mol

(c) It is easiest to work with moles in this calculation. The 250 mL of buffer solution contains

$$250 \text{ mL}\left(\frac{10^{-3} \text{ L}}{1 \text{ mL}}\right)\left(\frac{1.00 \text{ mol}}{1 \text{ L}}\right) = 0.250 \text{ mol NH}_3$$

$$250 \text{ mL}\left(\frac{10^{-3} \text{ L}}{1 \text{ mL}}\right)\left(\frac{0.654 \text{ mol}}{1 \text{ L}}\right) = 0.163 \text{ mol NH}_4^+$$

The amount of added acid is $5.0 \text{ mL}\left(\frac{10^{-3} \text{ L}}{1 \text{ mL}}\right)\left(\frac{12.0 \text{ mol}}{1 \text{ L}}\right) = 0.060 \text{ mol}$

The new amounts are

$(0.163 + 0.060) = 0.223 \text{ mol NH}_4^+$

$(0.250 - 0.060) = 0.190 \text{ mol NH}_3$

$$\text{pH} = \text{p}K_a + \log\left(\frac{n_{\text{NH}_3}}{n_{\text{NH}_4^+}}\right) = 9.25 + \log\left(\frac{0.190 \text{ mol}}{0.223 \text{ mol}}\right) = 9.25 - 0.07 = 9.18$$

16.69 To calculate an equilibrium constant when experimental data concerning concentrations are available, identify the reaction, and then complete a concentration table and substitute into the equilibrium constant expression.

The reaction is $\text{Ca}_3(\text{PO}_4)_2 \ (s) \ \rightleftharpoons \ 3 \text{ Ca}^{2+} \ (aq) + 2 \text{ PO}_4^{3-} \ (aq)$

The solubility in mol/L is $\left(\frac{3.5 \times 10^{-5} \text{ g}}{1 \text{ L}}\right)\left(\frac{1 \text{ mol}}{310.174 \text{ g}}\right) = 1.13 \times 10^{-7} \text{ M}$

Reaction:	$\text{Ca}_3(\text{PO}_4)_2 \ \rightleftharpoons$	$3 \text{ Ca}^{2+} \ +$	2 PO_4^{3-}
Initial concentration (M)	Solid	0	0
Change in concentration (M)	Solid	$3(1.13 \times 10^{-7})$	$2(1.13 \times 10^{-7})$
Equilibrium concentration (M)	Solid	3.39×10^{-7}	2.26×10^{-7}

Now substitute into the equilibrium constant expression and evaluate K_{sp}:

$$K_{\text{sp}} = [\text{Ca}^{2+}]_{\text{eq}}^3 [\text{PO}_4^{3-}]_{\text{eq}}^2 = (3.39 \times 10^{-7})^3 (2.26 \times 10^{-7})^2 = 2.0 \times 10^{-33}$$

16.71 The major species present are different at various points during a titration, so there are different dominant equilibria that must be identified before doing an equilibrium calculation to determine pH.

(a) Before titration begins, the major species are a weak base, formate (HCO_2^-) and H_2O, and the dominant equilibrium is

$$HCO_2^- \ (aq) + H_2O\ (l) \rightleftharpoons HCO_2H\ (aq) + OH^-\ (aq)$$

$$K_b = \frac{K_w}{K_a} = \frac{1.0 \times 10^{-14}}{1.8 \times 10^{-4}} = 5.6 \times 10^{-11}$$

Reaction: H_2O +	HCO_2^- $\rightleftharpoons$	HCO_2H +	OH^-
Initial concentration (M)	0.200	0	0
Change in concentration (M)	$-x$	$+x$	$+x$
Final concentration (M)	$0.200 - x$	x	x

Now substitute into the equilibrium constant expression and solve for x:

$$K_b = 5.6 \times 10^{-11} = \frac{[HCO_2H]_{eq}[OH^-]_{eq}}{[HCO_2^-]_{eq}} = \frac{x^2}{(0.200 - x)}\text{; assume that } x << 0.200\text{:}$$

$x^2 = 1.1 \times 10^{-11}$, so $x = 3.3 \times 10^{-6}$; the assumption is valid.

$[OH^-] = 3.3 \times 10^{-6}$ M

$pOH = -\log(3.3 \times 10^{-6}) = 5.48$ and $pH = 14.00 - 5.48 = 8.52$

(b) At the midpoint of the titration, the buffer equation applies and the concentrations of acid and conjugate base are equal, so pH = pK_a = 3.74.

(c) At the stoichiometric point, the major acid–base species present are HCO_2H and H_2O and the dominant equilibrium is HCO_2H (*aq*) + H_2O (*l*).

$$HCO_2H\ (aq) + H_2O\ (l) \rightleftharpoons HCO_2^-\ (aq) + H_3O+\ (aq) \quad K_a = 1.8 \times 10^{-4}$$

Correct the initial concentration of HCO_2H for the volume added during titration:

$$[HCO_2H] = \frac{0.0600\text{ mol}}{0.300\text{ L} + 0.010\text{ L}} = 0.193\text{ M}$$

Here is the concentration table:

Reaction: H_2O +	HCO_2H $\rightleftharpoons$	HCO_2^- +	H_3O^+
Initial concentration (M)	0.193	0	0
Change in concentration (M)	$-x$	$+x$	$+x$
Final concentration (M)	$0.193 - x$	x	x

Now substitute into the equilibrium constant expression and solve for x:

$$K_a = 1.8 \times 10^{-4} = \frac{[HCO_2^-]_{eq}[H_3O^+]_{eq}}{[HCO_2H]_{eq}} = \frac{x^2}{0.193 - x}$$

$$x^2 = (0.193 - x)(1.8 \times 10^{-4}) = (3.5 \times 10^{-5}) - (1.8 \times 10^{-4})x$$

$$x^2 + (1.8 \times 10^{-4})x - (3.5 \times 10^{-5}) = 0$$

$$x = \frac{-b \pm \sqrt{b^2 - 4ac}}{2a} = \frac{-(1.8 \times 10^{-4}) \pm \sqrt{(1.8 \times 10^{-4})^2 + 4(3.5 \times 10^{-5})}}{2} = 5.8 \times 10^{-3}$$

$[H_3O^+]_{eq} = 5.8 \times 10^{-3}$ M, and pH = 2.23

(d) A suitable indicator for this titration must change colour at around pH = 2. Thymol blue, $pK_{In} = 1.75$, would be the best choice.

16.73 When a precipitate forms upon mixing aqueous solutions, the equilibrium reaction is the reverse of the solubility reaction. Follow the standard procedure for a reaction with a large equilibrium constant.

Reaction is Mn^{2+} $(aq) + CO_3^{2-}$ (aq) ▢ $MnCO_3$ (s)

"Initial" concentrations are after mixing but before reaction:

$$[Mn^{2+}] = \left(\frac{0.750 \text{ L}}{0.750 + 0.150 \text{ L}}\right)(0.0250 \text{ M}) = 0.0208 \text{ M}$$

$$[CO_3^{2-}] = \left(\frac{0.150 \text{ L}}{0.750 + 0.150 \text{ L}}\right)(0.500 \text{ M}) = 0.0833 \text{ M}$$

Allow the reaction to go to completion:

$[Mn^{2+}] = 0.000$ M $[CO_3^{2-}] = 0.0625$ M

These are the initial concentrations to enter in the concentration table:

Reaction:	$MnCO_3$ (s) ▢	CO_3^{2-} +	Mn^{2+}
Initial concentration (M)	Solid	0.0625	0
Change in concentration (M)	Solid	$+x$	$+x$
Equilibrium concentration (M)	Solid	$0.0625 + x$	x

Assume that $x << 0.0625$:

$$K_{sp} = 2.2 \times 10^{-11} = (x)(0.0625) \qquad x = \frac{2.2 \times 10^{-11}}{0.0625} = 3.5 \times 10^{-10}$$

Thus, $[Mn^{2+}] = 3.5 \times 10^{-10}$ M.

To calculate the mass of precipitate, use standard stoichiometric methods. The reaction goes virtually to completion, so

$$n_{MnCO_3} = n_{Mn^{2+}} = cV = (0.0250 \text{ mol/L})(0.750 \text{ L}) = 0.01875 \text{ mol}$$

$$m = nM = (0.01875 \text{ mol})(114.95 \text{ g/mol}) = 2.16 \text{ g}$$

16.75 "What mass will dissolve" means "calculate the amount present in solution at equilibrium." Initial amounts are zero, so it is easy to complete a concentration table. Calculate molarity from the solubility product expression, and then convert to mass using standard stoichiometric methods. Let x be the number of moles of salt dissolving in 1 L of solution.

Reaction:	$Zn(OH)_2$ (s) ⇌	Zn^{2+} +	2 OH$^-$
Initial concentration (M)	Solid	0	0
Change in concentration (M)	Solid	+ x	+ $2x$
Equilibrium concentration (M)	Solid	x	$2x$

$$K_{sp} = 3.0 \times 10^{-17} = [Zn^{2+}]_{eq}[OH^-]^2_{eq} = 4x^3$$

$$x = 2.0 \times 10^{-6}$$

$$m = cVM = (2.0 \times 10^{-6} \text{ M})(1.00 \text{ L})(99.40 \text{ g/mol}) = 2.0 \times 10^{-4} \text{ g dissolves}$$

16.77 The best choice for a buffer solution generally is the weak acid whose pK_a is closest to the desired pH of the buffer solution. For a pH = 4.80 buffer solution, acetic acid–acetate (pK_a = 4.75) would be the best choice. To prepare the buffer solution, add the appropriate amounts of sodium acetate (NaAc) and acetic acid (HAc) solution to water and make up to 1.0 L with additional water. Calculate the acetate:acetic acid ratio using the buffer equation:

$$\log\left(\frac{[Ac^-]}{[HAc]}\right) = pH - pK_a = 4.80 - 4.75 = 0.05 \qquad \frac{[Ac^-]}{[HAc]} = 10^{0.05} = 1.12$$

Thus, $[Ac^-] = 1.12 [HAc]$

$[Ac^-] + [HAc] = 0.35 \text{ M}$ $[HAc] = 0.35 \text{ M} - [Ac^-]$

$[Ac^-] = 1.12(0.35 \text{ M} - [Ac^-]) = 0.39 - 1.12 [Ac^-]$

$2.12 [Ac^-] = 0.39$ and $[Ac^-] = 0.18 \text{ M}$ $n_{Ac^-} = (0.18 \text{ M})(1.0 \text{ L}) = 0.18 \text{ mol}$

$m_{NaAc} = nM = (0.18 \text{ mol})(82.0 \text{ g/mol}) = 15 \text{ g}$

$[HAc] = 0.35 - 0.18 = 0.17 \text{ M}$

$n_{HAc} = (0.17 \text{ M})(1.0 \text{ L}) = 0.17 \text{ mol}$

$$V_{HAc} = \frac{n}{c} = 0.17 \text{ mol} \left(\frac{1 \text{ L}}{1.00 \text{ mol}} \right) = 0.17 \text{ L}$$

Mix together 15 g of sodium acetate, 0.17 L of 1.0 M acetic acid, and enough water to make 1.0 L.

16.79 This is a complexation equilibrium problem. Because K_{eq} for complexation generally is large, we create a concentration table that takes the reaction to completion and then brings it back to equilibrium. The problem gives information about the amounts of both starting materials, so this is a limiting reactant situation. We must calculate the initial concentration of each species, construct a table of amounts, and use the results to determine the final solution concentrations.

$$Zn^{2+} (aq) + 3 \text{ } C_2O_4^{2-} (aq) \rightleftharpoons [Zn(C_2O_4)_3]^{4-} (aq) \qquad K_f = 1.4 \times 10^8$$

Calculate the initial concentration of metal cations:

$$n_{Zn^{2+}} = \frac{m}{M} = \frac{0.275 \text{ g}}{136.29 \text{ g/mol}} = 2.02 \times 10^{-3} \text{ mol}$$

$$c = \frac{n}{V} = \frac{2.02 \times 10^{-3} \text{ mol}}{0.450 \text{ L}} = 4.49 \times 10^{-3} \text{ mol/L}$$

Complete a concentration table after taking the reaction to completion:

Reaction:	$Zn^{2+} (aq) +$	$3 \text{ } C_2O_4^{2-} (aq) \rightleftharpoons$	$[Zn(C_2O_4)_3]^{4-} (aq)$
Start concentration (M)	4.49×10^{-3}	0.250	0
Change in concentration (M)	-4.49×10^{-3}	$-3(4.49 \times 10^{-3})$	$+4.49 \times 10^{-3}$
Completion concentration (M)	0	0.237	4.49×10^{-3}
Change in concentration (M)	$+x$	$+3x$	$-x$
Equilibrium concentration (M)	x	$0.237 + 3x$	$4.49 \times 10^{-3} - x$

$$K_f = 1.4 \times 10^8 = \frac{0.00449 - x}{x(0.237 + 3x)^3}; \text{ assume that } x \ll 0.00449:$$

$$x(0.237)^3 (1.4 \times 10^8) = 0.00449$$

$$x = \frac{0.00449}{(0.237)^3 (1.4 \times 10^8)} = 2.4 \times 10^{-9}; \text{ the assumption is valid.}$$

Here are the final concentrations:

$$[C_2O_4^{2-}] = 0.237 \text{ M} \qquad [Zn^{2+}] = 2.4 \times 10^{-9} \text{ M} \qquad [[Zn(C_2O_4)_3]^{4-}] = 4.5 \times 10^{-3} \text{ M}$$

16.81 "Saturated aqueous solution" identifies this as a solubility equilibrium.

(a) The reaction is $MgF_2\ (s) \rightleftharpoons Mg^{2+}\ (aq) + 2\ F^-\ (aq)$ $K_{sp} = [Mg^{2+}]_{eq}[F^-]_{eq}^2$

(b) If $[Mg^{2+}]_{eq} = 1.14 \times 10^{-3}$ M, by stoichiometry $[F^-]_{eq} = 2(1.14 \times 10^{-3}$ M$)$

$K_{sp} = (1.14 \times 10^{-3})[2(1.14 \times 10^{-3})]^2 = 5.93 \times 10^{-9}$

(c) To estimate the equilibrium constant at a temperature different from 298 K, calculate

$\Delta H^\circ_{reaction}$ and $\Delta S^\circ_{reaction}$ at 298 K and then use Equations 12-11 and 14-3:

$$\Delta G^\circ_{reaction} = \Delta H^\circ_{reaction} - T\Delta S^\circ_{reaction} \qquad \Delta G^\circ = -RT \ln K_{eq}$$

$\Delta H^\circ_{reaction} = 1\ mol(-467.0\ kJ/mol) + 2\ mol(-335.4\ kJ/mol) - 1\ mol(-1124.2$ kJ/mol$) = -13.6$ kJ

$\Delta S^\circ_{reaction} = 1\ mol(-137\ J\ mol^{-1}\ K^{-1}) + 2\ mol(-13.8\ J\ mol^{-1}\ K^{-1}) - 1\ mol(57.2\ J$ $mol^{-1}\ K^{-1}) = -222\ J\ mol^{-1}\ K^{-1}$

$$\Delta G^\circ_{reaction,\ 373\ K} = (-13.6\ kJ/mol) - (373\ K)(-222\ J\ mol^{-1}\ K^{-1})\left(\frac{10^{-3}\ kJ}{1\ J}\right)$$

$$= 69.2\ kJ/mol$$

$\ln K_{sp} = -\dfrac{6.92 \times 10^4\ J/mol}{(8.314\ J\ mol^{-1}\ K^{-1})(373\ K)} = -22.3$ $\qquad K_{sp} = e^{-22.3} = 2.1 \times 10^{-10}$

16.83 To calculate equilibrium concentrations, set up a concentration table:

Reaction:	$HgS\ (s) \rightleftharpoons$	$Hg^{2+}\ (aq) +$	$S^{2-}\ (aq)$
Initial concentration (M)	Solid	0	0
Change in concentration (M)	Solid	$+x$	$+x$
Equilibrium concentration (M)	Solid	x	x

$4.0 \times 10^{-53} = [Hg^{2+}]_{eq}[S^{2-}]_{eq} = x^2$, so $x = 6.3 \times 10^{-27}$

$[Hg^{2+}]_{eq} = 6.3 \times 10^{-27}$ M (a very small concentration)

To calculate the volume that would be expected to contain a single Hg^{2+} cation:

$$V = \frac{n}{c}, \text{ with } n = \frac{1 \text{ cation}}{N_A}$$

$$V/\text{cation} = \frac{1}{N_A c} = \left(\frac{1 \text{ mol}}{6.022 \times 10^{23} \text{ cations}}\right)\left(\frac{1 \text{ L}}{6.3 \times 10^{-27} \text{ mol}}\right) = 2.6 \times 10^2 \text{ L/cation}$$

16.85 (a) To determine the ion concentrations in a solution, follow the five-step procedure for solving an equilibrium problem:

1. The species in solution are Ca^{2+}, Cl^-, PO_4^{3-}, and H_2O.

2, 3. The reaction is formation of $Ca_3(PO_4)_2$ precipitate:

$$3 Ca^{2+} (aq) + 2 PO_4^{3-} (aq) \rightleftharpoons Ca_3(PO_4)_2 (s) \qquad K_{eq} = \frac{1}{K_{sp}} = \frac{1}{[Ca^{2+}]_{eq}^3 [PO_4^{3-}]_{eq}^2}$$

4. Initial concentrations are those present after adding solid but before reaction occurs:

$$[Ca^{2+}] = \frac{120 \text{ mol}}{3.00 \times 10^3 \text{ L}} = 4.0 \times 10^{-2} \text{ M} \qquad [PO_4^{3-}] = 2.2 \times 10^{-3} \text{ M}$$

Take the reaction to completion and then return to equilibrium. At completion

$$[Ca^{2+}] = 0.037 \text{ M} \qquad [PO_4^{3-}] = 0.000 \text{ M}$$

Reaction:	$Ca_3(PO_4)_2 (s) \rightleftharpoons$	$3 Ca^{2+} (aq) +$	$2 PO_4^{3-} (aq)$
Initial concentration (M)	Solid	0.037	0
Change in concentration (M)	Solid	$+ 3x$	$+ 2x$
Equilibrium concentration (M)	Solid	$0.037 + 3x$	$2x$

5. Assume that $3x \ll 3.7 \times 10^{-2}$, substitute into the solubility product expression, and solve:

$$K_{sp} = 2.0 \times 10^{-33} = [Ca^{2+}]_{eq}^3 [PO_4^{3-}]_{eq}^2 = (3.7 \times 10^{-2})^3 (2x)^2$$

$$x^2 = \frac{(2.0 \times 10^{-33})}{(3.7 \times 10^{-2})^3 (4)} = 9.9 \times 10^{-30}, \text{ so } x = 3.1 \times 10^{-15}$$

$$[PO_4^{3-}]_{eq} = 2x = 6.2 \times 10^{-15} \text{ M}$$

(b) Because this reaction goes essentially to completion, the calculation of the mass of precipitate can be done using standard stoichiometric methods:

For phosphate ions, $n = 3.00 \times 10^3 \text{ L} \left(\dfrac{2.2 \times 10^{-3} \text{ mol}}{1 \text{ L}} \right) = 6.6 \text{ mol}$

$n_{Ca_3(PO_4)_2} = \dfrac{1}{2} n_{PO_4^{3-}} = 3.3 .\text{mol}$

$m = nM = (3.3 \text{ mol})(310.18 \text{ g/mol}) = 1.0 \times 10^3 \text{ g}$

16.87 To calculate an equilibrium constant when experimental data concerning concentrations are available, identify the reaction, and then complete an amounts table and substitute into the equilibrium constant expression.

The reaction is $Sr(IO_3)_2$ (s) $\rightleftharpoons$ Sr^{2+} (aq) $+ 2 IO_3^-$ (aq)

At 25 °C, the solubility in mol/L is $\dfrac{0.030 \text{ g}}{(0.100 \text{ L})(437.42 \text{ g/mol})} = 6.86 \times 10^{-4} \text{ M}$

Use this value to complete a concentration table at 25 °C:

Reaction:	$Sr(IO_3)_2$ (s) $\rightleftharpoons$	Sr^{2+} (aq)	$2 IO_3^-$ (aq)
Initial concentration (M)	Solid	0	0
Change in concentration (M)	Solid	$+ 6.86 \times 10^{-4}$	$+ 2(6.86 \times 10^{-4})$
Equilibrium concentration (M)	Solid	6.86×10^{-4}	1.37×10^{-3}

Now substitute into the equilibrium constant expression and evaluate K_{sp}:

$K_{sp} = [Sr^{2+}]_{eq}[IO_3^-]_{eq}^2 = (6.86 \times 10^{-4})(1.37 \times 10^{-3})^2 = 1.29 \times 10^{-9}$

$\Delta G° = -RT \ln K_{eq} = -(8.314 \times 10^{-3} \text{ kJ mol}^{-1} \text{ K}^{-1})(298 \text{ K}) \ln(1.29 \times 10^{-9}) = 50.7 \text{ kJ/mol}$

At 100.0 °C, the solubility in mol/L is $\dfrac{0.80 \text{ g}}{(0.100 \text{ L})(437.42 \text{ g/mol})} = 1.83 \times 10^{-2} \text{ M}$.

Use this value to complete a concentration table at 100 °C:

Reaction:	$Sr(IO_3)_2$ (s) $\rightleftharpoons$	Sr^{2+} (aq)	$2 IO_3^-$ (aq)
Initial concentration (M)	Solid	0	0
Change in concentration (M)	Solid	$+ 1.83 \times 10^{-2}$	$+ 2(1.83 \times 10^{-2})$
Equilibrium concentration (M)	Solid	1.83×10^{-2}	3.66×10^{-2}

Now substitute into the equilibrium constant expression and evaluate K_{sp}:

$$K_{sp} = [Sr^{2+}]_{eq} [IO_3^-]_{eq}^2 = (1.83 \times 10^{-2})(3.66 \times 10^{-2})^2 = 2.5 \times 10^{-5}$$

$$\Delta G^\circ = -RT \ln K_{eq} = -(8.314 \times 10^{-3} \text{ J mol}^{-1} \text{ K}^{-1})(373 \text{ K}) \ln(2.45 \times 10^{-5}) = 33 \text{ kJ/mol}$$

16.89 To determine equilibrium concentrations, set up the appropriate concentration table. In this problem, one equilibrium concentration is given and the other two are stoichiometrically related, so the table can be relatively simple (note that the volume of the atmosphere is so large relative to the volume of ground water that the change in CO_2 pressure is negligible):

Reaction: $CaCO_3$ (s) + H_2O (l)	+ CO_2 (g) ⇌	Ca^{2+} (aq) +	2 HCO_3^- (aq)
Initial values	3.2×10^{-4} bar	0 M	0 M
Change in values	~0	+ x M	+ 2x M
Equilibrium values	3.2×10^{-4} bar	x M	2x M

Substitute equilibrium values into the equilibrium constant expression and solve for x:

$$K_{eq} = 1.56 \times 10^{-8} = \frac{[Ca^{2+}]_{eq}[HCO_3^-]_{eq}^2}{(pCO_2)_{eq}} = \frac{(x)(2x)^2}{3.2 \times 10^{-4}}$$

$(1.56 \times 10^{-8})(3.2 \times 10^{-4}) = 4\,x^3$, so $x^3 = 1.25 \times 10^{-12}$ and $x = [Ca^{2+}] = 1.1 \times 10^{-4}$ M.

16.91 Salts that are more soluble in acidic solution are those whose anions have weak conjugate acids. These are Ag_2CO_3, Ag_2SO_4, and Ag_2S. Those that are independent of pH have anions that are conjugate bases of strong acids, AgBr and AgCl.

16.93 Use the buffer equation to carry out calculations on buffer solutions:

(a) $pH = pK_a + \log\left(\dfrac{[TRIS]}{[TRISH^+]}\right) = (14.00 - 5.91) + \log\left(\dfrac{0.30 \text{ M}}{0.60 \text{ M}}\right) = 7.79$

(b) Adding HCl provides H_3O^+ ions, which react quantitatively with TRIS to generate $TRISH^+$.

The amount of added acid is $n = 5.0 \text{ mL}\left(\dfrac{10^{-3} \text{ L}}{1 \text{ mL}}\right)\left(\dfrac{12 \text{ mol}}{1 \text{ L}}\right) = 0.060 \text{ mol}$.

Using moles in the buffer equation saves us from calculating the dilution effect of adding 5.0 mL of the acid. If the initial solution is 1.0 L of buffer, then the amounts in moles are the same as the molarity of each species:

$$pH = (14.00 - 5.91) + \log\left(\frac{(0.30 - 0.06 \text{ mol})}{(0.60 + 0.06 \text{ mol})}\right) = 8.09 + \log\left(\frac{0.24 \text{ mol}}{0.66 \text{ mol}}\right) = 7.65$$

16.95 At different points during a titration, different major species are present, because the titration reaction consumes formic acid and produces formate:

$$HCO_2H\ (aq) + OH^-\ (aq) \ \rightleftharpoons\ HCO_2^-\ (aq) + H_2O\ (l)$$

Point A is before titration begins, when the major species are H_2O and HCO_2H.

Point B is near the midpoint of the titration, when the major species are H_2O, HCO_2H, and HCO_2^-; the dominant equilibrium for both points A and B is the acid reaction:

$$H_2O\ (l) + HCO_2H\ (aq) \ \rightleftharpoons\ HCO_2^-\ (aq) + H_3O^+\ (aq) \qquad K_a = \frac{[HCO_2^-][H_3O^+]}{[HCO_2H]}$$

Point C is the stoichiometric point where all formic acid has been consumed, so the major species are H_2O and HCO_2^-. The dominant equilibrium for point C is

$$H_2O\ (l) + HCO_2^-\ (aq) \ \rightleftharpoons\ HCO_2H\ (aq) + OH^-\ (aq) \qquad K_b = \frac{[HCO_2H][OH^-]}{[HCO_2^-]}$$

Point D is beyond the stoichiometric point, so excess hydroxide is present and the major species are H_2O, OH^-, and HCO_2^-. The dominant equilibrium for point D is

$$H_2O\ (l) + HCO_2^-\ (aq) \ \rightleftharpoons\ HCO_2H\ (aq) + OH^-\ (aq) \qquad K_b = \frac{[HCO_2H][OH^-]}{[HCO_2^-]}$$

The cation of the strong base will also be present as a major species at points B, C, and D.

16.97 Molecular pictures show the correct relative numbers of the various species in the solution. The starting condition shows nine molecules of acetic acid and three of acetate anions.

(a) Use the buffer equation to calculate the pH of this solution:

$$pH = pK_a + \log\left(\frac{[A^-]}{[HA]}\right) = 4.75 + \log\left(\frac{3}{9}\right) = 4.75 - 0.48 = 4.27$$

(b) A hydroxide ion reacts with an acetic acid molecule to form a water molecule and an acetate ion, so the new picture shows eight acetic acid molecules, four acetate ions, and one H_2O:

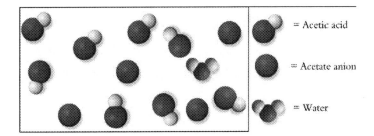

(c) HSO_4^- is a weak acid, so it transfers a proton to an acetate ion:

$$C_2H_3O_2^- \ (aq) + HSO_4^- \ (aq) \ \square \quad SO_4^{2-} \ (aq) + HC_2H_3O_2 \ (aq)$$

The new picture shows 10 molecules of acetic acid, two of acetate anions, and one sulphate:

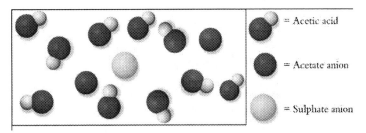

16.99 The key to working this problem is to recognize that the reaction can be written as the sum of two reactions for which K_{eq} values are available:

$$Ni^{2+} \ (aq) + 4 \ CN^- \ (aq) \ \square \quad [Ni(CN)_4]^{2-} \ (aq) \qquad K_f = 1 \times 10^{22}$$

$$\underline{Ni(OH)_2 \ (s) \ \square \quad Ni^{2+} \ (aq) + 2 \ OH^- \ (aq) \qquad K_{sp} = 5.5 \times 10^{-16}}$$

$$Ni(OH)_2 \ (s) + 4 \ CN^- \ (aq) \ \square \quad [Ni(CN)_4]^{2-} \ (aq) + 2 \ OH^- \ (aq)$$

$$K_{eq} = K_f K_{sp} = 5.5 \times 10^6$$

We can use this equilibrium to calculate the molarity of ions at equilibrium. The equilibrium constant is large, so take the reaction to completion and then return to equilibrium:

Reaction: $Ni(OH)_2 \ (s) +$	$4 \ CN^- \ (aq) \ \square$	$[Ni(CN)_4]^{2-} \ (aq) +$	$2 \ OH^- \ (aq)$
Initial concentration (M)	0.500	0	0
Change in concentration(M)	-0.500	$+0.125$	$+0.250$
Final concentration (M)	0	0.125	0.250
Change in concentration (M)	$+4x$	$-x$	$-2x$
Equilibrium concentration (M)	$4x$	$0.125 - x$	$0.250 - 2x$

Substitute into the equilibrium constant expression:

$$K_{eq} = 5.5 \times 10^6 = \frac{(0.125-x)(0.250-2x)^2}{(4x)^4}$$

Make the approximation that x is small relative to 0.125 and solve for x:

$$5.5 \times 10^6 = \frac{(0.125)(0.250)^2}{(4x)^4} \qquad 256x^4 = \frac{0.0078}{5.5 \times 10^6} \qquad x^4 = 5.5 \times 10^{-12}$$

$$x = 1.5 \times 10^{-3}$$

This is only 1.2% of 0.125, so the approximation is valid.

Because the value for K_f is only known to one significant figure, x must be rounded to one significant figure: $x = 0.002$.

$[CN^-] = 4x = 0.008$ M $[Ni(CN)_4]^{2-} = 0.125 - 0.002 = 0.123$ M

$[OH^-] = 0.250 - 0.004 = 0.246$ M

Use the concentration of the complex to calculate the amount of $Ni(OH)_2$ that will dissolve, and then use molar mass to convert to mass:

$$n = cV = (0.123 \text{ mol/L})(225 \text{ mL})(10^{-3} \text{ L/mL}) = 0.0277 \text{ mol}$$

$$m = nM = (0.0277 \text{ mol})(92.7 \text{ g/mol}) = 2.57 \text{ g}$$

16.101 (a) The best choice for a buffer solution is the weak acid whose pK_a is closest to the desired pH of the buffer solution. For a pH = 7.25 buffer solution, $H_2PO_4^-/HPO_4^{2-}$ ($pK_a = 7.21$) would be a good choice. To prepare the buffer solution, add the appropriate amounts of KH_2PO_4 and K_2HPO_4 to water and make up to 1.5 L with additional water.

(b) Calculate the $HPO_4^{2-}:H_2PO_4$ ratio using the buffer equation:

$$\log\left(\frac{[HPO_4^{2-}]}{[H_2PO_4^-]}\right) = pH - pK_a = 7.25 - 7.21 = 0.04 \qquad \frac{[HPO_4^{2-}]}{[H_2PO_4^-]} = 10^{0.04} = 1.10$$

$$[HPO_4^{2-}] = 1.10\left[H_2PO_4^-\right]$$

Also, $[HPO_4^{2-}] + [H_2PO_4^-] = 0.085$ M, so $[H_2PO_4^-] = 0.085$ M $-[HPO_4^{2-}]$

$$[HPO_4^{2-}] = 1.10(0.085 \text{ M} - [HPO_4^{2-}]) = 0.0935 - 1.10[HPO_4^{2-}]$$

$$2.10[HPO_4^{2-}] = 0.0935 \quad \text{and} \quad HPO_4^{2-} = 0.0445 \text{ M}$$

$$[H_2PO_4^-] = 0.085 - 0.0445 = 0.0405 \text{ M}$$

$$m_{KH_2PO_4} = 1.5 \text{ L}\left(\frac{0.0405 \text{ mol}}{1 \text{ L}}\right)\left(\frac{136 \text{ g}}{1 \text{ mol}}\right) = 8.3 \text{ g}$$

$$m_{K_2HPO_4} = 1.5\ L \left(\frac{0.0445\ mol}{1\ L} \right) \left(\frac{174\ g}{1\ mol} \right) = 12\ g$$

(c) Since hydronium ions are being generated in the solution, the pH will decrease. The enzyme will lose its activity once the pH reaches 7.1. Use the buffer equation to determine the ratio of the ions at this pH:

$$\log\left(\frac{[HPO_4^{2-}]}{[H_2PO_4^-]} \right) = pH - pK_a = 7.1 - 7.21 = -0.11 \qquad \frac{[HPO_4^{2-}]}{[H_2PO_4^-]} = 10^{-0.11} = 0.776$$

$$[HPO_4^{2-}] = 0.776[H_2PO_4^-]$$

The acid will react with the HPO_4^{2-} to form $H_2PO_4^-$, so their new equilibrium concentrations will be $[HPO_4^{2-}] = 0.0445 - x;$ $[H_2PO_4^-] = 0.0405 + x$. Substituting gives $0.0445 - x = 0.776(0.0405 + x)$, so $x = 0.00736$ M.

Convert to moles: 0.00736 M (1.5 L) = 0.011 moles H_3O^+ can be generated.

16.103 (a and b) To determine if cations can be separated by selective precipitation of their hydroxides, calculate the maximum concentration of hydroxide ions that can be present in the solution before each ion precipitates:

For $Zn(OH)_2$: $K_{sp} = 3 \times 10^{-17} = \left[Zn^{2+} \right]_{eq} \left[OH^- \right]_{eq}^2 = (0.300)\left[OH^- \right]_{eq}^2$

$$\left[OH^- \right]_{eq}^2 = 1 \times 10^{-16}, \text{ so } \left[OH^- \right]_{eq} = 1 \times 10^{-8}\ M$$

$$pOH = -\log\left(1 \times 10^{-8}\right) = 8.0 \qquad pH = 14.00 - 8.0 = 6.0$$

For $Fe(OH)_3$: $K_{sp} = 2.8 \times 10^{-39} = \left[Fe^{3+} \right]_{eq} \left[OH^- \right]_{eq}^3 = (0.100)\left[OH^- \right]_{eq}^3$

$$\left[OH^- \right]_{eq}^3 = 2.8 \times 10^{-38}, \text{ so} \left[OH^- \right]_{eq} = 3.0 \times 10^{-13}\ M$$

$$pOH = -\log\left(3.0 \times 10^{-13}\right) = 12.52 \qquad pH = 14.00 - 12.52 = 1.48$$

Thus, Fe(OH)$_3$ will precipitate from this acid solution when the pH rises above 1.48, but Zn(OH)$_2$ will not begin to precipitate until the pH reaches 6.0. The two ions can be separated by adjusting the solution pH to around 5.9.

(c) Zn(OH)$_2$ starts to precipitate when the hydroxide ion concentration reaches 1.0 $\times 10^{-8}$ M:

For $Fe(OH)_3$: $K_{sp} = 2.8 \times 10^{-39} = \left[Fe^{3+} \right]_{eq} \left[OH^- \right]_{eq}^3 = \left[Fe^{3+} \right]_{eq} \left(1.0 \times 10^{-8}\right)^3$

Thus, $\left[Fe^{3+} \right]_{eq} = \dfrac{2.8 \times 10^{-39}}{\left(1.0 \times 10^{-8}\right)^3} = 3 \times 10^{-15}$ M (which is negligible compared with

the zinc ion concentration).

16.105 Suppose we have an aqueous solution of two weak acids, HA and HB, which have acid hydrolysis constants K_{HA} and K_{HB} respectively. Derive an equation for $[H_3O^+(aq)]$ in this solution based only on constants, [HA], and [HB]. *Hint*: Use the fact that total positive charge = total negative charge.

The equilibria in this solution are:

$$HA + H_2O(l) \rightleftharpoons H_3O^+(aq) + A^-(aq)$$
$$HB + H_2O(l) \rightleftharpoons H_3O^+(aq) + B^-(aq)$$
$$2\,H_2O(l) \rightleftharpoons H_3O^+(aq) + OH^-(aq)$$

The equilibrium expressions are:

$$K_{HA} = \frac{[H_3O^+][A^-]}{[HA]}$$

$$K_{HB} = \frac{[H_3O^+][B^-]}{[HB]}$$

$$K_w = [H_3O^+][OH^-]$$

For a charge balance, it must be true that $[H_3O^+] = [OH^-] + [A^-] + [B^-]$

All three terms on the right side of this equation can be derived from the equilibrium expressions above:

$$[A^-] = \frac{K_{HA}[HA]}{[H_3O^+]}$$

$$[B^-] = \frac{K_{HB}[HB]}{[H_3O^+]}$$

$$[OH^-] = \frac{K_w}{[H_3O^+]}$$

Substituting these expressions into the charge balance equation, we have:

$$[H_3O^+] = \frac{K_{HA}[HA]}{[H_3O^+]} + \frac{K_{HB}[HB]}{[H_3O^+]} + \frac{K_w}{[H_3O^+]}$$

$$[H_3O^+]^2 = K_{HA}[HA] + K_{HB}[HB] + K_w$$

$$[H_3O^+] = (K_{HA}[HA] + K_{HB}[HB] + K_w)^{1/2}$$

Chapter 17 Electron Transfer Reactions

Solutions to Problems in Chapter 17

17.1 Oxidation numbers are determined by applying the rules given in your textbook:

(a) Ionic compound containing OH^- anions, with O –2, H +1, and Fe +3 to give overall neutrality

(b) F (most electronegative) is –1, so N must be +3 to give overall neutrality

(c) O is –2, H is +1, and C is –2 to give overall neutrality

(d) Ionic compound, with K^+ = +1 and CO_3^{2-}; in the anion, O (more electronegative) is –2 and C is +4 to give the –2 net charge on the ion

(e) Ionic compound; in NH_4^+, H is +1 and N is –3, and in NO_3^-, O is –2 and N is +5

(f) Cl (more electronegative) is –1, so Ti is +4 to give overall neutrality

(g) Ionic; Pb must be +2 to balance the –2 charge on sulphate; O is –2, so S is +6 to give the –2 charge on the anion

(h) When it is a pure element, P is 0

17.3 Identify redox reactions by determining whether or not oxidation numbers change:

(a) No change in oxidation numbers (Br remains –1), so not redox

(b) Redox, because Fe changes from +2 to +3 (O also changes)

(c) No change in oxidation numbers (Fe remains +2), so not redox

(d) Redox, because O in O_2 changes from 0 to –2 (C also changes)

(e) Redox, because N changes from 0 to –3 and H changes from 0 to +1

17.5 Oxidation numbers are determined by applying the rules given in your textbook:

(a) F is –1, so Cl is +5 to give overall neutrality

(b) Cl is more electronegative, so it is –1

(c) Ionic; ClO_4^- has net charge of –1, and O is –2, so Cl is +7 to give net charge of –1

(d) Cl is 0 because this is a pure element

(e) Ionic; ClO_2^- has net charge of –1, and O is –2, so Cl is +3 to give net charge of –1

(f) Ionic; ClO^- has net charge of –1, and O is –2, so Cl is +1 to give net charge of –1

17.7 Determine half-reactions by inspection, making use of oxidation numbers if necessary. Balance each half-reaction following the standard steps in order (balance all but H and O by inspection; balance O by adding H_2O; balance H by adding H_3O^+ cations to the side that is deficient in hydrogen and an equal number of H_2O to the other side; balance charge by adding e^-):

(a) The reactants are Na and H_2O, and one product is H_2; Na^+ and OH^- must be the other products:

$$Na\ (s) \rightarrow Na^+\ (aq) + e^- \text{ and } 2\ H_2O\ (l) + 2\ e^- \rightarrow H_2\ (g) + 2\ OH^-\ (aq)$$

(b) The reactants are Au and NO_3^-, and the products are $[AuCl_4]^-$ and NO; "aqua regia" also contains Cl^- anions:

$$Au\ (s) + 4\ Cl^-\ (aq) \rightarrow [AuCl_4]^-\ (aq) + 3\ e^- \text{ and } NO_3^-\ (aq) + 4\ H_3O^+\ (aq) + 3\ e^- \rightarrow NO\ (g) + 6\ H_2O\ (l)$$

(c) The reactants are MnO_4^- and $C_2O_4^{2-}$, and the products are Mn^{2+} and CO_2:

$$MnO_4^-\ (aq) + 8\ H_3O^+\ (aq) + 5\ e^- \rightarrow Mn^{2+}\ (aq) + 12\ H_2O\ (l) \text{ and } C_2O_4^{2-}\ (aq) \rightarrow 2\ CO_2\ (g) + 2\ e^-$$

(d) The reactants are C and H_2O, and the products are H_2 and CO:

$$C\ (s) + 3\ H_2O\ (l) \rightarrow CO\ (g) + 2\ H_3O^+\ (aq) + 2e^- \text{ and } 2\ H_3O^+\ (aq) + 2e^- \rightarrow H_2\ (g) + 2\ H_2O\ (l)$$

17.9 Balance each half-reaction following the standard steps in order: (1) Balance all but H and O by inspection; (2) balance O by adding H_2O; (3) balance H by adding H_3O^+ cations to the side that is deficient in hydrogen and an equal number of H_2O to the other side; if the solution is basic, add an H_2O for each deficient H to the side that is deficient in hydrogen and an equal number of OH^- anions to the other side; (4) balance charge by adding e^-:

(a) (1) Cu is balanced; (2) add H_2O on the left; (3) add 2 H_3O^+ on the right and 2 H_2O on the left; (4) add 1 e^- on the right: $Cu^+\ (aq) + 3\ H_2O\ (l) \rightarrow CuO\ (s) + 2\ H_3O^+\ (aq) + e^-$

(b) (1) S is balanced; (2) no O is present; (3) add 2 H_3O^+ on the left and 2 H_2O on the right; (4) add 2 e^- on the left: $S\ (s) + 2\ H_3O^+\ (aq) + 2\ e^- \rightarrow H_2S\ (g) + 2\ H_2O\ (l)$

(c) (1) Ag is balanced; add Cl^- on the right; (2 & 3) there are no H or O atoms; (4) add 1 e^- on the left: $AgCl + e^- \rightarrow Ag + Cl^-$

(d) (1) I is balanced; (2) add 3 H_2O on the left; (3) add 6 H_2O on the right and 6 OH^- on the left; (4) add 6 e^- on the right: $I^- + 3\ H_2O + 6\ OH^- \rightarrow IO_3^- + 6\ H_2O + 6\ e^-$

Cancel duplicated species: $I^-\ (aq) + 6\ OH^-\ (aq) \rightarrow IO_3^-\ (aq) + 3\ H_2O\ (l) + 6\ e^-$

(e) (1) I is balanced; (2) add 2 H_2O on the right; (3) add 4 H_2O on the left and 4

OH$^-$ on the right; (4) add 4 e$^-$ on the left: IO$_3^-$ + 4 H$_2$O + 4 e$^-$ → IO$^-$ + 2 H$_2$O + 4 OH$^-$

Cancel duplicated species: IO$_3^-$ (*aq*) + 2 H$_2$O (*l*) + 4 e$^-$ → IO$^-$ (*aq*) + 4 OH$^-$ (*aq*)

(f) (1) C is balanced; (2) add H$_2$O on the left; (3) add 4 H$_3$O$^+$ on the right and 4 H$_2$O on the left; (4) add 4 e$^-$ on the right:

H$_2$CO (*aq*) + 5 H$_2$O (*l*) → CO$_2$ (*g*) + 4 H$_3$O$^+$ (*aq*) + 4 e$^-$

17.11 To combine half-reactions into a net redox reaction, multiply by appropriate integers so that the electrons will cancel:

(a) 2 [Cu$^+$ + 3 H$_2$O → CuO + 2 H$_3$O$^+$ + e$^-$]

  + [S + 2 H$_3$O$^+$ + 2 e$^-$ → H$_2$S + 2 H$_2$O]
  _____
  2 Cu$^+$ + 4 H$_2$O + S → 2 CuO + 2 H$_3$O$^+$ + H$_2$S

(b) 6 [AgCl + e$^-$ → Ag + Cl$^-$]

  + [I$^-$ + 6 OH$^-$ → IO$_3^-$ + 3 H$_2$O + 6 e$^-$]
  _____
  6 AgCl + I$^-$ + 6 OH$^-$ → 6 Ag + 6 Cl$^-$ + 3 H$_2$O + IO$_3^-$

(c) 2 [I$^-$ + 6 OH$^-$ → IO$_3^-$ + 3 H$_2$O + 6 e$^-$]

  + 3 [IO$_3^-$ + 2 H$_2$O + 4 e$^-$ → IO$^-$ + 4 OH$^-$]
  _____
  2 I$^-$ + IO$_3^-$ → 3 IO$^-$

(d) 2 [S + 2 H$_3$O$^+$ + 2 e$^-$ → H$_2$S + 2 H$_2$O]

  + [H$_2$CO + 5 H$_2$O → CO$_2$ + 4 H$_3$O$^+$ + 4 e$^-$]
  _____
  2 S + H$_2$CO + H$_2$O → 2 H$_2$S + CO$_2$

17.13 The species that loses electrons is oxidized and acts as the reducing agent, whereas the species that gains electrons is reduced and acts as the oxidizing agent. In this reaction, Cl$^-$ loses electrons (oxidized, reducing agent) to become Cl$_2$, and Mn in MnO$_4^-$ gains electrons (reduced, oxidizing agent) to become Mn^{2+}.

reactions, balance each half-reaction using the stepwise technique, and then multiply by appropriate integers so that the electrons cancel:

(a) CN$^-$ → CNO$^-$

  (1) C and N are already balanced; (2) add 1 H$_2$O on the left; (3) add 2 H$_2$O on the right and 2 OH$^-$ on the left; (4) add 2 e$^-$ on the right:

  CN$^-$ + H$_2$O + 2 OH$^-$ → CNO$^-$ + 2 H$_2$O + 2 e$^-$

  Cancel duplicated species: CN$^-$ + 2 OH$^-$ → CNO$^-$ + H$_2$O + 2 e$^-$

$MnO_4^- \rightarrow MnO_2$

(1) Mn is balanced; (2) add 2 H_2O on the right; (3) add 4 H_2O on the left and 4 OH^- on the right; (4) add 3 e^- on the left:

$$MnO_4^- + 4\ H_2O + 3\ e^- \rightarrow MnO_2 + 2\ H_2O + 4\ OH^-$$

Cancel duplicated species: $MnO_4^- + 2\ H_2O + 3\ e^- \rightarrow MnO_2 + 4\ OH^-$

Multiply the first reaction by 3, the second by 2, and

$$3\ [CN^- + 2\ OH^- \rightarrow CNO^- + H_2O + 2\ e^-]$$

add: $\dfrac{+2\ [MnO_4^- + 2\ H_2O + 3\ e^- \rightarrow MnO_2 + 4\ OH^-]}{3\ CN^- + 2\ MnO_4^- + H_2O \rightarrow 3\ CNO^- + 2\ MnO_2 + 2\ OH^-}$

(b) $As \rightarrow HAsO_2$

(1) As is balanced; (2) add 2 H_2O on the left; (3) add 3 H_3O^+ on the right and 3 H_2O on the left; (4) add 3 e^- on the right: $As + 5\ H_2O \rightarrow HAsO_2 + 3\ H_3O^+ + 3\ e^-$

$O_2 \rightarrow H_2O$

(1) No elements other than O and H; (2) add 1 H_2O on the right; (3) add 4 H_3O^+ on left and 4 H_2O on the right; (4) add 4 e^- on the left: $O_2 + 4\ H_3O^+ + 4\ e^- \rightarrow 6\ H_2O$.

Multiply the first reaction by 4, the second by 3, and add:

$$4\ [As + 5\ H_2O \rightarrow HAsO_2 + 3\ H_3O^+ + 3\ e^-]$$
$$\underline{+3\ [O_2 + 4\ H_3O^+ + 4\ e^- \rightarrow 6\ H_2O]}$$
$$4\ As + 3\ O_2 + 2\ H_2O \rightarrow 4\ HAsO_2$$

(c) $Br^- \rightarrow BrO_3^-$

(1) Br is balanced; (2) add 3 H_2O on the left; (3) add 6 H_2O on the right and 6 OH^- on the left; (4) add 6 e^- on the right: $Br^- + 3\ H_2O + 6\ OH^- \rightarrow BrO_3^- + 6H_2O + 6\ e^-$

Cancel duplicated species: $Br^- + 6\ OH^- \rightarrow BrO_3^- + 3\ H_2O + 6\ e^-$

$MnO_4^- \rightarrow MnO_2$, same as the reaction in part (a):

$$MnO_4^- + 2\ H_2O + 3\ e^- \rightarrow MnO_2 + 4\ OH^-$$

Multiply the second reaction by 2 and add:

$$[Br^- + 6\,OH^- \rightarrow BrO_3^- + 3\,H_2O + 6\,e^-]$$

$$+2\,[MnO_4^- + 2\,H_2O + 3\,e^- \rightarrow MnO_2 + 4\,OH^-]$$

$$\overline{Br^- + 2\,MnO_4^- + H_2O \rightarrow BrO_3^- + 2\,MnO_2 + 2\,OH^-}$$

(d) $NO_2 \rightarrow NO_3^-$

(1) N is balanced; (2) add 1 H_2O on the left; (3) add 2 H_3O^+ on the right and 2 H_2O on the left; (4) add 1 e^- on the right: $NO_2 + 3\,H_2O \rightarrow NO_3^- + 2\,H_3O^+ + e^-$

$NO_2 \rightarrow NO$

(1) N is balanced; (2) add 1 H_2O on the right; (3) add 2 H_3O^+ on the left and 2 H_2O on the right; (4) add 2 e^- on the left: $NO_2 + 2\,H_3O^+ + 2\,e^- \rightarrow NO + 3\,H_2O$

Multiply the first reaction by 2 and add:

$$2\,[NO_2 + 3\,H_2O \rightarrow NO_3^- + 2\,H_3O^+ + e^-]$$

$$+[NO_2 + 2\,H_3O^+ + 2\,e^- \rightarrow NO + 3\,H_2O]$$

$$\overline{3\,NO_2 + 3\,H_2O \rightarrow 2\,NO_3^- + NO + 2\,H_3O^+}$$

(e) $ClO_4^- \rightarrow ClO^-$

(1) Cl is balanced; (2) add 3 H_2O on the right; (3) add 6 H_3O^+ on the left and 6 H_2O on the right; (4) add 6 e^- on the left: $ClO_4^- + 6\,H_3O^+ + 6\,e^- \rightarrow ClO^- + 9\,H_2O$

$Cl^- \rightarrow Cl_2$

(1) Multiply Cl^- by 2; (2 & 3) no H or O present; (4) add 2 e^- on the right:

$$2\,Cl^- \rightarrow Cl_2 + 2\,e^-$$

Multiply the second reaction by 3 and add:

$$[ClO_4^- + 6\,H_3O^+ + 6\,e^- \rightarrow ClO^- + 9\,H_2O]$$

$$+3\,[2\,Cl^- \rightarrow Cl_2 + 2\,e^-]$$

$$\overline{ClO_4^- + 6\,Cl^- + 6\,H_3O^+ \rightarrow ClO^- + 9\,H_2O + 3\,Cl_2}$$

(f) $AlH_4^- \rightarrow Al^{3+}$

(1) Al is balanced; (2) no O present; (3) add 4 H_2O on the right and 4 OH^- on the left; (4) add 8 e^- on the right: $AlH_4^- + 4\,OH^- \rightarrow Al^{3+} + 4\,H_2O + 8\,e^-$

$H_2CO \rightarrow CH_3OH$

(1) C is balanced; (2) O is balanced; (3) add 2 H_2O on the left and 2 OH^- on the right; (4) add 2 e^- on the left: $H_2CO + 2\,H_2O + 2\,e^- \rightarrow CH_3OH + 2\,OH^-$

Multiply the second reaction by 4 and add:

$$[AlH_4^- + 4\,OH^- \rightarrow Al^{3+} + 4\,H_2O + 8\,e^-]$$

$$+ 4[H_2CO + 2\,H_2O + 2\,e^- \rightarrow CH_3OH + 2\,OH^-]$$

$$\overline{AlH_4^- + 4\,H_2CO + 4\,H_2O \rightarrow Al^{3+} + 4\,CH_3OH + 4\,OH^-}$$

17.17 To determine spontaneity using standard thermodynamic values, calculate $\Delta G^\circ_{reaction}$ from tabulated values for ΔG°_f. The reaction is spontaneous if $\Delta G^\circ_{reaction}$ is negative:

(a) $\Delta G^\circ_{reaction}$ = 2 mol(–129.7 kJ/mol) – 0 kJ/mol = –259.4 kJ; spontaneous

(b) $\Delta G^\circ_{reaction}$ = 2 mol(–58.5 kJ/mol) – 0 kJ/mol = –117.0 kJ; spontaneous

(c) $\Delta G^\circ_{reaction}$ = 1 mol(–300.1 kJ/mol)

   – [1 mol(–53.6 kJ/mol) + 1 mol(0 kJ/mol)] = –246.5 kJ; spontaneous

(d) $\Delta G^\circ_{reaction}$ = [1 mol(–300.1 kJ/mol) + 1 mol(0 kJ/mol)]

   – [1 mol(–100.4 kJ/mol) + 1 mol(0 kJ/mol)] = –199.7 kJ; spontaneous

17.19 A molecular view of a process occurring at an electrode should show the species involved in charge transfer and indicate the direction of movement of electrons. See Figure 17-5 for an example. At a silver–silver chloride electrode undergoing reduction, solid AgCl gains an electron from the electrode to release chloride anions into the solution and produce solid Ag metal:

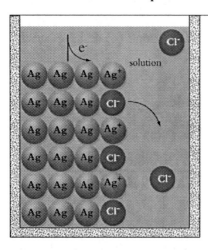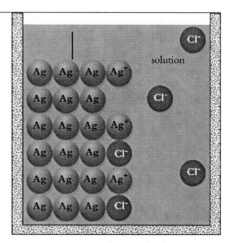

17.21 A passive electrode and a supply of gas are needed to study a redox reaction that includes gases. See Figure 16-8 for a sketch of a hydrogen electrode. A chlorine electrode is exactly analogous:

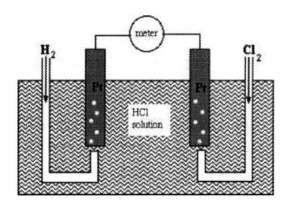

17.23 Active electrodes participate in redox reactions, whereas passive electrodes only provide or accept electrons. The Ag–AgCl electrode is active, whereas the Pt electrodes are passive.

17.25   The shorthand notation is always written as anode to cathode (left to right).

(a) Ag (*s*) | Ag$^+$ (*aq*) ‖ Fe^{2+} (*aq*) | Fe (*s*)

(b) Pt (*s*) | H$_2$O$_2$ (*aq*), H$_3$O$^+$ (*aq*) | O$_2$ (*g*) ‖ NO$_3^-$ (*aq*), H$_3$O$^+$ (*aq*) | NO (*g*) | Pt (*s*)

(c) Pt (*s*) | H$_2$ (*g*) | H$_3$O$^+$ (*aq*) ‖ Cl$_2$ (*g*) | Cl$^-$ (*aq*) | Pt (*s*)

17. 27 Standard electrode potentials are calculated using Equation 17-1:

$$E^\circ_{cell} = E^\circ_{cathode} - E^\circ_{anode}$$

17.15 (a) MnO$_4^-$ undergoes reduction to MnO$_2$ (cathode) $E^\circ_{cathode}$ = 0.595 V

CN$^-$ undergoes oxidation to CNO$^-$ (anode) E°_{anode} = –0.970 V

E°_{cell} = (0.595 V) – (–0.970 V) = 1.565 V

17.15 (b) O$_2$ undergoes reduction to H$_2$O (cathode) $E^\circ_{cathode}$ = 1.229 V

As undergoes oxidation to HAsO$_2$ (anode) E°_{anode} = 0.248 V

E°_{cell} = (1.229 V) – (0.248 V) = 0.981 V

17.15 (e) ClO$_4^-$ undergoes reduction to ClO$^-$ (cathode) $E^\circ_{cathode}$ = 1.36 V

Cl$^-$ undergoes oxidation to Cl$_2$ (anode) E°_{anode} = 1.35827 V

E°_{cell} = (1.36 V) – (1.35827 V) = 0.002 V, which rounds to 0.00 V

17.29 Balance redox reactions following the standard procedure. Break into half-reactions, balance each half-reaction using the stepwise technique, and then multiply by appropriate integers so that the electrons cancel:

$NO \rightarrow N_2O$

(1) Multiply NO by 2 to balance N; (2) add 1 H_2O on the right; (3) add 2 H_3O^+ on the left and 2 H_2O on the right; (4) add 2 e^- on the left:

$$2\,NO + 2\,H_3O^+ + 2\,e^- \rightarrow N_2O + 3\,H_2O$$

$NO \rightarrow NO_3^-$

(1) N is already balanced; (2) add 2 H_2O on the left; (3) add 4 H_3O^+ on the right and 4 H_2O on the left; (4) add 3 e^- on the right:

$$NO + 6\,H_2O \rightarrow NO_3^- + 4\,H_3O^+ + 3\,e^-$$

Multiply the first reaction by 3 and the second reaction by 2, and then add to obtain the overall reaction:

$$3[2\,NO + 2\,H_3O^+ + 2\,e^- \rightarrow N_2O + 3\,H_2O]$$
$$\underline{+\,2[NO + 6\,H_2O \rightarrow NO_3^- + 4\,H_3O^+ + 3\,e^-]}$$
$$8\,NO + 3\,H_2O \rightarrow 3\,N_2O + 2\,NO_3^- + 2\,H_3O^+$$

Standard electrode potentials are calculated using Equation 17-1:

$$E^\circ_{cell} = E^\circ_{cathode} - E^\circ_{anode}$$

NO undergoes reduction to N_2O (cathode) $E^\circ_{cathode} = 1.591$ V

NO undergoes oxidation to NO_3^- (anode) $E^\circ_{anode} = 0.957$ V

$E^\circ_{cell} = (1.591\ \text{V}) - (0.957\ \text{V}) = 0.634$ V

17.31 The direct way to measure a standard reduction potential is with a cell containing a standard hydrogen electrode as a reference. Because F^- is the conjugate base of a weak acid, the F_2/F^- portion of the cell must be separated from the H_2/H_3O^+ portion, requiring a porous plate. In addition, the fluoride solution could be made basic to prevent formation of HF. Set up a standard hydrogen electrode on one side and a Pt electrode immersed in a 1 M solution of NaF with F_2 bubbling over the electrode on the other side:

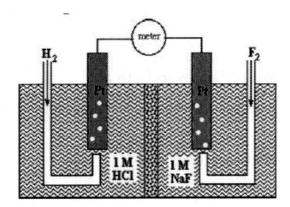

The standard reduction potential of F_2 is +2.866 V, so F_2 will be reduced in this cell and the H_2/H^+ electrode will be the anode.

17.33 To act as a reducing agent under standard conditions, a substance must have a more negative standard reduction potential than the substance that is to be reduced. The standard reduction potential of Be is –1.847 V. Scanning the table, we find the following metals with standard potentials more negative than this:

Ba (–2.912 V), Ca (–2.868 V), Cs (–3.026 V), Li (–3.0401 V), Mg (–2.37 V), K (–2.931 V), and Na (–2.71 V). All these metals lie in the *s* block of the periodic table and easily lose 1 or 2 electrons.

17.35 The standard free energy change of a reaction is related to the standard reduction potential through Equation 17-3: $\Delta G° = -nFE°$

The standard potentials for the reactions were calculated in Problem 16.27:

16.27 (a) $E°_{cell} = 1.565$ V, $n = 6$:

$$\Delta G° = -(6\text{ mol})(9.6485 \times 10^4 \text{ C/mol})(1.565\text{ V})\left(\frac{1\text{ J}}{1\text{ V C}}\right)\left(\frac{10^{-3}\text{ kJ}}{1\text{ J}}\right) = -906.0\text{ kJ}$$

16.27 (b) $E°_{cell} = 0.981$ V, $n = 12$:

$$\Delta G° = -(12\text{ mol})(9.6485 \times 10^4 \text{ C/mol})(0.981\text{ V})\left(\frac{1\text{ J}}{1\text{ V C}}\right)\left(\frac{10^{-3}\text{ kJ}}{1\text{ J}}\right) = -1.14 \times 10^3 \text{ kJ}$$

16.27 (e) $E°_{cell} = 0.002$ V (rounds to 0.00 V), $n = 6$:

$$\Delta G° = (6\text{ mol})(9.6485 \times 10^4 \text{ C/mol})(0.002\text{ V})\left(\frac{1\text{ J}}{1\text{ V C}}\right)\left(\frac{10^{-3}\text{ kJ}}{1\text{ J}}\right) = 1\text{ kJ (rounds to}$$

0 kJ)

17.37 Operating potentials are related to standard cell potentials through Equation 17-6:

$$E = E° - \frac{RT}{nF} \ln Q$$

The reaction for the nickel–cadmium battery is

$$2\ NiO(OH)\ (s) + 2\ H_2O\ (l) + Cd\ (s) \rightarrow 2\ Ni(OH)_2\ (s) + Cd(OH)_2\ (s)$$

Here, all the substances are solids or pure liquids, so $Q = 1$, $\ln Q = 0$, and $E = E° = 1.35$ V. Hence, E is independent of $[OH^-]$.

17.39 The connection between electric current and chemical amounts is provided by Equation 17-7, $n = \frac{It}{F}$ where n is the moles of electrons flowing.

$$n = 15\ s \left(\frac{5.9\ C}{1\ s}\right) \left(\frac{1\ mol}{9.6485 \times 10^4\ C}\right) = 9.2 \times 10^{-4}\ mol\ electrons$$

The half-reactions in a lead storage cell are

$$H_2O + Pb\ (s) + HSO_4^-\ (aq) \rightarrow PbSO_4\ (s) + H_3O^+\ (aq) + 2\ e^-$$

$$PbO_2\ (s) + HSO_4^-\ (aq) + 3\ H_3O^+\ (aq) + 2\ e^- \rightarrow PbSO_4\ (s) + 5\ H_2O\ (l)$$

$$n_{Pb} = n_{PbO_2} = 9.2 \times 10^{-4}\ mol \left(\frac{1\ mol\ Pb}{2\ mol\ e^-}\right) = 4.6 \times 10^{-4}\ mol$$

Finally, convert moles to mass using molar masses:

$$m_{Pb} = 4.6 \times 10^{-4}\ mol \left(\frac{207.5\ g}{1\ mol}\right) = 9.5 \times 10^{-2}\ g$$

$$m_{PbO_2} = 4.6 \times 10^{-4}\ mol \left(\frac{239.2\ g}{1\ mol}\right) = 0.11\ g$$

17.41 The connection between electric current and chemical amounts is provided by Equation 17-7, $n = \frac{It}{F}$, where n is the moles of electrons flowing. Thus, $t = \frac{nF}{I}$

The reaction being driven by the alternator is

$$PbSO_4\ (s) + H_3O^+\ (aq) + 2\ e^- \rightarrow H_2O\ (l) + Pb\ (s) + HSO_4^-\ (aq)$$

When 0.850 g of $PbSO_4$ is converted into Pb in each cell, the amount of electrons that flows is as follows:

$$n_e = 2n_{PbSO_4} = 2(0.850\ g) \left(\frac{1\ mol}{303.3\ g}\right) = 5.6 \times 10^{-3}\ mol$$

$$I = (1.750\ A) - (1.350\ A) = 0.400\ C/s$$

$$t = \frac{(5.6 \times 10^{-3}\ \text{mol})(9.6485 \times 10^{4}\ \text{C/mol})}{0.400\ \text{C/s}} = 1.35 \times 10^{3}\ \text{s}$$

17.43 Balance redox reactions following the standard procedure. Break into half-reactions, balance each half-reaction using the stepwise technique, and then multiply by appropriate integers so that the electrons cancel:

$$Cr_2O_7{}^{2-} \rightarrow Cr^{3+}$$

(1) Multiply Cr^{3+} by 2; (2) add 7 H_2O on the right; (3) add 14 H_3O^+ on the left and 14 H_2O on the right; (4) add 6 e$^-$ on the left:

$$Cr_2O_7{}^{2-} + 14\ H_3O^+ + 6\ e^- \rightarrow 2\ Cr^{3+} + 21\ H_2O$$

$$CH_3CHO \rightarrow CH_3CO_2H$$

(1) C is balanced; (2) add 1 H_2O on the left; (3) add 2 H_3O^+ on the right and 2 H_2O on the left; (4) add 2 e$^-$ on the right:

$$CH_3CHO + 3\ H_2O \rightarrow CH_3CO_2H + 2\ H_3O^+\ (aq) + 2\ e^-$$

Multiply the second reaction by 3 and add:

$$\begin{array}{l} [Cr_2O_7{}^{2-} + 14\ H_3O^+\ (aq) + 6\ e^- \rightarrow 2\ Cr^{3+} + 21\ H_2O] \\ \underline{+ 3[CH_3CHO + 3\ H_2O \rightarrow CH_3CO_2H + 2\ H_3O^+\ (aq) + 2\ e^-]} \\ Cr_2O_7{}^{2-} + 3\ CH_3CHO + 8\ H_3O^+\ (aq) \rightarrow 2\ Cr^{3+} + 3\ CH_3CO_2H + 12\ H_2O \end{array}$$

It takes 2 moles of electrons to oxidize 1 mole of CH_3CHO:

$$n = 1.00\ \text{g} \left(\frac{1\ \text{mol}}{44.1\ \text{g}} \right) \left(\frac{2\ \text{mol e}^-}{1\ \text{mol } CH_3CHO} \right) = 0.0454\ \text{mol of electrons needed}$$

Each mole of $Cr_2O_7{}^{2-}$ consumes 6 moles of electrons:

$$n_{\text{dichromate}} = 0.0454\ \text{mol} \left(\frac{1\ \text{mol } Cr_2O_7^{2-}}{6\ \text{mol e}^-} \right) = 7.57 \times 10^{-3}\ \text{mol}$$

$$m = 7.57 \times 10^{-3}\ \text{mol} \left(\frac{262\ \text{g}}{1\ \text{mol}} \right) = 1.98\ \text{g of sodium dichromate needed}$$

17.45 Use the Nernst equation, and identify the two half-reactions to find *n* and *Q*.

Oxidation: $Cr^{2+}\ (aq) \rightarrow Cr^{3+}\ (aq) + e^-$	$E° = +0.407$ V
Reduction: $Ni^{2+}\ (aq) + 2\ e^- \rightarrow Ni\ (s)$	$E° = -0.257$ V
Overall: $Ni^{2+}\ (aq) + 2\ Cr^{2+}\ (aq) \rightarrow Ni\ (s) + 2\ Cr^{3+}\ (aq)$	$E°_{\text{cell}} = +0.150$ V

Under non-standard state conditions, use the Nernst equation; $n = 2$ for this reaction.

$$Q = \frac{[Cr^{3+}]^2}{[Ni^{2+}][Cr^{2+}]^2} = \frac{(0.0015)^2}{(0.0001)(0.002)^2} = 5625$$

$$E = 0.150 - \frac{0.0592 \text{ V}}{2} \log 5625 = 0.039 \text{ V}$$

17.47   Use the Nernst equation. You need to find $E°$, n, and Q.

$$Sn^{2+} (aq) + 2 e^- \rightarrow Sn (s) \qquad E° = -0.137 \text{ V}$$

$$2 H^+ (aq) + 2 e^- \rightarrow H_2 (g) \qquad E° = 0.00 \text{ V}$$

(a) For standard state conditions, $E°_{cell} = E°_{cathode} - E°_{anode} = (0.00 \text{ V}) - (-0.137 \text{ V}) = 0.137 \text{ V}$

(b) For non-standard state conditions, use the Nernst equation, where $n = 2$ for this reaction.

$$Q = \frac{[Sn^{2+}]p_{H_2}}{[H^+]^2} = \frac{(1.0)(1.0)}{(1.0 \times 10^{-2})^2} = 1.0 \times 10^4$$

$$E = 0.137 - \frac{0.0592 \text{ V}}{2} \log(1.0 \times 10^4) = 0.0187 \text{ V}$$

(c) $Q = \dfrac{[Sn^{2+}]p_{H_2}}{[H^+]^2} = \dfrac{(1.0)(1.0)}{(1.0 \times 10^{-5})^2} = 1.0 \times 10^{10}$

$$E = 0.137 - \frac{0.0592 \text{ V}}{2} \log(1.0 \times 10^{10}) = -0.159 \text{ V}$$

17.49 The Nernst equation is Equation 17-6, $E = E° - \dfrac{RT}{nF} \ln Q$. To determine Q, obtain the balanced redox equation for the lead storage cell:

$$Pb (s) + 2 HSO_4^- (aq) + PbO_2 (s) + 2 H_3O^+ (aq) \rightarrow 2 PbSO_4 (s) + 4 H_2O (l)$$

$Q = \dfrac{1}{[H_3O^+]^2[HSO_4^-]^2}$; as the battery operates, these concentrations decrease, Q and $\ln Q$ increase, and the potential of the battery decreases with use.

17.51 When two metals are in contact, the one with the more negative standard reduction potential corrodes first. Here are the values for Zn and Fe, from Appendix F:

$$Zn^{2+} + 2 e^- \rightarrow Zn \qquad E° = -0.7618 \text{ V}$$

$$Fe^{2+} + 2 e^- \rightarrow Fe \qquad E° = -0.447 \text{ V}$$

Thus, Zn oxidizes more easily and will preferentially corrode, protecting the steel propeller. This is why zinc metal is called sacrificial.

17.53 Lead storage batteries are used in automobiles because they deliver large amounts of current and can be readily recharged, but they are heavy. For space flights, weight is a more important consideration than rechargeability, so lead batteries are not suitable.

17.55 The connection between electric current and chemical amounts is provided by Equation 17-7, $n = \dfrac{It}{F}$, where n is moles of electrons flowing. Examine the balanced half-reaction to determine the stoichiometric relationship between moles of electrons and moles of chemical species. Here, the reaction is $2\ Cl^- \rightarrow Cl_2 + 2\ e^-$, so 1 mole of Cl_2 is formed by 2 moles of electrons:

$$t = 200.0\ \text{min}\left(\frac{60\ \text{s}}{1\ \text{min}}\right) = 1.200 \times 10^4\ \text{s}$$

$$n = 1.200 \times 10^4\ \text{s}\left(\frac{4.50\ \text{C}}{1\ \text{s}}\right)\left(\frac{1\ \text{mol e}^-}{9.6485 \times 10^4\ \text{C}}\right)\left(\frac{1\ \text{mol Cl}_2}{2\ \text{mol e}^-}\right) = 0.280\ \text{mol Cl}_2$$

$$m = nM = 0.280\ \text{mol}\left(\frac{70.906\ \text{g}}{1\ \text{mol}}\right) = 19.9\ \text{g}$$

17.57 When an external potential is applied to a galvanic cell, the reduction that occurs is the one with the least negative reduction potential, and the oxidation that occurs is the one with the least positive reduction potential. The oxidation reaction during recharging is the reverse of the galvanic reduction:

$$Cu \rightarrow Cu^{2+} + 2\ e^-$$

However, instead of $Zn^{2+} + 2\ e^- \rightarrow Zn$ ($E° = -0.7618$ V), during recharging the reduction of H_3O^+ occurs:

$$2\ H_3O^+\ (aq) + 2\ e^- \rightarrow H_2 + 2\ H_2O \qquad (E° = 0\ V)$$

In neutral H_2O, the H_3O^+ ion concentration is only 10^{-7} M, but the resulting E is still less negative than $E°$ for Zn^{2+}/Zn.

17.59 (a) To determine where to attach the negative wire from the charger, examine the half-reactions of the battery when it is producing current:

$$\text{Cathode: NiO(OH)}\ (s) + H_2O\ (l) + e^- \rightarrow Ni(OH)_2\ (s) + OH^-\ (aq)$$

$$\text{Anode: Cd}\ (s) + 2\ OH^-\ (aq) \rightarrow Cd(OH)_2\ (s) + 2\ e^-$$

To recharge, electrons must be supplied to the cadmium electrode, so this is the electrode to which the negative wire (anode) from the charger should be attached.

The reaction is $Cd(OH)_2$ (s) + 2 e$^-$ → Cd (s) + 2 OH$^-$ (aq)

(b) The connection between electric current and chemical amounts is provided by Equation 17-7, $n = \dfrac{It}{F}$, where n is moles of electrons flowing. Thus $t = \dfrac{nF}{I}$

$$n_{Cd(OH)_2} = 1.55\ g\left(\frac{1\ mol}{146.4\ g}\right) = 0.0106\ mol$$

$$n_{e^-} = 2\,n_{Cd(OH)_2} = 0.0212\ mol$$

$$t = 0.0212\ mol\left(\frac{9.6485 \times 10^4\ C}{1\ mol}\right)\left(\frac{1\ s}{125 \times 10^{-3}\ C}\right) = 1.64 \times 10^4\ s$$

Convert to hours:

$$1.64 \times 10^4\ s\left(\frac{1\ min}{60\ s}\right)\left(\frac{1\ h}{60\ min}\right) = 4.56\ h$$

17.61 Quantity of charge can be calculated once the amount of electron flow has been determined. Use the half-reaction to determine the relationship between amount of chemical change and amount of electron flow: $Ag^+ + e^- \rightarrow Ag$

$$n_e = n_{Ag} = (12.89\ g - 10.77\ g)\left(\frac{1\ mol}{107.9\ g}\right) = 0.0196\ mol\ e^-$$

Charge = nF = (0.0196 mol)(9.6485 × 10^4 C/mol) = 1.89 × 10^3 C. Determine the current by dividing the charge by the time in seconds:

$$t = 15.0\ min\left(\frac{60\ s}{1\ min}\right) = 900\ s \qquad I = \frac{Charge}{t} = \frac{1.89 \times 10^3\ C}{900\ s} = 2.10\ A$$

17.63 A molecular picture must show the species undergoing redox reactions and the direction of electron flow. In a silver coulometer, both electrodes are silver. Ag is oxidized to Ag^+ at the anode, and Ag^+ is reduced to Ag at the cathode.

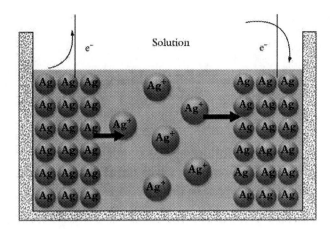

17.65 $2 \text{ Cl}^-(aq) \rightarrow \text{Cl}_2(g) + 2 \text{ e}^-$

1 million tonnes = 1 x 10^6 tonnes = 1 x 10^9 kg = 1 x 10^{12} g

$$\frac{1 \times 10^{12}\,\text{g}}{70.91\,\text{g mol}^{-1}} = 1.41 \times 10^{10}\,\text{mol Cl}_2$$

Charge required = 1.41 x 10^{10} mol Cl$_2$ $\times \dfrac{2 \text{ mol e}^-}{1 \text{ mol Cl}_2} \times \dfrac{96487 \text{ C}}{\text{mol e}^-} = 2.72 \times 10^{15}\,\text{C}$

1 V is 1 J C^{-1} so 1 J = 1 C x 1 V

Energy required = (2.72 x 10^{15} C)(4.5 V) = 1.22 x 10^{16} J

$$\text{kWh} = (1.22 \text{ x } 10^{16}\,\text{J}) \left(\frac{1\,\text{kWh}}{3.6 \times 10^6\,\text{J}} \right) = 3.4 \text{ x } 10^9\,\text{kWh}$$

17.67 An equilibrium constant can be calculated from standard reduction potentials if half-reactions can be combined to give the equilibrium reaction whose constant is desired.

The equilibrium is $\text{Zn}^{2+} + 4\,\text{NH}_3 \rightarrow [\text{Zn(NH}_3)]_4^{2+}$

Appendix F provides these standard potentials:

$\text{Zn}^{2+} + 2\,\text{e}^- \rightarrow \text{Zn}$ $\qquad\qquad\qquad E^\circ = -0.7618$ V

$[\text{Zn(NH}_3)]_4^{2+} + 2\,\text{e}^- \rightarrow \text{Zn} + 4\,\text{NH}_3 \qquad E^\circ = -1.04$ V

Subtract the second half-reaction from the first to give the standard reduction potential corresponding to the desired equilibrium:

$E^\circ = (-0.7618 \text{ V}) - (-1.04 \text{ V}) = +0.2782$ V

Now use the relationship between K_{eq} and $E°$ to determine the equilibrium constant:

$$\log K_{eq} = \frac{nE°}{0.0592 \text{ V}} = \frac{(2)(0.2782 \text{ V})}{0.0592 \text{ V}} = 9.405$$

$$K_{eq} = 10^{9.405} = 2.5 \times 10^9$$

17.69 (a) Balance redox reactions following the standard procedure. Break into half-reactions, balance each half-reaction using the stepwise technique, and then multiply by appropriate integers so the electrons cancel when the half-reactions are combined.

$$H_2 \rightarrow H_2O$$

(1) No elements except O and H; (2) add 1 H_2O on the left; (3) add 2 H_2O on the right and 2 OH^- on the left; (4) add 2 e^- on the right:

$$H_2 + 2 \, OH^- + H_2O \rightarrow 3 \, H_2O + 2 \, e^-$$

Cancel duplicated species: $H_2 + 2 \, OH^- \rightarrow 2 \, H_2O + 2 \, e^-$

$$Cr(OH)_3 \rightarrow Cr$$

(1) Cr is balanced; (2) add 3 H_2O on the right; (3) add 3 H_2O on the left and 3 OH^- on the right; (4) add 3 e^- on the left:

$$Cr(OH)_3 + 3 \, H_2O + 3 \, e^- \rightarrow Cr + 3 \, H_2O + 3 \, OH^-$$

Cancel duplicated species: $Cr(OH)_3 + 3 \, e^- \rightarrow Cr + 3 \, OH^-$

Multiply the first reaction by 3 and the second reaction by 2, and then add:

$$3 \, [H_2 + 2 \, OH^- \rightarrow 2 \, H_2O + 2 \, e^-]$$
$$+ \, 2 \, [Cr(OH)_3 + 3 \, e^- \rightarrow Cr + 3 \, OH^-]$$
$$\overline{3 \, H_2 + 2 \, Cr(OH)_3 \rightarrow 6 \, H_2O + 2 \, Cr}$$

(b) To determine the standard potential, consult values in Appendix F, and subtract $E°$ for the oxidation from $E°$ for the reduction:

$$2 \, H_2O + 2 \, e^- \rightarrow H_2 + 2 \, OH^- \qquad E° = -0.828 \text{ V}$$

$$Cr(OH)_3 + 3 \, e^- \rightarrow Cr + 3 \, OH^- \qquad E° = -1.48 \text{ V}$$

$$E°_{cell} = (-1.48 \text{ V}) - (-0.828 \text{ V}) = -0.65 \text{ V}$$

(c) Calculate $\Delta G°$ from $\Delta G° = -nFE°_{cell}$:

$$\Delta G^\circ = -(6 \text{ mol})(9.6485 \times 10^4 \text{ C/mol})(-0.65 \text{ V})(10^{-3} \text{ kJ/J}) = 3.8 \times 10^2 \text{ kJ}$$

17.71 (a) Balance the half-reaction using the stepwise technique:

$$Cr_2O_7^{2-} \rightarrow Cr^{3+}$$

(1) Multiply Cr^{3+} by 2; (2) add 7 H_2O on the right; (3) add 14 H_3O^+ on the left and 14 H_2O on the right; (4) add 6 e^- on the left:

$$Cr_2O_7^{2-} + 14 \text{ } H_3O^+ \text{ } (aq) + 6 \text{ } e^- \rightarrow 2 \text{ } Cr^{3+} + 21 \text{ } H_2O$$

(b) It takes 2 moles of K_2CrO_4 to generate 1 mole of $Cr_2O_7^{2-}$, which in turn consumes 6 moles of e^-, so 1 mole of K_2CrO_4 consumes 3 moles of e^-:

$$0.250 \text{ mol} \left(\frac{1 \text{ mol K}_2\text{CrO}_4)}{3 \text{ mol e}^-} \right) = 0.0833 \text{ mol K}_2\text{CrO}_4 \text{ required}$$

$$m = nM = 0.0833 \text{ mol} \left(\frac{194 \text{ g}}{1 \text{ mol}} \right) = 16.2 \text{ g of K}_2\text{CrO}_4$$

17.73 Identify what is taking place in a galvanic cell by identifying the species present:

(a) On the left: $Ni^{2+} + 2 \text{ } e^- \rightarrow Ni$ $E^\circ = -0.257 \text{ V}$

     On the right: $Fe^{2+} + 2 \text{ } e^- \rightarrow Fe$ $E^\circ = -0.447 \text{ V}$

(b) To generate a positive overall potential, the iron half-reaction operates as oxidation:

$$Fe + Ni^{2+} \rightarrow Ni + Fe^{2+} \qquad E^\circ = (-0.257 \text{ V}) - (-0.447 \text{ V}) = 0.190 \text{ V}$$

(c) Oxidation occurs at the anode, so the Fe electrode is the anode and Ni is the cathode.

(d) A molecular picture of an electrochemical cell must show the species undergoing redox reactions and the direction of electron flow:

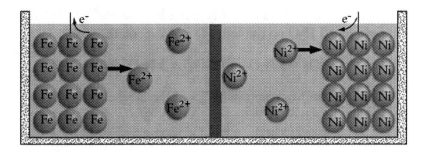

17.75 Use the Nernst equation to determine the concentration at which $E = 0.0 \text{ V}$:

$$Fe + Ni^{2+} \rightarrow Ni + Fe^{2+} \qquad E^\circ = (-0.257 \text{ V}) - (-0.447 \text{ V}) = 0.190 \text{ V}$$

$$E = 0.0 \text{ V} = E° - \frac{0.0592 \text{ V}}{n} \log\left(\frac{[\text{Fe}^{2+}]}{[\text{Ni}^{2+}]}\right) \qquad \log\left(\frac{[\text{Fe}^{2+}]}{[\text{Ni}^{2+}]}\right) = \frac{nE°}{0.05916 \text{ V}}$$

The concentration of reactant must be reduced to reach a potential of 0.0 V:

$$\log\left(\frac{1.00}{[\text{Ni}^{2+}]}\right) = \frac{(2)(0.190 \text{ V})}{0.0592 \text{ V}} = 6.42 \qquad [\text{Ni}^{2+}] = 10^{-6.42} = 3.8 \times 10^{-7} \text{ M}$$

17.77 Balance redox reactions following the standard procedure. Break into half-reactions, balance each half-reaction using the stepwise technique, and then multiply by appropriate integers so that the electrons cancel when the half-reactions are combined:

$$\text{MnO}_4^- \rightarrow \text{Mn}^{2+}$$

(1) Mn is already balanced; (2) add 4 H_2O on the right; (3) add 8 H_3O^+ on the left and 8 H_2O on the right; (4) add 5 e$^-$ on the left:

$$\text{MnO}_4^- + 8 \text{ H}_3\text{O}^+ (aq) + 5 \text{ e}^- \rightarrow \text{Mn}^{2+} + 12 \text{ H}_2\text{O}$$

(a) $H_2SO_3 \rightarrow HSO_4^-$

(1) S is already balanced; (2) add 1 H_2O on the left; (3) add 3 H_3O^+ on the right and 3 H_2O on the left; (4) add 2 e$^-$ on the right:

$$\text{H}_2\text{SO}_3 + 4 \text{ H}_2\text{O} \rightarrow \text{HSO}_4^- + 3 \text{ H}_3\text{O}^+ (aq) + 2 \text{ e}^-$$

Multiply the Mn reaction by 2 and the S reaction by 5, and add:

$$2[\text{MnO}_4^- + 8 \text{ H}_3\text{O}^+ (aq) + 5 \text{ e}^- \rightarrow \text{Mn}^{2+} + 12 \text{ H}_2\text{O}]$$
$$+ 5[\text{H}_2\text{SO}_3 + 4 \text{ H}_2\text{O} \rightarrow \text{HSO}_4^- + 3 \text{ H}_3\text{O}^+ (aq) + 2 \text{ e}^-]$$
$$\overline{2 \text{ MnO}_4^- + 5 \text{ H}_2\text{SO}_3 + \text{H}_3\text{O}^+ (aq) \rightarrow 2 \text{ Mn}^{2+} + 5 \text{ HSO}_4^- + 4 \text{ H}_2\text{O}}$$

(b) $SO_2 \rightarrow HSO_4^-$

(1) S is already balanced; (2) add 2 H_2O on the left; (3) add 3 H_3O^+ on the right and 3 H_2O on the left; (4) add 2 e$^-$ on the right:

$$\text{SO}_2 + 5 \text{ H}_2\text{O} \rightarrow \text{HSO}_4^- + 3 \text{ H}_3\text{O}^+ (aq) + 2 \text{ e}^-$$

Multiply the Mn reaction by 2 and the S reaction by 5, and add:

$$2[\text{MnO}_4^- + 8 \text{ H}_3\text{O}^+ (aq) + 5 \text{ e}^- \rightarrow \text{Mn}^{2+} + 12 \text{ H}_2\text{O}]$$
$$+ 5[\text{SO}_2 + 5 \text{ H}_2\text{O} \rightarrow \text{HSO}_4^- + 3 \text{ H}_3\text{O}^+ (aq) + 2 \text{ e}^-]$$
$$\overline{2 \text{ MnO}_4^- + 5 \text{ SO}_2 + \text{H}_2\text{O} + \text{H}_3\text{O}^+ (aq) \rightarrow 2 \text{ Mn}^{2+} + 5 \text{ HSO}_4^-}$$

(c) $H_2S \rightarrow HSO_4^-$

(1) S is already balanced; (2) add 4 H_2O on the left; (3) add 9 H_3O^+ on the right and 9 H_2O on the left; (4) add 8 e^- on the right:

$$H_2S + 13\,H_2O \rightarrow HSO_4^- + 9\,H_3O^+\,(aq) + 8\,e^-$$

Multiply the Mn reaction by 8 and the S reaction by 5, and add:

$$8[MnO_4^- + 8\,H_3O^+\,(aq) + 5\,e^- \rightarrow Mn^{2+} + 12\,H_2O]$$
$$+\,5[H_2S + 13\,H_2O \rightarrow HSO_4^- + 9\,H_3O^+\,(aq) + 8\,e^-]$$
$$\overline{8\,MnO_4^- + 5\,H_2S + 19\,H_3O^+\,(aq) \rightarrow 8\,Mn^{2+} + 5\,HSO_4^- + 31\,H_2O}$$

(d) $H_2S_2O_3 \rightarrow HSO_4^-$

(1) Multiply HSO_4^- by 2; (2) add 5 H_2O on the left; (3) add 10 H_3O^+ on the right and 10 H_2O on the left; (4) add 8 e^- on the right:

$$H_2S_2O_3 + 15\,H_2O \rightarrow 2\,HSO_4^- + 10\,H_3O^+\,(aq) + 8\,e^-$$

Multiply the Mn reaction by 8 and the S reaction by 5, and add:

$$8[MnO_4^- + 8\,H_3O^+\,(aq) + 5\,e^- \rightarrow Mn^{2+} + 12\,H_2O]$$
$$+\,5[H_2S_2O_3 + 15\,H_2O \rightarrow 2\,HSO_4^- + 10\,H_3O^+\,(aq) + 8\,e^-]$$
$$\overline{8\,MnO_4^- + 5\,H_2S_2O_3 + 14\,H_3O^+\,(aq) \rightarrow 8\,Mn^{2+} + 10\,HSO_4^- + 21\,H_2O}$$

17.79 The connection between electric current and chemical amounts is provided by Equation 16-7, $n = \dfrac{It}{F}$, where n is moles of electrons flowing. Thus, $t = \dfrac{nF}{I}$.

The reaction is $Cu^{2+} + 2\,e^- \rightarrow Cu$

$$n_{Cu} = 0.250\,L\left(\frac{0.245\,mol}{1\,L}\right) = 0.06125\,mol \qquad n_{electron} = 2\,n_{Cu} = 0.1225\,mol$$

$$t = 0.1225\,mol\left(\frac{9.6485 \times 10^4\,C}{1\,mol}\right)\left(\frac{1\,s}{2.45\,C}\right)\left(\frac{1\,min}{60\,s}\right) = 80.4\,min$$

17.81 The process in Problem 16.79 is electrodeposition, presumably with Cu serving as both electrodes. Reduction of Cu^{2+} to Cu occurs at the cathode, and oxidation of Cu to Cu^{2+} occurs at the anode:

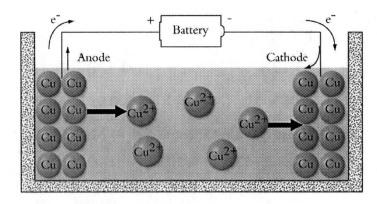

17.83 The better oxidizing agent under standard conditions is the reactant having the more positive reduction potential:

(a) MnO_4^- (1.679 V vs. 1.232 V)

(b) O_2 (0.401 V vs. 0.109 V)

(c) Sn^{2+} (–0.137 V vs. –0.447 V)

17.85 Under standard conditions, a substance can oxidize any other substance that appears on the RIGHT side of a half-reaction whose reduction potential is less positive: For O_2, $E° = 0.401$ V; all the substances listed in Problem 17.84 (except for Co^{2+} and H_2) have less positive reduction potentials, so O_2 can oxidize Cu, Ag, Fe^{2+}, and I^- (not surprisingly, oxygen is a good oxidizing agent!).

17.87 Although the anions are different in the two solutions, the redox process involves only Zn metal and Zn^{2+} cations, so this is a concentration cell:

(a) In the $Zn(NO_3)_2$ solution, $Zn^{2+} + 2\,e^- \rightarrow Zn$

In the $ZnCl_2$ solution, $Zn \rightarrow Zn^{2+} + 2\,e^-$

(b) A molecular picture shows Zn^{2+} cations depositing on the electrode from the 1.25 M solution and dissolving off the electrode into the 0.250 M solution:

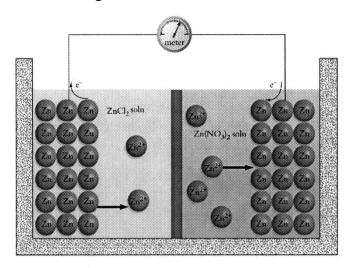

(c) To calculate a cell potential from concentration measurements, use the Nernst equation. For a concentration cell, $E° = 0$ V, so $E = -\dfrac{0.0592 \text{ V}}{n} \log Q$.

Both reactions are $Zn^{2+} + 2 \text{ e}^- \rightarrow Zn$, and the concentrations are 0.250 M and 1.25 M:

$$Zn^{2+}{}_{(1.25 \text{ M})} \rightarrow Zn^{2+}{}_{(0.250 \text{ M})} \qquad n = 2 \qquad Q = \dfrac{0.250 \text{ M}}{1.25 \text{ M}}$$

$$E = -\left(\dfrac{0.0592 \text{ V}}{2}\right) \log\left(\dfrac{0.250 \text{ M}}{1.25 \text{ M}}\right) = 0.0207 \text{ V}$$

17.89 The connection between time and amount of electrons is provided by Equation 17-7:

$$n = \dfrac{It}{F} \qquad\qquad t = \dfrac{nF}{I}$$

From the half-reaction, $n_{\text{electron}} = n_{MnO_2}$. Only 90% of the MnO_2 is available before the battery fails:

$$n_{MnO_2} = 4.0 \text{ g}\left(\dfrac{1 \text{ mol}}{86.94 \text{ g}}\right)\left(\dfrac{90\%}{100\%}\right) = 0.0414 \text{ mol}$$

$$t = 0.0414 \text{ mol}\left(\dfrac{9.6485 \times 10^4 \text{ C}}{1 \text{ mol}}\right)\left(\dfrac{1 \text{ s}}{0.0048 \text{ C}}\right)\left(\dfrac{1 \text{ min}}{60 \text{ s}}\right)\left(\dfrac{1 \text{ h}}{60 \text{ min}}\right) = 2.3 \times 10^2 \text{ h}$$

17.91 Use the Nernst equation to determine a concentration from cell voltages. First determine the net reaction in order to find n, $E°$, and Q:

Reactions: $\quad Fe^{3+} + \text{e}^- \rightarrow Fe^{2+} \qquad\qquad E° = 0.771 \text{ V}$

$\qquad\qquad\quad Cu^{2+} + 2 \text{ e}^- \rightarrow Cu \qquad\qquad E° = 0.3419 \text{ V}$

Multiply the first reaction by 2 and subtract the second:

$$2 Fe^{3+} + Cu \rightarrow 2 Fe^{2+} + Cu^{2+} \qquad E° = (0.771 \text{ V}) - (0.3419 \text{ V}) = 0.429 \text{ V}$$

$$E = E° - \dfrac{0.0592 \text{ V}}{n} \log Q = E° - \left(\dfrac{0.0592 \text{ V}}{n}\right) \log\left(\dfrac{[Fe^{2+}]^2[Cu^{2+}]}{[Fe^{3+}]^2}\right)$$

Substitute values and solve for the unknown concentration:

$$0.00 \text{ V} = 0.429 \text{ V} - \left(\dfrac{0.0592 \text{ V}}{2}\right) \log\left(\dfrac{[1.00]^2[1.00]}{[Fe^{3+}]^2}\right)$$

$$-\log\left([Fe^{3+}]^2\right) = \frac{(2)(0.429\ V)}{0.0592\ V} = 14.50 \quad [Fe^{3+}]^2 = 3.2 \times 10^{-15} \quad [Fe^{3+}] = 5.6 \times 10^{-8}\ M$$

17.93 (a) In molten NaCl, the only species present are Na^+ cations and Cl^- anions; the Pt electrodes are passive, so the reactions are $Na^+ + e^- \rightarrow Na$, and $2\ Cl^- \rightarrow Cl_2 + 2\ e^-$.

(b) Reduction is driven by electrons supplied at the negative terminal, so the cathode is the Pt electrode connected to the negative pole of the battery, and the anode is the Pt electrode connected to the positive pole of the battery.

(c)

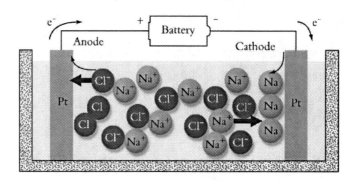

17.95 A sketch of a cell that shows molecular processes should show the molecular species undergoing redox reactions and the direction of electron flow:

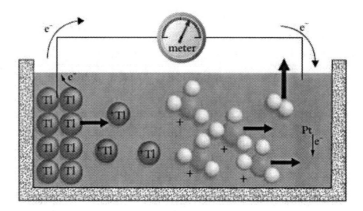

17.97 The standard potential for a redox reaction is easily calculated from the equilibrium constant for the reaction by using Equation 17-5, $E^\circ = \dfrac{0.0592\ V}{n} \log K_{eq}$:

$$E^\circ = \frac{1}{2}(0.0592\ V) \log (2.7 \times 10^{12}) = (0.02958)(12.43) = 0.368\ V$$

17.99 To determine the redox chemistry taking place in an electrochemical cell, start with the species present in each compartment. The first compartment contains an Ag/AgCl electrode and 0.500 M HCl, and the second compartment contains an Mg electrode and 1.00 M MgCl$_2$ solution.
Compartment 1: $AgCl + e^- \ \square \quad Ag + Cl^-$

Compartment 2: $Mg^{2+} + 2\,e^- \;\square\;\; Mg$

(a) Net reaction (Mg^{2+}/Mg has a more negative $E°$):

$Mg + 2\,AgCl \;\square\;\; Mg^{2+} + 2\,Ag + 2\,Cl^-$

(b) Use the Nernst equation to calculate the cell emf:

$$E = E° - \left(\frac{0.0592}{n}\right)\log([Cl^-]^2[Mg^{2+}])$$

$$E° = (0.22233\ \text{V}) - (-2.37\ \text{V}) = 2.59\ \text{V}$$

$$E = 2.59\ \text{V} - \left(\frac{0.0592}{n}\right)\log[(0.500)^2(1.00)] = 2.59\ \text{V} + 0.018\ \text{V} = 2.61\ \text{V}$$

(c) A molecular picture of an electrochemical cell must show the species undergoing redox reactions and the direction of electron flow:

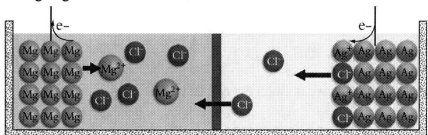

17.101   The most negative half-reaction potentials are around –3 V, so to generate a net cell potential greater than 5 V requires a half-reaction with a positive potential greater than +2 V. There is only one half-reaction in Table 17-1 that meets these requirements:

$F_2 + 2\,e^- \;\square\;\; 2\,\text{F-}$ $\hspace{5cm}$ $E° = 2.866$ V

This half-reaction could be coupled with any of the following:

$Mg^{2+} + 2\,e^- \;\square\;\; Mg$ $\hspace{4cm}$ $E° = -2.37$ V

$Ca^{2+} + 2\,e^- \;\square\;\; Ca$ $\hspace{4.1cm}$ $E° = -2.868$ V

$Na^+ + e^- \;\square\;\; Na$ $\hspace{4.6cm}$ $E° = -2.71$ V

$Li^+ + e^- \;\square\;\; Li$ $\hspace{4.8cm}$ $E° = -3.0401$ V

$Ca(OH)_2 + 2\,e^- \;\square\;\; Ca + 2\,OH^-$ $\hspace{2.4cm}$ $E° = -3.02$ V

The problem with any of these batteries is that they would have to contain F_2 gas, which would require a large, heavy container. In addition, fluorine is corrosive and highly toxic.

17.103   Standard electrode potentials allow calculation of equilibrium constants if the half-reactions can be combined to give the equilibrium reaction whose constant is desired. In this case, subtract the second reaction from the first and divide by 4:

$$[O_2 + 2\,H_2O + 4\,e^- \rightarrow 4\,OH^-] \qquad\qquad E° = 0.401\ V$$

$$-[O_2\,(g) + 4\,H_3O^+\,(aq) + 4\,e^- \rightarrow 6\,H_2O] \qquad E° = 1.229\ V$$

$$\text{Net}:\ 8\,H_2O\,(l) \rightarrow 4\,OH^-\,(aq) + 4\,H_3O^+\,(aq)$$

$$E°_{reaction} = (0.401\ V) - (1.229\ V) = -0.828\ V$$

Dividing by 4 changes n_e but not $E°$:

$$2\,H_2O\,(l) \rightarrow OH^-\,(aq) + H_3O^+\,(aq) \qquad n_e = 1 \qquad E° = -0.828\ V$$

Use Equation 17-5, rearranged:

$$\log K_w = \frac{nE°}{0.0592\ V} = \frac{(1)(-0.828V)}{0.0592\ V} = -14.00$$

$$K_w = 1.0 \times 10^{-14} = [OH^-][H_3O^+]$$

17.105 (a) The charge that flows through the cell is:

$$q_{cell} = it$$

$$= 6.00\ C\,s^{-1} \times 10.0\ min \times \left(\frac{60\ s}{min}\right) = 3{,}600\ C$$

The charge consumed by the desired reduction reaction is calculated from the mass of copper deposited:

$$\frac{0.80\ g\ Cu}{63.55\ g\ mol^{-1}} = 0.0126\ mol\ Cu$$

Reduction of one mole of copper ions requires two moles of electrons:

$$Cu^{2+}_{(aq)} + 2\ e^- \rightarrow Cu_{(s)}$$

Thus, n = 2(0.0126 mol) = 0.0252 mol electrons

The charge used to reduce the copper is therefore:

$$q_{Cu} = nF$$

$$= 0.0252\ mol(96{,}487\ C\ mol^{-1})$$

$$= 2{,}431\ C$$

And the current efficiency is:

$$\text{Current efficiency} = \frac{2{,}431\ C}{3{,}600\ C} = 0.68$$

(b) The other reduction reactions are:

$$2\,H_3O^+\,(aq) + 2\ e^- \rightarrow H_2\,(aq) + 2\,H_2O\,(l)$$

NO_3^- *(aq)* $+ 4 H_3O^+$ *(aq)* $+ 3 e^- \rightarrow NO$ *(g)* $+ 6 H_2O$ *(l)*

From these reactions, we see that 5 moles of electrons produce one mole of hydrogen gas and one mole of nitric oxide. The total charge that is used in these reduction reactions in a ten minute period is (3,600 – 2,431) = 1,169 C. This corresponds to a number of moles of electrons equal to

$$\frac{1,169 \text{ C}}{96,487 \text{ C mol}^{-1}} = 0.0121 \text{ mol}$$

The number of moles of each gas produced in ten minutes is therefore 0.0121 / 5 = 0.00242 mol, which is 0.0145 mol per hour.

At 1 bar (= 10^5 Pa) and 30°C, the volumes of each gas produced per hour are:

$$V = \frac{nRT}{p}$$

$$= \frac{0.0145 \text{ mol}(0.08314 \text{ L bar K}^{-1}\text{mol}^{-1})(30 + 273.15)\text{K}}{1 \text{ bar}}$$

$$= 0.37 \text{ L}$$

Chapter 18 Macromolecules

Solutions to Problems in Chapter 18

18.1 Convert line drawings to structural formulas following the standard rules, adding a C atom at each vertex and line end and adding –H bonds until each C atom has four bonds. Identify functional groups from memory or from Table 18-1 in your textbook:

18.3 Identify functional groups from memory and convert structures to line drawings following the standard rules:

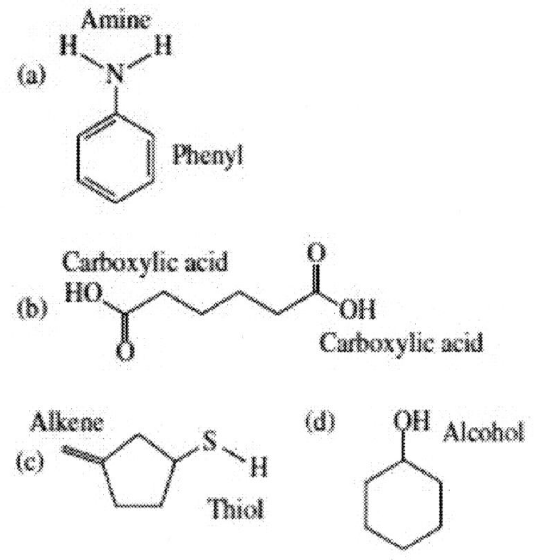

(a) Amine, Phenyl

(b) Carboxylic acid, Carboxylic acid

(c) Alkene, Thiol

(d) Alcohol

18.5 Convert line and ball-and-stick drawings to structural formulas following the standard rules, adding a C atom at each vertex and line end and adding –H bonds until each C atom has four bonds. Identify linkage groups from memory or Table 17-1 in your textbook:

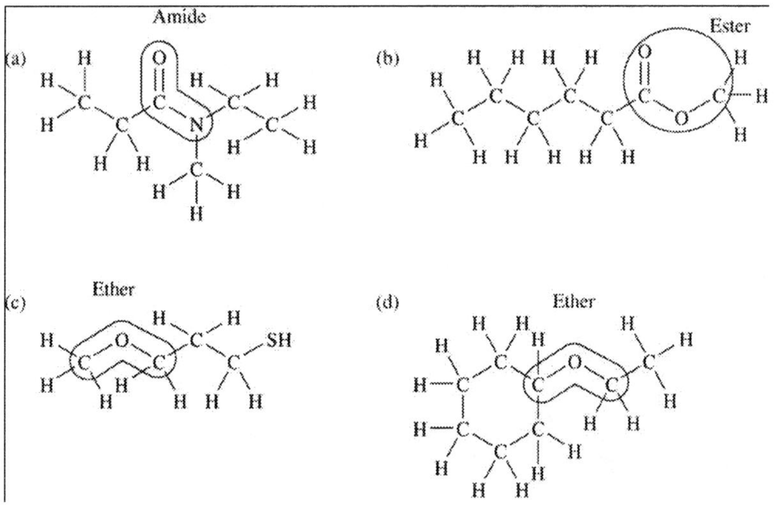

(a) Amide

(b) Ester

(c) Ether

(d) Ether

18.7 To draw examples of compound types, first identify the chemical formulas of its functional groups. Remember that carbon always forms four bonds. There are many possible correct answers to this problem. We provide just one example for each.

(a) An amine contains N bonded to C. One example is the tertiary amine diethylmethylamine:

Other isomers are also possible.

(b) An ester contains C–CO$_2$–C. There must be five additional C atoms, arranged in any fashion:

Other isomers are also possible.

(c) The functional group in an aldehyde is O=C–H, M = 29 g/mol. To have a molar mass greater than 80, an aldehyde must contain at least four additional C atoms:

Others are also possible.

(d) The ether linkage is –O–, and phenyl is the benzene ring:

Others are also possible.

18.9 In a condensation reaction, two monomers react to form a larger unit, eliminating a small molecule such as H$_2$O in the process.

(a) The amide linkage in the centre of this molecule forms in a condensation reaction between a carboxylic acid and an amine:

(b) The ester linkage in this molecule forms in a condensation reaction between a carboxylic acid and an alcohol:

(c) The ether linkage in this molecule forms in a condensation reaction between two alcohols:

18.11 All amino acids can condense with one another in two ways. In addition, the extra carboxylic acid of aspartic acid could undergo condensation, so there are three possible products:

18.13 Polymers made from substituted ethylenes form by breakage of the C=C double bond to form two new single C–C bonds to other monomers. In a copolymer, each monomer occurs randomly:

18.15 To deconstruct an alkane polymer into its monomers, break C–C single bonds and form double bonds:

18.17 Polybutadiene forms from butadiene by a combination of double-bond breakage and migration. The resulting polymer differs from polyethylene in having one C=C double bond in each repeat unit, whereas polyethylene is entirely CH_2 units connected by C–C single bonds (each arrow represents shifting one electron):

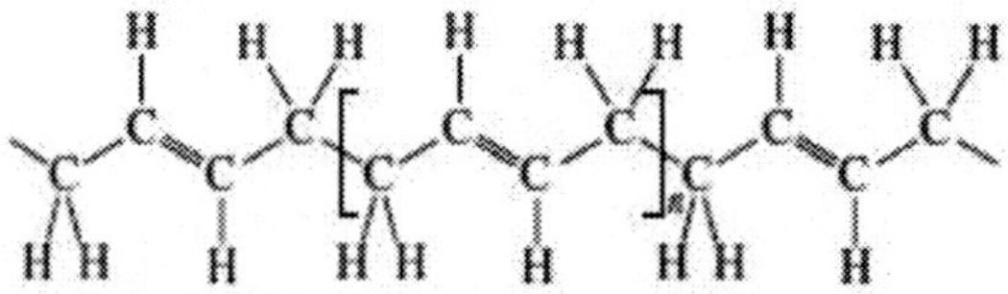

18.19 To determine the monomers from which a condensation polymer has formed, decompose the polymer into its monomeric parts and add components of a small molecule, for example –H and –OH. Here, the structure shows a single component whose repeat unit is $(CH_2)_{10}$, and the linkage is an amide, which breaks down into an amine and a carboxylic acid:

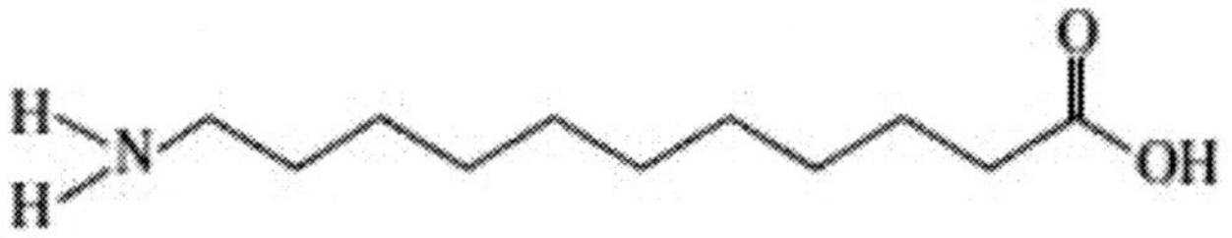

18.21 Construct polyethylene oxide by breaking a C–O bond and linking the fragments:

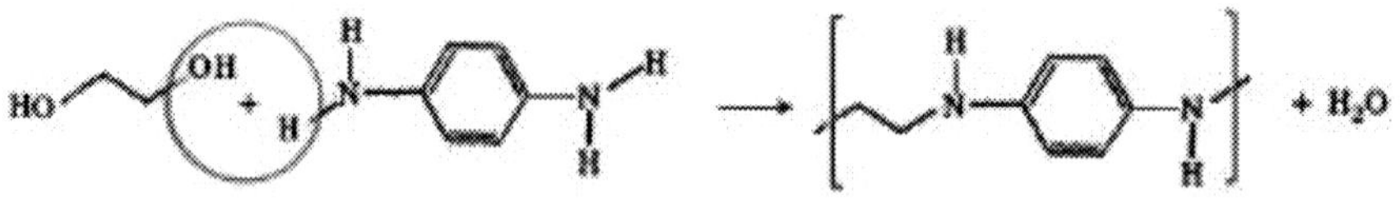

18.23 Cross-linking leads to increased rigidity, because chemical bonds between polymer chains restrict the ability of polymer chains to slide past one another. Thus, relatively rigid tires have more extensive cross-linking than flexible surgeon's gloves.

18.25 The categories of polymers and their characteristic properties are as follows: plastics, which exist as blocks or sheets; fibres, which can be drawn into long threads; and elastomers, which can be stretched without breaking.

(a) Balloons must stretch, so they are made of elastomers.

(b) Rope is made of fibres.

(c) Camera cases are rigid, so they are made of plastics.

18.27 Dioctylphthalate is an example of a liquid plasticizer, which reduces the amount of cross-linking in a polymer as well as adding a fluid component; both these changes result in improved flexibility of the polymer.

18.29 To draw the structures of sugars that are related to glucose, start with the glucose structure and make modifications as needed. For α-talose, move the OH group on carbons 2 and 4 from down positions to up positions.

α-Glucose   α-Talose

18.31 To switch between α and β isomers, move the OH group on carbon 1 from a down position to an up position (or vice versa when going from β to α). To draw β-talose, start with the structure of α-talose from Problem 18.29 and move the OH on carbon 1 from a down position to an up position.

β-talose

18.33 As described in your textbook, polymers of α-glucose coil upon themselves. Glycogen is this type of polymer. Polymers of β-glucose, of which cellulose is an example, form planar sheets. It is easier to see the distinction between the two polymers by looking at them from a side-on view (as shown). In the drawings, the dark solid lines indicate the direction of the continuing chain:

Cellulose   Glycogen

18.35 The complementary strands of DNA form from hydrogen-bond linkages between specific nucleic acids. A pairs with T, and G pairs with C, so the complementary sequence of A-A-T-G-C-A-C-T-G is T-T-A-C-G-T-G-A-C.

18.37 The structure of DNA consists of a phosphate–sugar–nucleotide-base trio bonded to others through its phosphate and sugar groups. Apart from the structure of the nucleotide base, each unit is identical:

18.39 The strands of DNA fit together as a result of the hydrogen-bonding interactions illustrated in Figure 18-27. The complementary sequence for A-T-C is T-A-G. The backbones of complementary strands run in opposite directions. For clarity in showing the hydrogen-bonding interactions, we represent the backbone with a solid line rather than showing its details:

18.41 Consult Figure 18-31 for the structures of the different amino acids, all of which have the same amino acid backbone.

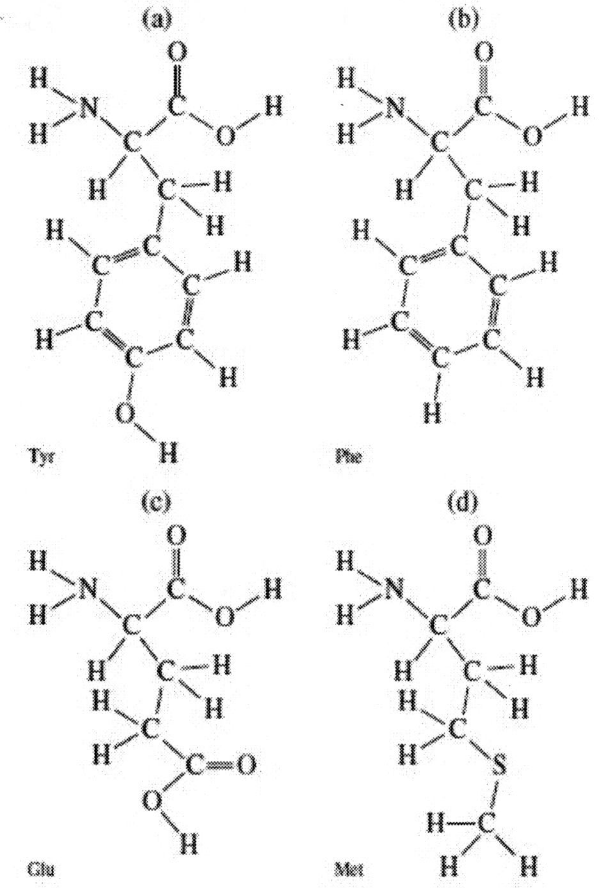

18.43 Hydrophilic side chains are characterized by the presence of N or O atoms that generate polar bonds and hydrogen-bonding capability, or an S–H bond that is polar. Among the structures shown in Problem 18.41, Tyr (O–H bond) and Glu (CO_2H group) are hydrophilic, whereas Phe and Met are hydrophobic.

18.45 To identify an amino acid, examine the side chain attached to the carbon atom between the N atom and the C=O bond in the amino acid backbone: (a) The side chain is just –H (hydrophobic), making this glycine, the simplest amino acid. (b) The side chain is –CH_2OH (hydrophilic side chain), so this is serine. (c) The side chain is –CH_2SH (hydrophilic side chain), so this is cysteine.

18.47 When two amino acids condense, the amino end of either molecule can link to the carboxylic acid end of the other. A water molecule is eliminated, creating a C–N bond. Three amino acids can combine in six different ways: A–B, B–A, A–C, C–A, B–C, and C–B.

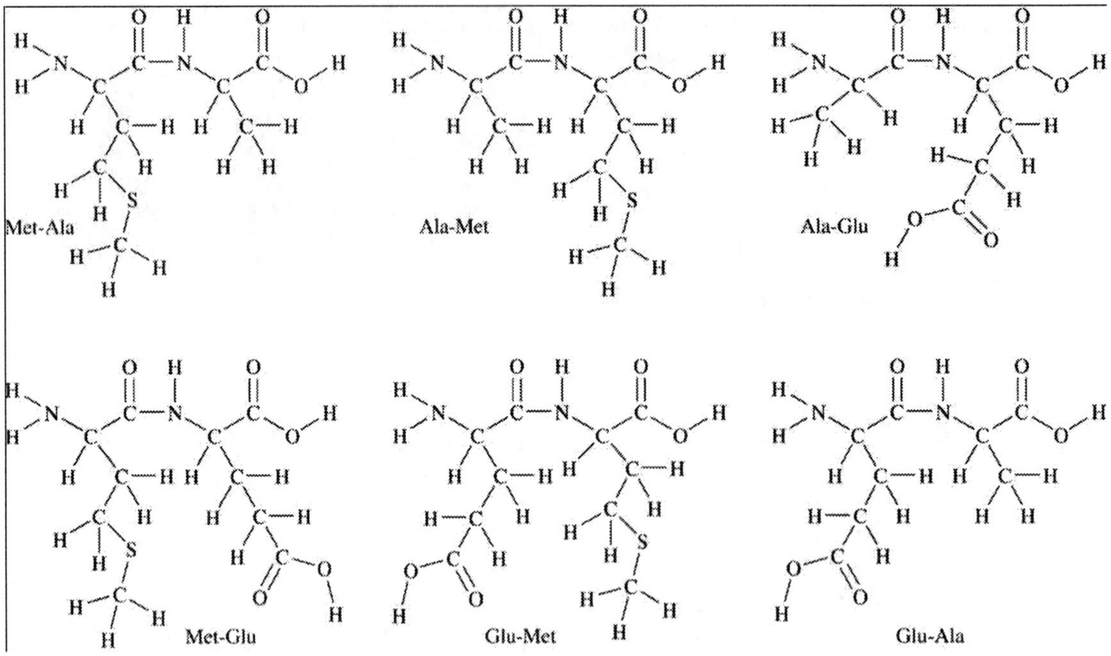

Met-Ala   Ala-Met   Ala-Glu

Met-Glu   Glu-Met   Glu-Ala

18.49 The repeat unit in polyethylene is the ethylene molecule, C_2H_4, $M = 28.1$ g/mol, so a polymer with 744 repeat units has

$$M = (744)(28.1 \text{ g/mol}) = 2.09 \times 10^4 \text{ g/mol.}$$

18.51 There are four possible choices for each base in a DNA strand, so the number of ways to connect 12 bases is $4^{12} = 16\ 777\ 216$ (this is an exact number, because 4 is an exact number).

18.53 (a) The monomers from which hair spray is made are substituted ethylenes, which polymerize in a free radical process in which each ethylene has an equal probability of adding to the growing chain. Thus, this polymer will have a random arrangement.

(b) The backbone of a polyethylene is a carbon chain with the substituents connected to every other carbon atom. Here is a line structure of six monomer units:

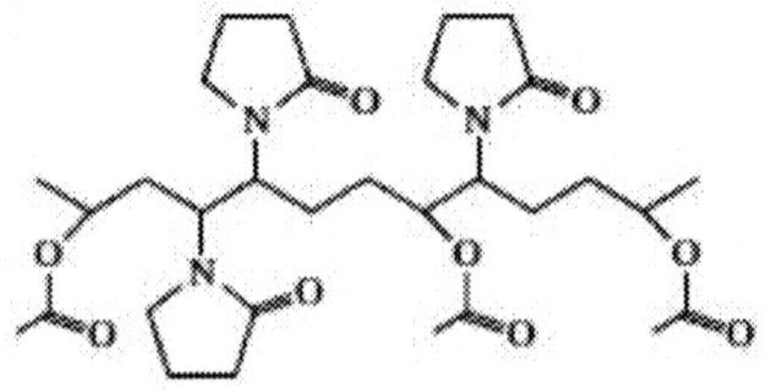

(c) The backbone of this polymer is hydrophobic, but its side chains contain highly polar C=O groups that interact with one another and with polar and hydrogen-bonding groups on hair.

18.55 A nucleotide is a combination of one base, one sugar, and one phosphate group. A duplex is a pair of nucleotides bound together by hydrogen-bonding interactions.

The second nucleotide in a guanine duplex is cytosine.

18.57 The three steps of free radical polymerization are (1) initiation by a free radical initiator, which generates a new free radical; (2) propagation, in which monomer units add to the free radical end of the polymer chain; and (3) termination, in which two free radical chains link together. The functional group on an ethylene monomer does not participate in any of these processes:

18.59 All sugar molecules have multiple –OH functional groups, each of which can participate in hydrogen-bond formation with water molecules. α-Glucose has five –OH groups; in addition, its ring O atom can participate in formation of hydrogen bonds.

18.61 Nylons and proteins share one common feature: they form by condensation reactions between carboxylic acids and amines, so their linkages are amide groups. Otherwise, they are quite different. Nylons contain one or at most two different monomers, each of which typically contains several carbon atoms that form part of the backbone of the polymer. Proteins contain backbones that are absolutely regular repetitions of amide–C–amide bonding, but the carbon atoms in the backbone have a variety of substituent groups attached to them, generating the immense variety of different proteins (compared with only a few different nylons).

18.63 Consult your textbook for the structures of the polymers, which indicate the monomers from which they are made: (a) Kevlar is made from terephthalic acid and phenylenediamine. (b) PET is made from ethylene glycol and terephthalic acid. (c) Styrofoam is the common name for polystyrene, so it is made from styrene.

18.65 All polyethylenes have the same empirical formula, $(CH_2)_n$, but whereas high-density polyethylene has all straight chains that nest together readily, low-density polyethylene has many side chains that cannot nest easily. Thus, low-density polyethylene has more open space, accounting for the lower amount of CH_2 groups per unit volume. See Figure 18-9 for a visual representation.

18.67 The Watson–Crick model of DNA requires that there be a complementary base for each base in a given strand: one G for every C, one C for every G, one A for every T, and one T for every A. This requires that the molar ratios of A to T and G to C be 1.0. Chargoff's observations indicate that this relationship holds even though DNA from different sources has different sequences, some A–T rich and others G–C rich. These differences give rise to differences in the relative amounts of A, T vs. G, C, but pairing always results in 1:1 mole ratios of A to T and G to C.

18.69 The bases of the mRNA molecule must be complementary to the bases of the template DNA on which they are modelled, meaning A generates U, C generates G, G generates C, and T generates A (remember that in RNA, U appears rather than T). Thus, the sequence generated by this strand is the same as that of Strand B except that U replaces T:

1 = cytosine, 2 = uracil, 3 = adenine, and 4 = guanine.

18.71 All proteins contain both hydrophobic and hydrophilic amino acids. When a protein is in contact with a hydrophilic medium such as aqueous solution, its hydrophilic amino acids are most stable when facing outward, in contact with the solvent, whereas its hydrophobic amino acids are most stable when facing inward, in contact with the protein backbone. The opposite is the case when a protein is immersed in a hydrophobic medium such as a cell wall. The tertiary structure of a protein is the manner in which the individual amino acids orient themselves, so solvent interactions are the primary determinants of this tertiary structure.

18.73 To draw the structure of a condensation product, start with the structures of the two molecules undergoing condensation and then connect them together, eliminating a small molecule such as water:

18.75 Polystyrene has the polyethylene backbone of carbon atoms, with a benzene ring attached to every second carbon atom. A divinylbenzene monomer reacts with the growing chain by means of one of its C=C double bonds, leaving the second C=C double bond available to cross-link by becoming incorporated into another polystyrene chain. Thus, the benzene rings form "bridges" between polystyrene chains:

18.77 A cyclic product forms when one end of a monomer molecule can condense with the other end of the same molecule. The monomer illustrated in this problem contains five carbon atoms in addition to its terminal functional groups, so it can easily form a ring containing six carbon atoms and one oxygen atom:

18.79 To determine which amino acids are used to make a dipeptide, examine the structure of the dipeptide and see what *R* groups it contains. The amine end of aspartame contains an extra –CH₂COOH group, characteristic of aspartic acid; the carboxylic acid end of the molecule contains a benzene side group, characteristic of phenylalanine.

18.81 The structure of lactic acid shows that it is bifunctional, with a carboxylic acid and an alcohol group. These groups can condense to form an ester linkage, so the lactic acid polymer is a polyester:

18.83 When amino acids condense to form polypeptides, each added amino acid reacts to eliminate a water molecule. The chemical formula of alanine is $C_3H_7NO_2$, and the net reaction to form poly-Ala is $n\ C_3H_7NO_2 \rightarrow n\ H_2O + (C_3H_5NO)_n$.

Thus, the empirical formula of this polymer is C_3H_5NO, $M = 71$ g/mol.

Divide the molar mass of the polypeptide by the molar mass of the empirical formula unit to obtain the number of repeat units:

$$\frac{1.20 \times 10^3 \text{ g/mol}}{71 \text{ g/mol}} = 17 \text{ repeat units of alanine.}$$

18.85 The monomers from which this *alkyd* forms are an alcohol with three –OH groups and a carboxylic acid with two –COOH groups. These can condense together, eliminating water molecules to form ester linkages. The third –OH allows the possibility of branching in addition to linear structures, and the 2:3 stoichiometry of alcohol to carboxylic acid indicates that all the functional groups undergo condensation. Here is a small part of the structure, built around one starting alcohol molecule:

18.87 Polyethylene is a carbon backbone (with two C–H bonds to each C atom). The butadiene–styrene polymer, while also containing an all-carbon backbone, has benzene rings attached at intervals of approximately six carbon atoms, as well as a double bond in the backbone at intervals of approximately six carbon atoms. The extra bulk of the benzene rings and the geometrical changes where the double bonds exist (trigonal planar as opposed to tetrahedral) make it impossible for the carbon chains to stack as closely together in butadiene–styrene as in polyethylene. With less close stacking, the dispersion forces holding the chains together are weaker, so butadiene–styrene is not as rigid as polyethylene.

Chapter 19 The Transition Metals

Solutions to Problems in Chapter 19

19.1 Oxidation states are determined by applying the rules given in Section 16.1 (Chapter 16). The procedure can be simplified when a polyatomic ion of known charge is present:

(a) CO_3^{2-} has a charge of –2, so Mn must be +2 to give overall neutrality.

(b) Cl (more electronegative) is –1, so Mo must be +5 to give overall neutrality.

(c) Na is +1, so VO_4 has overall charge –3; each O is –2, so V is +5.

(d) O is –2, so Au must be +3 to give overall neutrality.

(e) H_2O has zero charge, and SO_4^{2-} has a charge of –2, so Fe must be +3 to give overall neutrality.

19.3 Use the periodic table to locate and identify elements from their valence configurations:

(a) There are six valence electrons, so the element is in Column 6; $3d$ orbital is filling: Cr

(b) There are full s and d blocks, so the element is in Column 12; $4d$ has just filled: Cd

(c) There are 11 valence electrons, so the element is in Column 11; $3d$ orbital is full: Cu

19.5 Use the periodic table to locate a transition metal and determine the principal quantum numbers of its valence electrons. Then remove electrons to give the appropriate cation, remembering that when cations form, the valence s electrons are the first ones removed:

(a) $3d^5$; (b) $5d^6$; (c) $3d^8$; and (d) $4d^4$.

19.7 The properties of transition metals vary regularly with their valence configurations, so predict relative properties based on locations in the d block:

(a) Pd is in Column 10, Cd is in Column 12, both in the $n = 4$ row. Beyond the middle of the d block, melting point decreases with Z because electrons are placed in antibonding orbitals, so Pd melts at higher temperature.

(b) Cu and Au are both in Column 11, but Au has a higher molar mass, so Au has higher density.

(c) Cr is in Column 6, Co is in Column 9, both in the $n = 3$ row. IE_1 increases with Z across a row because Z_{eff} increases, so Co has the higher IE_1.

19.9 Use the charges of ligands and ions to determine the oxidation states of transition metals in coordination complexes. Because s electrons are always removed first, the count of d electrons is given by the number of valence electrons — the oxidation state:

(a) Each Cl is –1, and NH_3 is neutral, so Ru has oxidation state +2. Ru is in Column 8 (eight valence electrons), giving d^6.

(b) Each I is –1, and en is neutral, so Cr has oxidation state +3. Cr is in Column 6 (six valence electrons), giving d^3.

(c) Each Cl is –1, and trimethylphosphine is neutral, so Pd has oxidation state +2. Pd is in Column 10 (10 valence electrons), giving d^8.

(d) Each Cl is –1, and NH_3 is neutral, so Ir has oxidation state +3. Ir is in Column 9 (nine valence electrons), giving d^6.

(e) CO is a neutral molecule, so Ni has oxidation state 0. Ni is in Column 10 (10 valence electrons), giving $s^2 d^8$.

19.11 Compounds that contain coordination complexes are named following the six rules stated in your textbook: name the cation first, name ligands in alphabetical order, name the metal, add "o" for anions, use Greek prefixes, add "-ate" for anionic complexes, give the oxidation number:

(a) Hexaammineruthenium(II) chloride

(b) *trans-bis*(Ethylenediamine)diiodochromium(III) iodide

(c) *cis*-Dichloro*bis*(trimethylphosphine)palladium(II)

(d) *fac*-Triamminetrichloroiridium(III)

(e) Tetracarbonylnickel(0)

19.13 The structure of a metal complex usually is octahedral (six ligands), tetrahedral (four ligands), or square planar (four ligands):

(a) Six NH_3 in an octahedron around the central Ru:

(b) Two I at opposite ends of one axis, two en in a square plane around the central Cr:

(c) Square planar arrangement about Pd, with two Cl adjacent to each other:

(d) Octahedral arrangement around Ir, with three Cl in a triangular face:

(e) Tetrahedral arrangement of C≡O about a central Ni:

19.15 The name of a complex contains the information needed to determine its chemical formula. Determine the charge of the complex from charges on the ligands and the oxidation number:

(a) *cis*-[Co(NH$_3$)$_4$ClNO$_2$]$^+$; (b) [PtNH$_3$Cl$_3$]$^-$; (c) *trans*-[Cu(en)$_2$(H$_2$O)$_2$]$^{2+}$; (d) [FeCl$_4$]$^-$.

19.17 (a) *cis*-Tetraamminechloronitrocobalt(III) has six ligands and octahedral geometry. The *cis* indicates that the chlorine and nitro ligands will have a 90° angle between them:

(b) In amminetrichloroplatinate(II), platinum has eight *d* electrons in its valence shell, resulting in square planar geometry:

(c) In *trans*-diaqua*bis*(ethylenediamine)copper(II), ethylenediamine is a bidentate ligand, giving a coordination number of 6 and octahedral geometry. The *trans* term means that the water ligands are opposite each other on the complex:

(d) In tetrachloroferrate(III), iron(III) is a 3*d* metal with five valence electrons, resulting in tetrahedral geometry:

19.19 The crystal field diagram for weak and strong octahedral fields is always the same, with the populations changing depending on how many *d* electrons must be accommodated. The valence configuration provides information about the number of *d* electrons:

(a) Ti^{2+} (Column 4, two electrons) is d^2; (b) Cr^{3+} (Column 6, three electrons) is d^3. For these two ions, the low-field and high-field configurations are the same:

(c) Mn^{2+} (Column 7, five electrons) is d^5.

(d) Fe^{3+} (Column 8, five electrons) is d^5. These two ions have identical diagrams, with high-spin when the splitting is small and low-spin when the splitting is large:

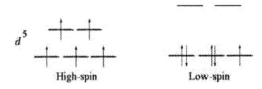

19.21 The magnetic properties of a complex are determined by its number of d electrons and the extent of its crystal field splitting energy:

(a) Ir^{3+} (Column 9, six electrons) is d^6, and NH_3 generates relatively large splitting. All the electrons will be paired, making the complex diamagnetic.

(b) Cr^{2+} (Column 6, four electrons) is d^4, and water generates relatively small splitting. The complex is paramagnetic with four unpaired electrons.

(c) Pt^{2+} (Column 10, eight electrons) is d^8, so regardless of the splitting energy, this square planar complex is paramagnetic with two unpaired electrons.

(d) Pd has d^{10} configuration, so all orbitals are filled and this complex is diamagnetic.

19.23 The colours of transition metal complexes are generally determined by d–d transitions, but Zr^{4+} (Column 4, four electrons) has all its valence electrons removed, so there is no valence electron that can undergo a transition involving the absorption of visible light.

19.25 (a) The coordination number is the number of bonds formed between metal and ligands. The en ligand is bidentate, so three of them form six bonds, CN = 6.

(b) The oxidation number is determined by examining the net charge and correcting for charges on all species other than the transition metal. Here, en is neutral. Each Cl^- anion contributes –1 charge, so the oxidation number of Fe is +3. Fe is in Column 8 of the periodic table, so its valence configuration is $3d^5$.

(c) Coordination number 6 means octahedral geometry.

(d) With an odd number of electrons, it is not possible for all of them to be paired, so this complex is paramagnetic.

(e) The complex is low-spin, so all but one of the electrons is paired.

19.27 Complexes are high-spin and paramagnetic when the crystal field splitting is small, and they become low-spin if the crystal field splitting is larger than the pairing energy. According to the spectrochemical series, CN^- produces greater splitting than H_2O. Replacing the H_2O ligands with CN^- ligands increases the splitting, to the point where it takes less energy to pair the electrons than to promote them to the less stable *d* orbitals. Cr is in Column 6 of the periodic table, so Cr^{2+} has d^4 valence configuration. There are four unpaired electrons in the high-spin configuration but only two in the low-spin configuration:

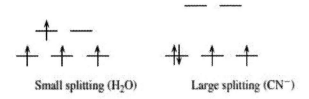

19.29 The wavelength of light that is absorbed provides a measure of the crystal field splitting energy:

$$\Delta = E = \frac{hcN_A}{\lambda} = \left(\frac{(6.626 \times 10^{-34} \text{ J s})(2.998 \times 10^8 \text{ m/s})(6.022 \times 10^{23} \text{ mol}^{-1})}{465 \times 10^{-9} \text{ m}} \right)\left(\frac{10^{-3} \text{ kJ}}{1 \text{ J}} \right)$$

$\Delta = 257$ kJ/mol
The complex absorbs visible light around 465 nm, in the blue. The colour of the complex will be the complementary colour to blue. Consult Table 19-5 to determine that this colour is orange.

19.31 The structural difference between haemoglobin and myoglobin is that the former has four subunits, whereas the latter has just one. As a consequence, haemoglobin has much more complex cooperative chemical behaviour than myoglobin does.

19.33 An iron ion bonded to four sulphur atoms from cysteines is in a tetrahedral environment, so the splitting pattern is the 2–3 pattern characteristic of tetrahedral complexes. The iron cation loses one electron and is converted from Fe^{2+} to Fe^{3+}:

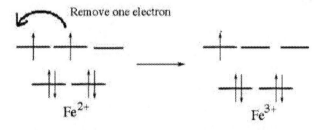

19.35 Consult your textbook for the chemical reactions of various metallurgical processes:

(a) $CuFeS_2 + 3 O_2 \rightarrow CuO + FeO + 2 SO_2$

(b) $Si + O_2 + CaO \rightarrow CaSiO_3$

(c) $TiCl_4 + 4 Na \rightarrow 4 NaCl + Ti$

19.37 This is a standard stoichiometry problem. Begin by analyzing the chemistry. The reactants are Cu_2S and air and the given products are Cu metal and SO_2 gas.

Balanced reaction: $Cu_2S + O_2 \rightarrow 2\,Cu + SO_2$.

Start by computing the number of moles of Cu_2S, and then use the appropriate mass–mole–mass and p–V–T calculations to determine the amounts of the products:

$$m_{Cu_2S} = 5.60 \times 10^4\ kg\left(\frac{10^3\ kg}{1\ kg}\right)\left(\frac{2.37\%}{100\%}\right) = 1.327 \times 10^6\ g$$

$$n_{Cu_2S} = \frac{1.327 \times 10^6\ g}{159.2\ g/mol} = 8.34 \times 10^3\ mol = n_{SO_2}$$

$$m_{Cu} = 8.34 \times 10^3\ mol\left(\frac{2\ mol\ Cu}{1\ mol\ Cu_2S}\right)(63.55\ g/mol) = 1.06 \times 10^6\ g$$

$$p_{total} = 755\ Torr\left(\frac{1\ bar}{750.06\ Torr}\right) = 1.006\ bar$$

$$V_{SO_2} = \frac{nRT}{p} = \frac{(8.34 \times 10^3\,mol)(0.08314\ L\ bar\ mol^{-1}\ K^{-1})(273.15 + 23.5\ K)}{1.006\ bar} = 2.04 \times 10^5\ L$$

19.39 Standard free energy changes are calculated using standard free energies of formation, found in Appendix D. $\Delta G^{\circ}_{reaction} = \Delta G^{\circ}_{products} - \Delta G^{\circ}_{reactants}$

$ZnO\ (s) + C\ (s) \rightarrow Zn\ (s) + CO\ (g)$

$\Delta G^{\circ}_{reaction} = [1\ mol(-137.2\ kJ/mol) + 0] - [1\ mol(-320.5\ kJ/mol) + 0] = 183.3\ kJ$

$ZnO\ (s) + CO\ (g) \rightarrow Zn\ (s) + CO_2\ (g)$

$\Delta G^{\circ}_{reaction} = [1\ mol(-394.4\ kJ/mol) + 1\ mol(0)] - [1\ mol(-320.5\ kJ/mol) + 1\ mol(-137.2\ kJ/mol)] = 63.3\ kJ$

19.41 The coinage metals are those that have been used since antiquity for coins: copper, silver, and gold. All are in Column 11 of the periodic table. They are characterized by high electrical conductivity, good ductility, and low chemical reactivity, in particular resistance to oxidation. Hence, they are used for money (a vanishing use in technologically advanced countries), for electrical wire, for jewellery, and for other decorative objects. See your text for special examples of uses for compounds of these elements.

19.43 Titanium is used as an engineering metal because of its relatively low density, high bond strength, resistance to corrosion, and ability to withstand high temperatures, all of which make it a favoured structural material.

19.45 The number of possible isomers of a complex is determined by its geometry and the number of ligands of each type:

(a) [Ir complex] mer   [Ir complex] fac

(b) [Pd complex] trans   [Pd complex] cis

(c) [Cr complex] trans   [Cr complex] cis

(d) [Cr complexes]

19.47 Bidentate ligands form complexes with two links. Each Fe ion forms an octahedral complex with three oxalate anions. When oxalate is applied to rust, it complexes and dissolves the Fe^{3+} cations. Here is a sketch showing the ligand–metal orientations:

19.49 To determine electron configurations, start from the position of the element in the periodic table and remove *s* electrons preferentially.

(a) Cr is in Column 6, configuration $[Ar] 4s^1 3d^5$; Cr^{2+} is $[Ar] 3d^4$; Cr^{3+} is $[Ar] 3d^3$

(b) V is in Column 5, configuration [Ar] $4s^2\, 3d^3$; V^{2+} is [Ar] $3d^3$; V^{3+} is [Ar] $3d^2$; V^{4+} is [Ar] $3d^1$; V^{5+} is [Ar]

(c) Ti is in Column 4, configuration [Ar] $4s^2\, 3d^2$; Ti^{2+} is [Ar] $3d^2$; Ti^{4+} is [Ar]

19.51 Compounds that contain coordination complexes are named following the six rules stated in your textbook: name the cation first, name ligands in alphabetical order, name the metal, add "o" for anions, use Greek prefixes, add "-ate" for anionic complexes, list the oxidation number: (a) *cis*-tetraaquadichlorochromium(III) chloride; (b) bromopentacarbonylmanganese(I); (c) *cis*-diamminedichloroplatinum(II).

19.53 Superoxide dismutase catalyzes the conversion of superoxide into molecular oxygen and hydrogen peroxide. This reaction occurs at a metal site that contains one Zn^{2+} ion and one Cu^{2+} ion, linked by a histidine ligand that bonds to both metal ions. An O atom binds to the Zn^{2+} ion:

$$2\, O_2^- + 2\, H_3O^+ \xrightarrow{\text{SOD}} O_2 + H_2O_2 + 2\, H_2O$$

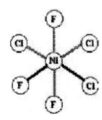

19.55 Cu(II) in water forms an aqua complex. Addition of fluoride produces the insoluble green salt, CuF_2, whereas addition of chloride produces the bright green tetrachlorocopper(II) complex, $[CuCl_4]^{2-}$:

$$Cu^{2+}\,(aq) + 2\, F^-\,(aq) \rightarrow CuF_2\,(s)$$

$$Cu^{2+}\,(aq) + 4\, Cl^-\,(aq) \rightarrow [CuCl_4]^{2-}\,(aq)$$

19.57 Silver (Column 11 of the periodic table) has a filled set of *d* orbitals, making the neutral metal difficult to oxidize. This gives silver good resistance to corrosion and makes it suitable for jewellery. Vanadium (Column 5), with d^3 configuration, is readily oxidized, so it corrodes rapidly and is unsuited to jewellery.

19.59 The *mer* isomer of an octahedral complex has three like-ligands arranged in a meridian plane:

19.61 Silver tarnish is silver sulphide, Ag_2S, formed by reaction with trace amounts of H_2S in the atmosphere:

$$4\,Ag\,(s) + 2\,H_2S\,(g) + O_2\,(g) \rightarrow 2\,Ag_2S\,(s) + 2\,H_2O\,(g)$$

19.63 In these complexes, chromium is in the +3 oxidation state. From its location in Column 6 of the periodic table, we deduce that is has d^3 valence electron configuration. The three electrons occupy the three different t_{2g} orbitals, regardless of the magnitude of the splitting energy.

19.65 Orbital sketches show that the two orbitals experience quite different electron–electron repulsion in a square planar environment:

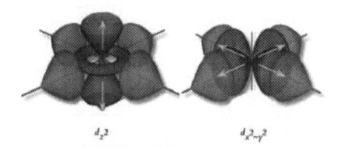

d_{z^2} $d_{x^2-y^2}$

19.67 (a) The charge on a complex is the charge on the transition metal cation, modified by any charges on the ligands. Ammonia and water are neutral, so when L is either of these, $n^+ = 3^+$; but chloride is –1, so for $L = Cl^-$, $n^+ = 2^+$.

(b) The colour of each complex is the colour that is complementary to the one corresponding to the wavelength of light that it absorbs. Consult Table 19-5 of your textbook to determine these: $L = Cl^-$, red to purple; $L = H_2O$, orange; and $L = NH_3$, yellow-orange.

(c) The wavelength of light that is absorbed provides a measure of the crystal field splitting energy:

$$\Delta = E = \frac{hcN_A}{\lambda}$$

$L = Cl^-$:

$$\Delta = \left(\frac{(6.626 \times 10^{-34}\text{ J s})(2.998 \times 10^8\text{ m/s})(6.022 \times 10^{23}\text{ mol}^{-1})}{515 \times 10^{-9}\text{ m}} \right)\left(\frac{10^{-3}\text{ kJ}}{1\text{ J}} \right)$$

$$= 232\text{ kJ/mol}$$

$L = H_2O$:

$$\Delta = \left(\frac{(6.626 \times 10^{-34}\text{ J s})(2.998 \times 10^8\text{ m/s})(6.022 \times 10^{23}\text{ mol}^{-1})}{480 \times 10^{-9}\text{ m}} \right)\left(\frac{10^{-3}\text{ kJ}}{1\text{ J}} \right)$$

$$= 249\text{ kJ/mol}$$

$L = NH_3$:

$$\Delta = \left(\frac{(6.626 \times 10^{-34} \text{ J s})(2.998 \times 10^8 \text{ m/s})(6.022 \times 10^{23} \text{ mol}^{-1})}{465 \times 10^{-9} \text{ m}} \right)\left(\frac{10^{-3} \text{ kJ}}{1 \text{ J}} \right)$$

$$= 257 \text{ kJ/mol}$$

The trend matches exactly the positions of these three ligands in the spectrochemical series: $Cl^- < H_2O < NH_3$.

19.69 Tetracarbonylnickel(0), $[Ni(CO)_4]$, has tetrahedral geometry since Ni is a first-row transition element. Because of the strong field ligand, CO, Δ is large, which will result in the compound absorbing UV light and appearing colourless.

Tetracyanozinc(II), $[Zn(CN)_4]^{2-}$, has tetrahedral geometry (Zn is a first-row transition metal) where Zn has a +2 charge (d^{10}). This complex is colourless because there are no possible d–d transitions.

19.71 Brass is an alloy of zinc and copper, and superoxide dismutase contains zinc and copper ions in its reaction centre.

19.73 Four-coordinate complexes may be either tetrahedral or square planar. The splitting patterns for these two geometries show that the d^8 configuration can have zero spin in the square planar case but not in the tetrahedral case. Thus, the magnetic behaviour indicates that $[Ni(CN)_4]^{2-}$ is square planar and $[NiCl_4]^{2-}$ is tetrahedral:

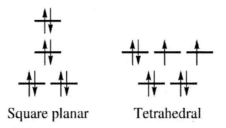

Square planar   Tetrahedral

19.75 When ferritin is neither empty nor filled to capacity, the protein has the capacity to provide iron as needed for haemoglobin synthesis, or to store iron if an excess is absorbed by the body.

19.77 Visible spectroscopy is useful when a compound has an energy gap between the highest occupied and lowest unoccupied orbital that matches the energy of visible light. Because Zn^{2+} has d^{10} configuration, its d orbitals are completely filled, and the lowest unoccupied orbital is quite high in energy. In contrast, Co^{2+} has d^7 configuration, giving this cation unfilled d orbitals. Consequently, metalloproteins that contain Co^{2+} absorb visible light, making it possible to study them with visible spectroscopy.

19.79 Use standard reduction potentials from Appendix F to determine which metals can be displaced by Zn. Any metal with a less negative standard potential can be displaced:

$$Zn^{2+} + 2 e^- \rightleftharpoons Zn \qquad E° = -0.7618 \text{ V}$$

The following are a few examples of metals that can be displaced by Zn:

$$Co^{2+} + 2\,e^- \; \epsilon \; Co \qquad\qquad E° = -0.28 \text{ V}$$

$$Cu^{2+} + 2\,e^- \; \epsilon \; Cu \qquad\qquad E° = +0.3419 \text{ V}$$

$$Fe^{2+} + 2\,e^- \; \epsilon \; Fe \qquad\qquad E° = -0.447 \text{ V}$$

$$Ni^{2+} + 2\,e^- \; \epsilon \; Ni \qquad\qquad E° = -0.257 \text{ V}$$

When combined with the Zn half-reaction, the overall cell voltage is positive and the process is spontaneous. For example:

$$Zn + Co^{2+} \; \epsilon \; Co + Zn^{2+} \qquad E° = (-0.28 \text{ V}) - (-0.7618 \text{ V}) = 0.48 \text{ V}$$

19.81 The colours of substances depend on what wavelengths of light they absorb. According to Table 19-5, blue colour results when a substance absorbs 610 nm light (196 kJ/mol), and red colour results when a substance absorbs 500 nm light (239 kJ/mol). Thus, deoxyhaemoglobin has a smaller value for Δ than oxyhaemoglobin, suggesting that O_2 causes larger energy-level splitting than H_2O. The hypothesis could be tested by replacing the H_2O ligand in deoxyhaemoglobin by a ligand that causes smaller splitting, such as Cl^-, which should push the colour toward the green. In addition, CO, which generates the largest energy-level splitting, should push the colour to red-orange or orange.

19.83 The densities of transition metals increase with **Z** across each row of the *d* block, up to about the middle of the block, as illustrated by Figure 19-2*b* in your textbook. Thus, the upper-left-most transition metal in the periodic table, scandium, would be the best replacement for Al. The density of scandium is only about 10% greater than that of aluminum. The figure shows that osmium, in the middle of the 5*d* block, is the densest transition metal. Its density is actually twice that of lead.

19.85 The chloride ions that form silver chloride precipitate are not bound to the complex and should not be included in the chemical formula of the complex ion:

$PtCl_4 \cdot 2NH_3$ is $[PtCl_4(NH_3)_2]$, diamminetetrachloroplatinum(IV), *cis* and *trans*:

$PtCl_4 \cdot 3NH_3$ is $[PtCl_3(NH_3)_3]Cl$, the complex ion being triamminetrichloroplatinum(IV), *mer* and *fac*:

$$\text{mer}\quad \left[\begin{array}{c} \text{Cl} \\ \text{Cl}_{\prime\prime\prime\prime}\!\!\underset{\text{Pt}}{\overset{|}{\longleftarrow}}\!\!_{\prime\prime\prime\prime}\text{NH}_3 \\ \text{H}_3\text{N}\overset{|}{\nearrow}\,\,{}^{\searrow}\text{Cl} \\ \text{Cl} \end{array}\right]^{+} \qquad \text{fac}\quad \left[\begin{array}{c} \text{Cl} \\ \text{H}_3\text{N}_{\prime\prime\prime\prime}\!\!\underset{\text{Pt}}{\overset{|}{\longleftarrow}}\!\!_{\prime\prime\prime\prime}\text{Cl} \\ \text{Cl}\overset{\nearrow}{}\,\,{}^{\searrow}\text{Cl} \\ \text{NH}_3 \end{array}\right]^{+}$$

$PtCl_4 \cdot 4NH_3$ is $[PtCl_2(NH_3)_4]Cl_2$, the complex ion being tetraamminedichloroplatinum(IV), *cis* and *trans:*

$$\text{trans}\quad \left[\begin{array}{c} \text{Cl} \\ \text{H}_3\text{N}_{\prime\prime\prime\prime}\!\!\underset{\text{Pt}}{\overset{|}{\longleftarrow}}\!\!_{\prime\prime\prime\prime}\text{NH}_3 \\ \text{H}_3\text{N}\overset{\nearrow}{}\,\,{}^{\searrow}\text{NH}_3 \\ \text{Cl} \end{array}\right]^{2+} \qquad \text{cis}\quad \left[\begin{array}{c} \text{Cl} \\ \text{H}_3\text{N}_{\prime\prime\prime\prime}\!\!\underset{\text{Pt}}{\overset{|}{\longleftarrow}}\!\!_{\prime\prime\prime\prime}\text{NH}_3 \\ \text{H}_3\text{N}\overset{\nearrow}{}\,\,{}^{\searrow}\text{Cl} \\ \text{NH}_3 \end{array}\right]^{2+}$$

$PtCl_4 \cdot 5NH_3$ is $[PtCl(NH_3)_5]Cl_3$, the complex ion being pentaamminechloroplatinum(IV), and $PtCl_4 \cdot 6NH_3$ is $[Pt(NH_3)_6]Cl_4$, the complex ion being hexaammineplatinum(IV):

$$\left[\begin{array}{c} \text{Cl} \\ \text{H}_3\text{N}_{\prime\prime\prime\prime}\!\!\underset{\text{Pt}}{\overset{|}{\longleftarrow}}\!\!_{\prime\prime\prime\prime}\text{NH}_3 \\ \text{H}_3\text{N}\overset{\nearrow}{}\,\,{}^{\searrow}\text{NH}_3 \\ \text{NH}_3 \end{array}\right]^{3+} \qquad \left[\begin{array}{c} \text{NH}_3 \\ \text{H}_3\text{N}_{\prime\prime\prime\prime}\!\!\underset{\text{Pt}}{\overset{|}{\longleftarrow}}\!\!_{\prime\prime\prime\prime}\text{NH}_3 \\ \text{H}_3\text{N}\overset{\nearrow}{}\,\,{}^{\searrow}\text{NH}_3 \\ \text{NH}_3 \end{array}\right]^{4+}$$

19.87 Since the d_{z^2} orbital is perpendicular to the trigonal plane of the molecule, it must lie along the axis containing two of the ligands. The repulsion between the d_{z^2} electrons and the bonding electrons would therefore be large, and we expect the d_{z^2} orbital to be at a very high energy in the complex. The $d_{x^2-y^2}$ and d_{xy} orbital lie in the trigonal plane of the molecule and would also be at a high energy, but not as high as that of the d_{z^2} orbital. This is because the $d_{x^2-y^2}$ and d_{xy} orbitals do not point directly at the three ligands in the trigonal plane. Finally, the d_{xz} and d_{yz} orbitals will be at the lowest energy because they do not point at the ligands in the trigonal plane nor along the z-axis. The crystal field splitting energy diagram therefore looks like:

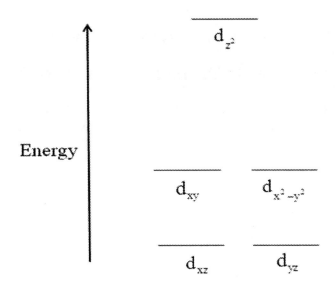

Chapter 20 The Main Group Elements

Solutions to Problems in Chapter 20

20.1 Lewis acids are electron-pair acceptors, and Lewis bases are electron-pair donors. Thus, Lewis acids are electron-deficient, whereas Lewis bases have lone pairs to donate:

(a) Lewis acid is Ni, Lewis base is CO.

(b) Lewis acid is $SbCl_3$, Lewis base is Cl^-.

(c) Lewis acid is $AlBr_3$, Lewis base is $P(CH_3)_3$.

(d) Lewis acid is BF_3, Lewis base is ClF_3.

20.3 Construct Lewis structures and determine steric numbers to identify the three-dimensional structure of a molecule.

(a) AlF_3

   1. There are $3 + 3(7) = 24$ valence electrons.

   2. Three electron pairs are needed for the bonding framework.

   3. The remaining electrons are used to place three pairs on each outer F atom.

The Lewis structure shows SN = 3 for Al, so the molecule has trigonal planar geometry and there is a vacant $3p$ orbital perpendicular to the molecular plane that can accept a pair of electrons:

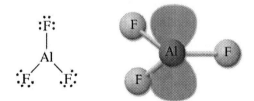

(b) SbF_5

   1. There are $5 + 5(7) = 40$ valence electrons.

   2. Five pairs are used in the bonding framework, leaving $40 - 2(5) = 30$ electrons.

   3. The remaining electrons are used to place three electron pairs on each F atom. The Lewis structure shows SN = 5 for Sb, so the molecule has trigonal bipyramidal geometry, indicating $sp^3 d$ hybridization. There are vacant $3d$ orbitals that do not participate in the $sp^3 d$ hybridization and can accept electrons:

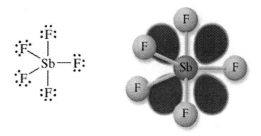

(c) SO_2

1. There are $6 + 2(6) = 18$ valence electrons.

2. Two electron pairs are needed for the bonding framework, leaving $18 - 2(2) = 14$.

3. Place three pairs of electrons on each outer O atom.

4. Place the remaining two electrons on the inner S atom.

5. The resulting structure has $FC_S = 6 - 2 - 2 = 2$; move two lone pairs from the outer

O atoms to form two double bonds and reduce the formal charge to 0.

The Lewis structure shows $SN = 3$ for S (there is a lone pair on the central sulphur atom, so the molecule has a bent geometry), and there is a delocalized π orbital to which an electron pair can be added:

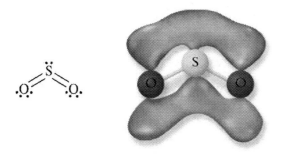

20.5 In the first step, an electron pair from the O atom of H_2O displaces a π bond in Lewis acid–base adduct formation. Then, a proton from H_2O migrates to a C–O oxygen atom:

Adduct formation   Proton tranfer

20.7 Polarizability of cations increases substantially with the value of n and decreases, within the same row of the periodic table, with Z and the charge on the ion:

Fe^{3+} (high charge) $<$ V^{3+} (lowest Z, high charge) $<$ Fe^{2+} $<$ Pb^{2+} (large n)

20.9 Hard acids have low polarizability, and soft acids have high polarizability:

(a) The hardest acid is BF_3 (both elements from Row 2), and then BCl_3, and $AlCl_3$ is the softest (both elements from Row 3).

(b) The hardest acid is Al^{3+} (Row 3), and then Tl^{3+} (Row 6), and Tl^+ (low charge) is the softest.

(c) Polarizability increases with *n*, so the hardest is $AlCl_3$, and then $AlBr_3$, and AlI_3 is the softest.

20.11 When an electronegative O atom bonds to a less electronegative S atom, it withdraws electron density from S, decreasing the polarizability about S and increasing the hardness of the base.

20.13 A metathesis reaction will occur if exchange of partners couples the harder Lewis acid with the harder Lewis base, and the softer Lewis acid with the softer Lewis base:

(a) Al^{3+} is harder than Na^+, and Cl^- is harder than I^-, so metathesis occurs, giving $AlCl_3$ (hard–hard) and NaI (soft–soft).

(b) The Lewis acid is Ti^{4+} in each substance, so no reaction occurs.

(c) Ca^{2+} is softer than H^+, and S^{2-} is softer than O^{2-}, so metathesis occurs, giving H_2O (hard–hard) and CaS (soft–soft).

(d) C (in CH_3^-) is a soft base, and Li^+ is a hard acid, so metathesis occurs, giving $(CH_3)_3P$ (soft–soft) and $LiCl$ (hard–hard).

(e) Ag^+ and I^- are both soft, and Si^{4+} and Cl^- are both hard, so no reaction occurs.

20.15 Descriptions of bonding always begin with a Lewis structure. As your textbook describes, Al_2Cl_6 contains two "bridging" chlorine atoms. Standard procedures would predict tetrahedral geometry about all inner atoms, but the Al–Cl–Al bond angles of 91° indicate that Cl uses *p* orbitals. Each Al atom can be described as using sp^3 hybrids to form four σ bonds to four different Cl atoms.

In Lewis acid–base terms, the bridged molecule forms from two $AlCl_3$ units, linking together in double adduct formation between the Al Lewis-acid atoms and two Cl Lewis-base atoms.

20.17 Thallium lies below indium and gallium, so its properties should be similar to those metals: valence of 3, soft Lewis acid. Like its neighbour, Pb, it is toxic.

20.19 Determine the Lewis structure using standard procedures:

SnCl$_4$

1. There are $4 + 4(7) = 32$ valence electrons.

2. Four electron pairs are needed for the bonding framework, leaving $32 - 4(2) = 24$.

3. Place three pairs of electrons on each outer Cl atom, leaving $24 - 4(6) = 0$.

4. No remaining electrons.

5. The resulting structure has $FC_{Sn} = 4 - 4 = 0$, so the structure is complete.

SnCl$_4$ can function as a Lewis acid because the Sn atom has empty *d* orbitals that can accept electrons to form more bonds.

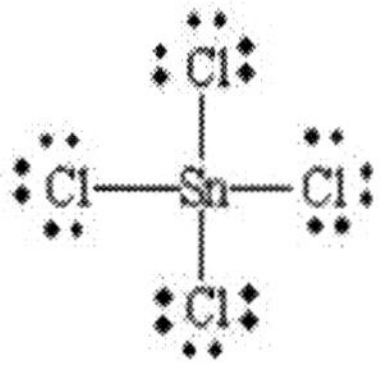

20.21 A BN pair has the same number of valence electrons as a pair of C atoms, so the structure and bonding of B$_3$N$_3$H$_6$ is just like that of benzene, a six-membered planar ring with a delocalized set of π bonds. There are $3(3) + 3(5) + 6(1) = 30$ valence electrons, all of which are involved in the bonding network:

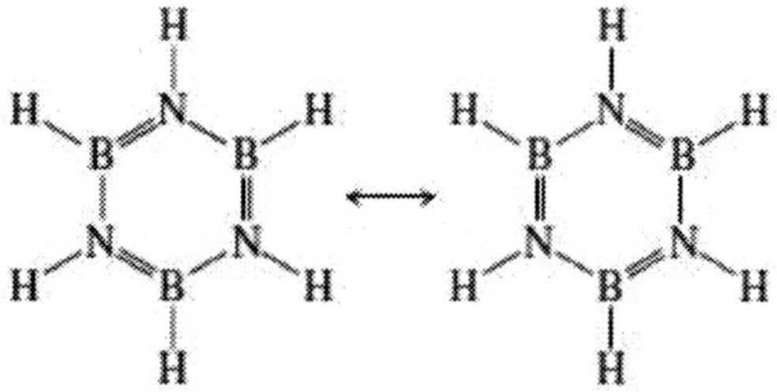

The B and N atoms have bonding and geometry that can be described using sp^2 hybrid orbitals, resulting in three σ bonds around each ring atom. In addition, there is a delocalized π-bonding network encompassing all six ring atoms and containing six electrons.

20.23 The band gap decreases from top to bottom of each column of the periodic table, so Ge has a smaller band gap than Si. In orbital terms, this is because the principal quantum number of the valence orbitals increases. This makes the valence orbitals larger, leading to less effective overlap and smaller energy difference between bonding and antibonding orbitals.

20.25 Follow the example in Section 20.4 of your text on polymer formation.

$$2\ C_2H_5Cl + Si(Cu) \rightarrow (C_2H_5)_2SiCl_2 + Cu$$

$$(C_2H_5)_2SiCl_2 + 2\ H_2O \rightarrow (C_2H_5)_2Si(OH)_2 + 2\ HCl$$

Condensation will eliminate water to give the polymer:

$$2\ (C_2H_5)_2Si(OH)_2 \rightarrow \text{polymer-linkage} + H_2O$$

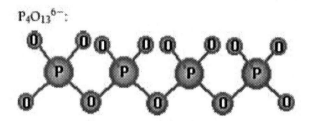

20.27 Polyphosphates form by sequential condensation of PO_3^- units, so the tetraphosphate has chemical formula $P_4O_{13}^{6-}$:

$P_4O_{13}^{6-}$:

20.29 Phosphoric acid is produced directly from apatite by reaction with sulphuric acid:

$$Ca_5(PO_4)_3X\ (s) + 5\ H_2SO_4\ (aq) \rightarrow 3\ H_3PO_4\ (aq) + 5\ CaSO_4\ (s) + HX\ (aq)$$

$(X = F^-, OH^-, \text{or } Cl^-)$

This is a Brønsted acid–base reaction in which protons are transferred from sulphuric acid to the phosphate anions. There are no redox reactions in this process.

20.31 White phosphorus consists of P_4 tetrahedra. To convert this form to the red form, break one bond in each tetrahedron and use the electrons to form bonds between P_4 units:

20.33 There are two industrial reactions in Section 20.6 in which sulphuric acid acts as a Brønsted acid:

$$CaF_2\,(s) + H_2SO_4\,(l) \rightarrow 2\,HF\,(g) + CaSO_4\,(s)$$

$$FeTiO_3\,(s) + 2\,H_2SO_4\,(aq) + 5\,H_2O\,(l) \xrightarrow{150\text{-}180\ ^\circ C} TiOSO_4\,(aq) + FeSO_4{\cdot}7H_2O\,(s)$$

20.35 The repeat structure of polyvinylchloride, in common with all polyethylene-type polymers, has an all-carbon backbone. There is a chlorine atom on every other carbon atom:

20.37 The reaction forming $TiCl_4$ from TiO_2 is as follows:

$$TiO_2\,(s) + C\,(s) + 2\,Cl_2\,(g) \rightarrow TiCl_4\,(l) + CO_2\,(g)$$

C goes from 0 to +4 oxidation state, so it is oxidized and serves as the reducing agent.

Cl goes from 0 to –1 oxidation state, so it is reduced and serves as the oxidizing agent.

20.39 This is a stoichiometry problem that can be worked by determining the percentage composition of bauxite:

Bauxite is AlOOH, $M = 60.00$ g/mol $\quad$ % Al $= 100\% \left(\dfrac{26.982\ \text{g/mol}}{60.00\ \text{g/mol}} \right) = 44.97\%$

1 kg of bauxite rock contains $1.00\ \text{kg} \left(\dfrac{85\%}{100\%} \right) \left(\dfrac{44.97\%}{100\%} \right) = 0.382\ \text{kg Al}$

If the processing is 75% efficient, each 1.00 kg of bauxite rock yields the following:

$$(0.382\ \text{kg})(0.75) = 0.287\ \text{kg Al}$$

To produce 2500 kg of Al requires the following:

$$2500\ \text{kg Al} \left(\dfrac{1\ \text{kg bauxite rock}}{0.287\ \text{kg Al}} \right) = 8.7 \times 10^3\ \text{kg bauxite rock}$$

20.41 Sulphur has *d* orbitals available for bond formation, allowing the formation of SF_6, in which the bonding can be described using $sp^3\,d^2$ hybrid orbitals on the S atom. Oxygen has no valence *d* orbitals available. In principle, SBr_6 could also form, but the Br atom is too large for six Br atoms to be accommodated around a central S atom.

20.43 Lewis acids are electron-pair acceptors, and Lewis bases are electron-pair donors. The As atom in $AsCl_3$ has a lone pair that it donates, giving this compound Lewis-base character. The As atoms also can accommodate additional electron pairs by using valence *d* orbitals, giving this compound Lewis-acid character:

$$AsCl_3 + BF_3 \rightarrow Cl_3As-BF_3 \text{ (} AsCl_3 \text{ acts as Lewis base)}$$

$$AsCl_3 + Cl^- \rightarrow AsCl_4^- \text{ (} AsCl_3 \text{ acts as Lewis acid)}$$

20.45 Balance a redox half-reaction following the standard steps outlined in Chapter 17 (balance all but H and O by inspection, balance O by adding H_2O, balance H by adding H_3O^+/H_2O or OH^-/H_2O, balance charge by adding electrons): begin by balancing the Al reaction:

$Al + 2\,H_2O \rightarrow AlO_2^-$; add 4 OH^- on the reactant side and 4 H_2O to the product side to balance H:

$$Al + 2\,H_2O + 4\,OH^- \rightarrow AlO_2^- + 4\,H_2O$$

Cancel 2 H_2O on each side and add three electrons to balance charge:

$$Al + 4\,OH^- \rightarrow AlO_2^- + 2\,H_2O + 3\,e^-$$

(a) $NO_3^- \rightarrow NH_3 + 3\,H_2O$; add 9 OH^- on the product side and 9 H_2O on the reactant to balance H:

$$NO_3^- + 9\,H_2O \rightarrow NH_3 + 3\,H_2O + 9\,OH^-$$

Cancel 3 H_2O on each side and add eight electrons to balance the charge:

$$NO_3^- + 6\,H_2O + 8\,e^- \rightarrow NH_3 + 9\,OH^-$$

Multiply this half-reaction by 3 and the Al half-reaction by 8 to balance the electrons, and add:

$$8\,Al + 32\,OH^- + 3\,NO_3^- + 18\,H_2O + 24\,e^- \rightarrow 3\,NH_3 + 27\,OH^- + 8\,AlO_2^- + 16\,H_2O + 24\,e^-$$

Cancelling duplicated species yields:

$$8\,Al + 5\,OH^- + 3\,NO_3^- + 2\,H_2O \rightarrow 3\,NH_3 + 8\,AlO_2^-$$

(b) $2\,H_2O + 2\,e^- \rightarrow H_2 + 2\,OH^-$

Multiply this half-reaction by 3 and the Al half-reaction by 2, and add:

$$2\,Al + 8\,OH^- + 6\,H_2O + 6\,e^- \rightarrow 3\,H_2 + 6\,OH^- + 2\,AlO_2^- + 4\,H_2O + 6\,e^-$$

Cancel duplicated species:

$$2\,Al + 2\,OH^- + 2\,H_2O \rightarrow 3\,H_2 + 2\,AlO_2^-$$

(c) $SnO_3^{2-} \rightarrow Sn + 3\ H_2O$

Add 6 OH^- on the product side and 6 H_2O on the reactant side to balance H:

$$SnO_3^{2-} + 6\ H_2O \rightarrow Sn + 3\ H_2O + 6\ OH^-$$

Cancel 3 H_2O on each side and add four electrons to balance charge:

$$SnO_3^{2-} + 3\ H_2O + 4\ e^- \rightarrow Sn + 6\ OH^-$$

Multiply this half-reaction by 3 and the Al half-reaction by 4, and add:

$$4\ Al + 16\ OH^- + 3\ SnO_3^{2-} + 9\ H_2O + 12\ e^- \rightarrow 3\ Sn + 18\ OH^- + 4\ AlO_2^- + 8\ H_2O + 12\ e^-$$

Cancelling duplicated species yields

$$4\ Al + 3\ SnO_3^{2-} + H_2O \rightarrow 3\ Sn + 2\ OH^- + 4\ AlO_2^-$$

20.47 Nitrogen, at the top of Group 15, is a non-metal showing a valence of 3 that forms polar bonds most readily with other non-metals (B, C, O, the halogens). Phosphorus has a similar pattern of reactivity but can also involve *d* orbitals in its bonding. Arsenic and antimony are metalloid with useful semiconductor properties, and bismuth is metallic. Interestingly, elemental nitrogen, N_2 (*g*), is very stable, whereas elemental phosphorus, P_4, is highly reactive.

20.49 The production of aluminum from its ore is described in Section 20.3. First, bauxite ore is treated with strong base to produce soluble $[Al(OH)_4]^-$:

$$Al(O)OH\ (s) + OH^-\ (aq) + H_2O\ (l) \rightarrow [Al(OH)_4]^-\ (aq)$$

When the solution is diluted with water, aluminum hydroxide precipitates:

$$[Al(OH)_4]^-\ (aq) \rightarrow Al(OH)_3\ (s) + OH^-\ (aq)$$

Strong heating drives off water, leaving aluminum oxide:

$$2\ Al(OH)_3 \xrightarrow{\ 1250\ ^\circ C\ } Al_2O_3\ (s) + 3\ H_2O\ (g)$$

Finally, electrolysis of aluminum oxide dissolved in cryolite reduces Al to pure metal:

$$2\ Al_2O_3\ (melt) + 3\ C\ (s) \rightarrow 3\ CO_2\ (g) + 4\ Al\ (l)$$

20.51 Silicon dioxide is first converted into silicon tetrachloride by reaction with molecular chlorine:

$$SiO_2\ (s) + 2\ C\ (s) + 2\ Cl_2\ (g) \xrightarrow{\ high\ T\ } SiCl_4\ (g) + 2\ CO\ (g)$$

The $SiCl_4$ is purified by distillation and then reduced by reaction with Mg metal:

$$SiCl_4\ (l) + 2\ Mg\ (s) \rightarrow Si\ (s) + 2\ MgCl_2\ (s)$$

20.53 Metal cations are Lewis acids, and the higher the principal quantum number of the valence electrons, the softer the acid. Hence, Hg is a very soft Lewis acid, whereas Zn is relatively hard. Among anions, S^{2-} is considerably softer than O-containing anions. By the HSAB principle, Hg forms compounds preferentially with S^{2-}, whereas Zn combines with harder bases that contain O.

20.55 Gaseous Al_2Cl_6 is in equilibrium with gaseous $AlCl_3$:

$Al_2Cl_6 (g)$ ⇌ $2\ AlCl_3 (g)$. Le Châtelier's principle predicts that because the forward reaction is endothermic (bonds must be broken), an increase in temperature shifts the position of the equilibrium to the right. Thus, as temperature increases, the number of moles of gaseous substance increases, so the pressure increases faster than would be predicted by the ideal gas equation.

20.57 The reaction can be balanced by recognizing the chemical formulas of the substances involved. The reactants are phosphoric acid (H_3PO_4) and fluoroapatite ($Ca_5(PO_4)_3F$); the products are calcium dihydrogen phosphate ($Ca(H_2PO_4)_2$) and HF: The unbalanced reaction is

$$Ca_5(PO_4)_3F\ (s) + H_3PO_4\ (aq) \rightarrow Ca(H_2PO_4)_2\ (s) + HF\ (aq)$$

$$5\ Ca + 4\ P + 16\ O + F + 3\ H \rightarrow Ca + 2\ P + 8\ O + F + 5\ H$$

Balance Ca by giving $Ca(H_2PO_4)_2$ a coefficient of 5:

$$Ca_5(PO_4)_3F\ (s) + H_3PO_4\ (aq) \rightarrow 5\ Ca(H_2PO_4)_2\ (s) + HF\ (aq)$$

$$5\ Ca + 4\ P + 16\ O + F + 3\ H \rightarrow 5\ Ca + 10\ P + 40\ O + F + 21\ H$$

Next, balance phosphorus by giving H_3PO_4 a coefficient of 7:

$$Ca_5(PO_4)_3F\ (s) + 7\ H_3PO_4\ (aq) \rightarrow 5\ Ca(H_2PO_4)_2\ (s) + HF\ (aq)$$

$$5\ Ca + 10\ P + 40\ O + F + 21\ H \rightarrow 5\ Ca + 10\ P + 40\ O + F + 21\ H$$

Note that now all atoms are balanced:

$$Ca_5(PO_4)_3F\ (s) + 7\ H_3PO_4\ (aq) \rightarrow 5\ Ca(H_2PO_4)_2\ (s) + HF\ (aq)$$

Calculate the mass percent of phosphorus:

$$M_{compound} = 40.078 + 2[(2)(1.008) + 30.974 + (4)(15.999)] = 234.05 \text{ g/mol}$$

$$\% \text{ P} = (100\%)\frac{2M_P}{M_{compound}} = 100\% \left(\frac{(2)(30.974 \text{ g/mol})}{234.05 \text{ g/mol}} \right) = 26.47\ \%$$

20.59 (a) Determine the Lewis structure of $SOCl_2$ following the usual procedure. There are 26 valence electrons, which give a provisional structure in which an inner S atom has +1 formal charge. Make one π bond to minimize formal charges:

(b) Balance the chemical reaction by inspection. Each H_2O requires one $SOCl_2$:

$$6 \; SOCl_2 \; (l) + FeCl_3 \cdot 6H_2O \; (s) \rightarrow 6 \; SO_2 \; (g) + 12 \; HCl \; (g) + FeCl_3 \; (s)$$

20.61 Polarizability increases with atomic size, and greater polarizability leads to larger dispersion forces. Thus, the larger the atoms in a molecule, the larger the intermolecular forces and the easier it is to condense the substance. In the sequence BCl_3, BBr_3, BI_3, the halogens increase in size, accounting for the different stable phases at room temperature.

20.63 The phosphate anion is the conjugate base of a weak acid, HPO_4^{2-}:

$$PO_4^{3-} + H_2O \rightarrow HPO_4^{2-} + OH^- \qquad pK_b = 14.00 - pK_a \; (HPO_4^{2-}) = 14.00 - 12.32 = 1.68$$

20.65 The reactions can be balanced by inspection:

$$2 \; Al_2O_3 + 6 \; Cl_2 \rightarrow 4 \; AlCl_3 + 3 \; O_2$$

$$AlCl_3 + 3 \; Na \rightarrow Al + 3 \; NaCl$$

20.67 Compounds with Lewis acid–base properties will undergo metathesis reactions if a transfer of bonding partners associates harder acids with harder bases.

(a) Al^{3+} is quite hard because of its +3 charge, and Cl^- is softer than CH_3^-:

$$AlCl_3 + 3 \; LiCH_3 \rightarrow Al(CH_3)_3 + 3 \; LiCl$$

(b) Sulphur trioxide is a Lewis acid and forms an adduct with H_2O, a Lewis base:

$$SO_3 + H_2O \rightarrow H_2SO_4 \; (aq)$$

(c) The antimony atom can bond to one additional F^- anion: $SbF_5 + LiF \rightarrow LiSbF_6$.

(d) S is harder than As, and F is harder than Cl, so there is no reaction.

20.69 The reaction consumes 12 H atoms for 4 P atoms, or 3 H atoms for every P atom. Thus, the product is H_3PO_4:

$$P_4O_{10} + 6 \; H_2O \rightarrow 4 \; H_3PO_4$$

20.71 Either position about phosphate detergents can be defended. Phosphates lead to eutrophication of lakes and rivers when used in excess, but other cleaning agents also have undesirable environmental consequences, and phosphates are biodegradable.

20.73 The line structures of dioxin and DDT reveal their similarities and differences. Both contain benzene rings with chlorine atoms attached. Dioxin is a planar three-ring molecule with two O atoms bridging between two benzene rings, whereas the benzene rings in DDT are linked by a tetrahedral carbon atom, to which another tetrahedral carbon atom is attached:

Dioxin   DDT

20.75 The Lewis acid–base adduct between tetrahydrofuran and BF₃ requires that the boron atom approach the oxygen atom closely. As the structural formulas show, this is much easier to accomplish in the case of tetrahydrofuran than in the case of dimethyltetrahydrofuran, because methyl groups are much bulkier than hydrogen atoms:

Tetrahydrofuran   Dimethyltetrahydrofuran

Chapter 21 Nuclear Chemistry and Radiochemistry

Solutions to Problems in Chapter 21

21.1 The elemental symbol identifies the value of Z, and the left superscript is A. $Z + N = A$:

Part:	(a)	(b)	(c)
Z	10	82	40
A	20	205	90
N	10	123	50

21.3 Nuclides in the "belt of stability" are stable. Instability occurs if a nuclide has too few neutrons, too many neutrons, or $Z > 83$. In addition, most odd–odd nuclides are unstable. Mn has $Z = 25$. In this region of the periodic table, the belt of stability has $N{:}Z \sim 1.2$.

A	$N{:}Z$	Stability	Reason
53	1.12	Unstable	Too few neutrons
54	1.16	Unstable	Odd–odd
55	1.20	Stable	In belt of stability
56	1.24	Unstable	Too many neutrons, odd–odd

21.5 Energy releases are calculated using Equation 21-3: $\Delta E = (\Delta m)(8.988 \times 10^{10} \text{ kJ/g})$:

$$\Delta m = 1.00 \text{ metric tonne}\left(\frac{10^3 \text{ kg}}{1 \text{ metric tonne}}\right)\left(\frac{10^3 \text{ g}}{1 \text{ kg}}\right) = 1.00 \times 10^6 \text{ g}$$

$$\Delta E = (1.00 \times 10^6 \text{ g})(8.988 \times 10^{10} \text{ kJ/g}) = 8.99 \times 10^{16} \text{ kJ}$$

21.7 Binding energy is calculated by determining the mass defect (difference between the mass of the atom and the sum of the masses of its individual components) and converting to energy. When an element has only one stable isotope, its molar mass is the molar mass of that isotope:

$M_{Cs} = 132.91$ g/mol

The stable isotope has 55 protons and electrons, and $133 - 55 = 78$ neutrons:

$m_{components} = 55(1.007276 + 0.0005486) + 78(1.008665) = 134.106223$ g/mol

$\Delta m = 132.91$ g/mol $- 134.106223$ g/mol $= -1.20$ g/mol

$\Delta E = (-1.20 \text{ g/mol})(8.988 \times 10^{10} \text{ kJ/g}) = -1.079 \times 10^{11} \text{ kJ/mol}$

$$\Delta E_{\text{per nucleon}} = \frac{-1.079 \times 10^{11} \text{ kJ/mol}}{133 \text{ nucleons}} = -8.11 \times 10^8 \text{ kJ mol}^{-1} \text{ nucleon}^{-1}$$

21.9 The repulsive barrier for fusion depends on the product of the nuclear charges and on the radii of the two nuclides. The stability of the product nuclide depends on its proton:neutron ratio: ^{1}H + ^{6}Li → ^{7}Be, charge product = (1)(3) = 3, and the product has N:Z = 0.75; ^{4}He + ^{4}He → ^{8}Be, charge product = (2)(2) = 4 and product has N:Z = 1.00; the sum of the two nuclear radii is about the same for both reactions. Thus, the repulsive barrier is greater for the second reaction, but the product nuclide is more stable.

21.11 Symbols and names for nuclear particles appear in Table 20-3 and should be memorized:

	Description	Symbol	Name
(a)	High-energy photon	γ	Gamma ray
(b)	positive particle with mass number 4	α	Alpha particle
(c)	Positive particle with m_e	β^+	Positron

21.13 To identify the products of nuclear decay processes, make use of the principles of conservation of mass number and charge:

(a) No change in mass number or charge, product is $^{125}_{52}$Te

(b) Electron capture changes nuclear charge by –1, no change in mass number, product is $^{123}_{51}$Sb

(c) Beta decay changes nuclear charge by +1, no change in mass number, gamma decay does not change mass number or charge, product is $^{127}_{53}$I

21.15 Nuclides with too many neutrons decay by emitting β particles, thereby increasing Z while decreasing N. Nuclides with too few neutrons decay by emitting positrons or capturing electrons. Odd–odd nuclides may decay by any of these three processes.

^{53}Mn has too few neutrons, may emit a positron or undergo electron capture (observed mode is EC)

^{54}Mn is odd–odd, could decay by any of the three modes (observed mode is EC)

^{55}Mn is stable

^{56}Mn has too many neutrons, decays by β emission

21.17 Equation 21-4 is used to determine half-lives from radioactive decay data:

$$\text{Rate} = \frac{dN}{dt} = \frac{-N \ln 2}{t_{1/2}}$$

Summarize the known data: rate = –242 nuclei decays/s; $m = 1.33 \times 10^{-12}$ g.

Convert mass to number of nuclei, using $M \cong A = Z + N = 26 + 33 = 59$ g/mol.

$$N = 1.33 \times 10^{-12} \text{ g} \left(\frac{1 \text{ mol}}{59 \text{ g}} \right) \left(\frac{6.022 \times 10^{23} \text{ nuclei}}{1 \text{ mol}} \right) = 1.36 \times 10^{10} \text{ nuclei}$$

$$t_{1/2} = -\frac{(1.36 \times 10^{10} \text{ nuclei})(\ln 2)}{-242 \text{ nuclei/s}} = 3.9 \times 10^7 \text{ s}$$

Convert to a more convenient time unit:

$$3.9 \times 10^7 \text{ s} \left(\frac{1 \text{ min}}{60 \text{ s}} \right) \left(\frac{1 \text{ h}}{60 \text{ min}} \right) \left(\frac{1 \text{ day}}{24 \text{ h}} \right) \left(\frac{1 \text{ yr}}{365 \text{ day}} \right) = 1.2 \text{ yr}$$

21.19 Identify products of nuclear decay by noting that α decay changes A by –4 and Z by –2, whereas β decay changes Z by +1 while leaving A unchanged:

$$^{232}_{90}\text{Th} \to \alpha + {}^{228}_{88}\text{Ra} \to \beta + {}^{228}_{89}\text{Ac} \to \beta + {}^{228}_{90}\text{Th} \to \alpha + {}^{224}_{88}\text{Ra} \to \alpha + {}^{220}_{86}\text{Rn}$$

$$^{220}_{86}\text{Rn} \to \alpha + {}^{216}_{84}\text{Po} \to \beta + {}^{216}_{85}\text{At} \to \alpha + {}^{212}_{83}\text{Bi} \to \beta + {}^{212}_{84}\text{Po} \to \alpha + {}^{208}_{82}\text{Pb}$$

21.21 Identify the compound nucleus and the final products of a nuclear reaction by applying the principles of conservation of mass number and charge:

(a) $^{12}_{6}\text{C} + {}^{1}_{0}\text{n} \to \left({}^{13m}_{6}\text{C} \right) \to {}^{12}_{5}\text{B} + {}^{1}_{1}\text{p}$

(b) $^{16}_{8}\text{O} + {}^{4}_{2}\alpha \to \left({}^{20m}_{10}\text{Ne} \right) \to {}^{20}_{10}\text{Ne} + \gamma$

(c) $^{247}_{96}\text{Cm} + {}^{11}_{5}\text{B} \to \left({}^{258m}_{101}\text{Md} \right) \to {}^{255}_{101}\text{Md} + 3\,{}^{1}_{0}\text{n}$

21.23 Pictures of nuclear reactions should show each nuclide with the appropriate numbers of protons and neutrons. The reaction in this problem is

$$^{14}_{7}\text{N} + {}^{4}_{2}\alpha \to \left({}^{18m}_{9}\text{F} \right) \to {}^{18}_{8}\text{O} + {}^{0}_{1}\beta^+$$

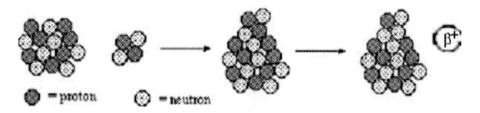

● = proton   ◉ = neutron

21.25 Stable elements with mass numbers around 135 have *N:Z* ratios in the range of 1.4. Use $A = N + Z$ to calculate what *Z* value is likely to be stable:

$$N = A - Z \qquad \frac{135 - Z}{Z} = 1.4 \qquad (2.4)Z = 135 \qquad Z = 56$$

Thus, the most likely high-mass elements resulting from U-235 fission are Cs, Ba, La, and Ce.

21.27 Calculate the mass defect from the masses of the individual reactants and products, and then convert to energy released:

$$\Delta m = [80.9199 + 151.9233 + 3(1.0087)] - [235.0439 + 1.0087] = -0.1833 \text{ g/mol}$$

$$\Delta E = (-0.1833 \text{ g/mol})(8.988 \times 10^{10} \text{ kJ/g}) = -1.648 \times 10^{10} \text{ kJ/mol}$$

This result is somewhat less than the result of the general calculation of Section Exercise 21.4.1 for a net change of two neutrons.

21.29 Your description should feature the fact that the core of a nuclear reactor generates radiation that converts materials in the vicinity of the core into radioactive substances. Thus, the heat exchanger in immediate contact with the core becomes radioactive and must be separated from the turbine that generates electricity. The primary heat exchanger transfers energy to a secondary heat exchanger, which does not become radioactive and can safely drive the turbine.

21.31 The amount of energy released in a fusion reaction can be calculated from the energy per event and the total amount of matter undergoing fusion. The reactions in this example are

$$^2\text{H} + {}^3\text{H} \rightarrow {}^4\text{He} + n \qquad \Delta E = -1.7 \times 10^9 \text{ kJ/mol}$$

$$^6\text{Li} + n \rightarrow {}^4\text{He} + {}^3\text{H} \qquad \Delta E = -4.6 \times 10^8 \text{ kJ/mol}$$

$$n_\text{H} = \frac{2.50 \text{ g}}{2.014 \text{ g/mol}} = 1.241 \text{ mol}$$

$$\Delta E = (1.241 \text{ mol})(-1.7 \times 10^9 \text{ kJ/mol}) = -2.11 \times 10^9 \text{ kJ}$$

In addition, 0.6207 mol Li react:

$$\Delta E = (0.6207 \text{ mol})(-4.6 \times 10^8 \text{ kJ/mol}) = -2.86 \times 10^8 \text{ kJ}$$

$$\Delta E_\text{total} = (-2.11 \times 10^9 \text{ kJ}) + (-2.86 \times 10^8 \text{ kJ}) = -2.4 \times 10^9 \text{ kJ}$$

(There are two significant figures because ΔE has two.)

21.33 To determine the speed of a nucleus that can fuse with another, first determine the energy needed to surmount the repulsive barrier, as described by Equation 3-1:

$$E_{electrical} = 1.389 \times 10^5 \text{ kJ pm mol}^{-1} \left(\frac{Z_1 Z_2}{r} \right)$$

Then calculate the speed from the equation for kinetic energy:

$$E_{kinetic} = \frac{1}{2}mv^2 \qquad\qquad v = \sqrt{\frac{2E_{kinetic}}{m}}$$

Determine the nuclide radii using the equation given in the problem:

$$r_{tritium\ nucleus} = 1.2(3)^{1/3} \text{ fm } (10^{-3} \text{ pm/fm}) = 1.7 \times 10^{-3} \text{ pm}$$

$$r_{deuterium\ nucleus} = 1.2(2)^{1/3} \text{ fm } (10^{-3} \text{ pm/fm}) = 1.5 \times 10^{-3} \text{ pm}$$

Here are the data:

$Z = 1$ for both nuclides, and $m = 3.0$ g/mol for tritium

$$E = 1.389 \times 10^5 \text{ kJ pm mol}^{-1} \left(\frac{(1)(1)}{(1.7 + 1.5) \times 10^{-3} \text{ pm}} \right) = 4.3 \times 10^7 \text{ kJ/mol}$$

$$v = \sqrt{\frac{(2)(4.3 \times 10^7 \text{ kJ/mol})(10^3 \text{ J/kJ})}{(3.0 \text{ g/mol})(10^{-3} \text{ kg/g})}} = 5.3 \times 10^6 \text{ m/s}$$

(Remember that $1 \text{ J} = 1 \text{ kg m}^2 \text{ s}^{-2}$, so the units in this calculation cancel to give m/s.)

21.35 Your description should include the extremely high energies required to initiate fusion and the difficulties in containing the fusion components at the temperature required for nuclei to have these high energies.

21.37 The characteristics of first-generation stars are described in your text:

Stage	Temperature	Composition
H-burning	4×10^7 K	H, He, e^-
He-burning	10^8 K	He, Be, C, O (H depleted)
C-burning	10^9 K	All nuclides from $Z = 6$ (C) up to $Z = 26$ (Fe)

21.39 Your description should include the fact that elements beyond $Z = 26$ are less stable than Fe, so they cannot be generated by fusion of lighter elements.

21.41 The problem asks for the energy released by a radioactive isotope. To determine this, it is necessary to calculate how much material decays during the time period. Determine the amount of the radioactive nuclide that decays in one day using

Equation 21-5:

$$\ln\left(\frac{N_0}{N}\right) = \frac{t \ln 2}{t_{1/2}}$$

$$\ln\left(\frac{N_0}{N}\right) = \frac{(1\ \text{day})(0.693)}{8.07\ \text{day}} = 0.0859 \qquad \frac{N_0}{N} = e^{0.0859} = 1.0897$$

$$N = \frac{7.45\ \text{pg}}{1.0897} = 6.84\ \text{pg} \qquad dN = 7.45\ \text{pg} - 6.84\ \text{pg} = 0.61\ \text{pg}$$

Now convert to moles and multiply by the energy released per mole to obtain the amount of energy captured by the gland:

$$n = 0.61\ \text{pg}\left(\frac{10^{-12}\ \text{g}}{1\ \text{pg}}\right)\left(\frac{1\ \text{mol}}{131\ \text{g}}\right) = 4.7 \times 10^{-15}\ \text{mol}$$

$$E = 4.7 \times 10^{-15}\ \text{mol}\left(\frac{9.36 \times 10^7\ \text{kJ}}{1\ \text{mol}}\right)\left(\frac{10^3\ \text{J}}{1\ \text{kJ}}\right) = 4.4 \times 10^{-4}\ \text{J}$$

21.43 Use Equation 21-4 to determine the rate of emission of a radioactive nuclide:

$$N = 7.45\ \text{pg}\left(\frac{10^{-12}\ \text{g}}{1\ \text{pg}}\right)\left(\frac{1\ \text{mol}}{131\ \text{g}}\right)\left(\frac{6.022 \times 10^{23}\ \text{nuclei}}{1\ \text{mol}}\right) = 3.425 \times 10^{10}\ \text{nuclei}$$

$$t_{1/2} = 8.07\ \text{day}\left(\frac{24\ \text{h}}{1\ \text{day}}\right)\left(\frac{60\ \text{min}}{1\ \text{h}}\right)\left(\frac{60\ \text{s}}{1\ \text{min}}\right) = 6.97 \times 10^5\ \text{s}$$

$$\frac{dN}{dt} = \frac{-N \ln 2}{t_{1/2}} = -\frac{(3.425 \times 10^{10}\ \text{decays})(0.693)}{6.97 \times 10^5\ \text{s}} = -3.41 \times 10^4\ \text{decays/s}$$

21.45 Exposure to radiation results first in damage to those cells that reproduce most quickly, including the white blood cells that are responsible for fighting infection and the mucous membrane lining of the intestinal tract. Thus, the early symptoms of radiation exposure include reduced resistance to infection and nausea due to disruption of the digestive tract.

21.47 Dating techniques using radioisotopes are based on Equation 21-5:

$$\ln\left(\frac{N_0}{N}\right) = \frac{t \ln 2}{t_{1/2}}.$$ To obtain an age estimate, the ratio $N_0{:}N$ must be determined.

Convert the mass ratio into the desired numerical ratio. Because both isotopes have the same value for A, their mole ratio and number ratio are the same as their mass ratio:

$$N_0 = N_{Sr} + N_{Rb} \qquad \frac{N_0}{N} = \frac{N_{Sr} + N_{Rb}}{N_{Rb}} = 1 + \frac{N_{Sr}}{N_{Rb}} = 1 + 0.0050 = 1.0050$$

$$t = \left(\frac{t_{1/2}}{\ln 2}\right)\ln\left(\frac{N_0}{N}\right) = \left(\frac{4.9 \times 10^{11} \text{ yr}}{0.693}\right)\ln(1.0050) = 3.5 \times 10^9 \text{ yr}$$

21.49 Add a small amount of a radioactive iron isotope to the manufacturer's iron waste stream before treatment, and monitor the iron content of the river downstream. If radioactivity appears in the river water, the source is the manufacturer; if not, the iron in the river comes from some other source.

21.51 Run the hydrolysis in water that is enriched with a radioactive isotope of oxygen. Isolate the two products and analyze them for radioactivity. Whichever product shows radioactivity is the one whose added oxygen atom comes from water.

21.53 Binding energy is calculated by determining the mass defect (difference between the mass of the atom and the sum of the masses of its individual components) and converting to energy. When an element has only one stable isotope, its molar mass is the mass of that isotope:

$M_{Bi} = 208.980$ g/mol

The isotope has 83 protons and electrons, $209 - 83 = 126$ neutrons.

$m_{components} = 83(1.007276 + 0.0005486) + 126(1.008665) = 210.741232$ g/mol

$\Delta m = 208.980 - 210.741232 = -1.761$ g/mol

$\Delta E = (-1.761 \text{ g/mol})(8.988 \times 10^{10} \text{ kJ/g}) = -1.583 \times 10^{11}$ kJ/mol

$$\Delta E_{per\ nucleon} = \frac{-1.583 \times 10^{11} \text{ kJ/mol}}{209 \text{ nucleons}} = -7.574 \times 10^8 \text{ kJ mol}^{-1} \text{ nucleon}^{-1}$$

21.55 Information about nuclear decays can be obtained using Equations 21-4 and 21-5.

$$\ln\left(\frac{N_0}{N}\right) = \frac{t \ln 2}{t_{1/2}} \qquad \ln N = \ln N_0 - \frac{t \ln 2}{t_{1/2}}$$

$N_0 = 5.0$ mg $t_{1/2} = 138.4$ day $t = 365$ days

$$\ln N = \ln(5.0) - \frac{(365 \text{ days})(\ln 2)}{138.4 \text{ days}} = -0.2182 \qquad N = e^{-0.2182} = 0.80 \text{ mg remain}$$

$$\frac{dN}{dt} = \frac{-N \ln 2}{t_{1/2}} \qquad t_{1/2} = 138.4 \text{ days}\left(\frac{24 \text{ h}}{1 \text{ day}}\right)\left(\frac{60 \text{ min}}{1 \text{ h}}\right)\left(\frac{60 \text{ s}}{1 \text{ min}}\right) = 1.196 \times 10^7 \text{ s}$$

$$N = 0.80 \text{ mg} \left(\frac{10^{-3} \text{ g}}{1 \text{ mg}} \right) \left(\frac{1 \text{ mol}}{210 \text{ g}} \right) \left(\frac{6.022 \times 10^{23} \text{ nuclei}}{1 \text{ mol}} \right) = 2.3 \times 10^{18} \text{ nuclei}$$

$$\frac{dN}{dt} = \frac{(2.3 \times 10^{18} \text{ decays})(\ln 2)}{1.196 \times 10^{7} \text{s}} = 1.3 \times 10^{11} \text{ decays/s}$$

21.57 Determine the products of nuclear decay by applying conservation of charge and mass number: the other product of neutron decay must be a particle with −1 charge and 0 mass, which is an electron: n → p + e.

Calculate the decay energy from the mass defect between reactant and products, using masses found in Table 21-1:

$\Delta m = (1.007276 \text{ g/mol}) + (5.486 \times 10^{-4} \text{ g/mol}) - (1.008665 \text{ g/mol}) = -0.0008404$ g/mol

$$\Delta E = (-0.0008404 \text{ g/mol})(8.988 \times 10^{10} \text{ kJ/g}) \left(\frac{10^{3} \text{ J}}{1 \text{ kJ}} \right) = -7.554 \times 10^{10} \text{ J/mol}$$

$$E_{\text{kinetic, electron}} = \frac{7.554 \times 10^{10} \text{ J/mol}}{6.022 \times 10^{23} \text{ mol}^{-1}} = 1.25 \times 10^{-13} \text{ J}$$

21.59 Calculations of decay times make use of Equation 21-5:

$$\ln \left(\frac{N_0}{N} \right) = \frac{t \ln 2}{t_{1/2}} \qquad\qquad t = \left(\frac{t_{1/2}}{\ln 2} \right) \ln \left(\frac{N_0}{N} \right)$$

When 1% has decayed, $t = \left(\dfrac{1622 \text{ yr}}{\ln 2} \right) \ln \left(\dfrac{N_0}{0.99 \, N_0} \right) = 24$ yr.

When 1% remains, $t = \left(\dfrac{1622 \text{ yr}}{\ln 2} \right) \ln \left(\dfrac{N_0}{0.01 \, N_0} \right) = 1.1 \times 10^{4}$ yr.

21.61 The *N:Z* ratio of an isotope determines where it lies with respect to the belt of stability. ^{11}C has six protons and $(11 - 6) = 5$ neutrons, *N:Z* = 0.833; ^{15}O has eight protons and $(15 - 8) = 7$ neutrons, *N:Z* = 0.875. Both isotopes are neutron-deficient and lie below (to the right of) the belt of stability. Isotopes that are neutron-deficient decay by positron emission, which increases their *N:Z* ratios:
$^{11}_{6}\text{C} \rightarrow\, ^{0}_{1}\beta^{+} +\, ^{11}_{5}\text{B}$ and $^{15}_{8}\text{O} \rightarrow\, ^{0}_{1}\beta^{+} +\, ^{15}_{7}\text{N}$

21.63 Calculations of decay times make use of Equation 21-5:

$$\ln \left(\frac{N_0}{N} \right) = \frac{t \ln 2}{t_{1/2}}, \text{ from which } t = \left(\frac{t_{1/2}}{\ln 2} \right) \ln \left(\frac{N_0}{N} \right)$$

When 1 μg is left, $t = \left(\dfrac{15\ \text{h}}{\ln 2}\right)\ln\left(\dfrac{25\ \mu\text{g}}{1\ \mu\text{g}}\right) = 70\ \text{h}.$

21.65 To identify the products of nuclear decay processes, make use of the principles of conservation of mass number and charge:

(a) Alpha emission changes nuclear charge by –2, mass number by –4, product is $_{90}^{234}\text{Th}$:

$$_{92}^{238}\text{U} \rightarrow\ _{2}^{4}\alpha +\ _{90}^{234}\text{Th}$$

(b) (n,p) changes nuclear charge by –1, no change in mass number, product is $_{27}^{60}\text{Co}$:

$$_{28}^{60}\text{Ni} +\ _{0}^{1}\text{n} \rightarrow\ _{1}^{1}\text{p} +\ _{27}^{60}\text{Co}$$

(c) This process changes nuclear charge by +6, mass number by (+12 – 3), product is $_{99}^{248}\text{Es}$:

$$_{93}^{239}\text{Np} +\ _{6}^{12}\text{C} \rightarrow 3_{0}^{1}\text{n} +\ _{99}^{248}\text{Es}$$

(d) (p, α) changes nuclear charge by –1, mass number by –3, product is $_{16}^{32}\text{S}$:

$$_{1}^{1}\text{p} +\ _{17}^{35}\text{Cl} \rightarrow\ _{2}^{4}\alpha +\ _{16}^{32}\text{S}$$

(e) Beta decay changes nuclear charge by +1, no change in mass number, product is $_{28}^{60}\text{Ni}$:

$$_{27}^{60}\text{Co} \rightarrow\ _{-1}^{0}\beta +\ _{28}^{60}\text{Ni}$$

21.67 Determine the expected product using conservation of mass number and charge number: $_{82}^{208}\text{Pb} +\ _{22}^{48}\text{Ti} \rightarrow\ _{104}^{256}\text{Rf}$. (This nucleus probably would decay by emitting neutrons and/or β particles.)

21.69 Energy releases are calculated using Equation 21-3: $\Delta E = (\Delta m)(8.988 \times 10^{10}\ \text{kJ/g})$:

$\Delta m = (147.9146) + (4.00260) - (151.9205) = -0.0033\ \text{g/mol}$

$\Delta E = (-0.0033\ \text{g/mol})(8.988 \times 10^{10}\ \text{kJ/g}) = -2.97 \times 10^{8}\ \text{kJ/mol}$

$$\Delta E_{\text{per nucleus}} = \left(\frac{-2.97 \times 10^{8}\ \text{J/mol}}{6.022 \times 10^{23}\ \text{nuclei/mol}}\right)\left(\frac{10^{3}\ \text{J}}{1\ \text{kJ}}\right) = -4.9 \times 10^{-13}\ \text{J/nucleus}$$

Fraction carried off by the α particle $= \dfrac{3.59 \times 10^{-13}\ \text{J}}{4.93 \times 10^{-13}\ \text{J}} = 0.73.$

21.71 Nuclides in the "belt of stability" are stable. Instability occurs if a nuclide has too few neutrons, too many neutrons, or $Z > 83$. In addition, most odd–odd nuclides are unstable: (a) too many neutrons; (b) $Z > 83$; (c) odd–odd; (d) too many neutrons.

21.73 In a nuclear reactor, the moderator serves to slow down fast neutrons, so they are more efficiently captured by the nuclear fuel. This reduces the amount of fuel required to sustain the reaction. The control rods serve to absorb some of the neutrons, allowing the reactor to operate just below its critical point and generate large quantities of heat without heating up beyond control.

21.75 Information about nuclear decays can be obtained using Equation 21-5.

$$\ln\left(\frac{N_0}{N}\right) = \frac{t \ln 2}{t_{1/2}} \qquad\qquad \ln N = \ln N_0 - \frac{t \ln 2}{t_{1/2}}$$

$$N_0 = (10^5 \text{ s}^{-1})(30 \text{ s}) = 3.0 \times 10^6$$

$$t = 5 \text{ h}\left(\frac{60 \text{ min}}{1 \text{ h}}\right)\left(\frac{60 \text{ s}}{1 \text{ min}}\right) = 1.8 \times 10^4 \text{ s}$$

If $t_{1/2} = 1100$ s, $\ln N = \ln(3.0 \times 10^6) - \left(\dfrac{(1.8 \times 10^4 \text{ s})(\ln 2)}{1100 \text{ s}}\right) = 3.57$

$$N = e^{3.57} = 36 \text{ nuclei remaining}$$

If $t_{1/2} = 876$ s, $\ln N = \ln(3.0 \times 10^6) - \left(\dfrac{(1.8 \times 10^4 \text{ s})(\ln 2)}{876 \text{ s}}\right) = 0.67$

$$N = e^{0.67} = 2 \text{ nuclei remaining}$$

21.77 Determine the nuclear reaction by applying conservation of mass number and charge number:

$${}^{10}_{5}\text{B} + \text{n} \rightarrow \left({}^{11}_{5}\text{B}\right) \rightarrow {}^{7}_{3}\text{Li} + \alpha$$. This reaction does not pose a significant health hazard, because Li-7 is a stable isotope and the α particles are easily stopped using an appropriate shield.

21.79 Dating techniques using radioisotopes are based on Equation 21-5:

$$\ln\left(\frac{N_0}{N}\right) = \frac{t \ln 2}{t_{1/2}}$$

Calculate the N_0:N ratio by taking the ratio of counts g^{-1} h^{-1} for the new and old samples:

$$\text{Fresh sample: count rate} = \frac{18\ 400 \text{ counts}}{(1.00 \text{ g})(20 \text{ h})} = 920 \text{ counts g}^{-1} \text{ h}^{-1}$$

$$\text{Old sample: count rate} = \frac{1020 \text{ counts}}{(0.250 \text{ g})(24 \text{ h})} = 170 \text{ counts g}^{-1} \text{ h}^{-1}$$

$$N_0/N = \frac{920}{170} = 5.4$$

$$t = \left(\frac{t_{1/2}}{\ln 2}\right)\ln\left(\frac{N_0}{N}\right) = \left(\frac{5730 \text{ yr}}{0.693}\right)\ln(5.4) = 1.4 \times 10^4 \text{ yr}$$

21.81 Use Equation 21-5 to calculate the useful lifetime of the dating technique:

$$\ln\left(\frac{N_0}{N}\right) = \frac{t \ln 2}{t_{1/2}} \qquad t = \left(\frac{t_{1/2}}{\ln 2}\right)\ln\left(\frac{N_0}{N}\right)$$

$$t = \left(\frac{5730 \text{ yr}}{0.693}\right)\ln\left(\frac{15.3 \text{ counts g}^{-1} \text{ min}^{-1}}{0.03 \text{ counts g}^{-1} \text{ min}^{-1}}\right) = 5.2 \times 10^4 \text{ yr}$$

21.83 Determine the products of nuclear decay using conservation of charge number and mass number. Use mass–energy equivalence to determine the mass of a product, given the mass of a reactant and the energy given off in the process:

(a) Sr has $Z = 38$ and Zr has $Z = 40$, so the decay requires two β particles (0 mass number, –1 charge number): $^{90}_{38}\text{Sr} \rightarrow {}^{90}_{39}\text{Y} + {}^{0}_{-1}\beta$; $^{90}_{39}\text{Y} \rightarrow {}^{90}_{40}\text{Zr} + {}^{0}_{-1}\beta$

(b) These nuclear decay processes involve emission of electrons and are exothermic, so there is a decrease in mass of the isotope in the process. Thus, Sr has the larger mass.

(c) $\Delta E = (\Delta m)(8.988 \times 10^{10} \text{ kJ/g})$; net reaction is $^{90}_{38}\text{Sr} \rightarrow {}^{90}_{40}\text{Zr} + 2{}^{0}_{-1}\beta$

$\Delta m = [89.9043 + 2(0.0005486)] - 89.9073 = -0.0019$ g/mol

$\Delta E = (-0.0019 \text{ g/mol})(8.988 \times 10^{10} \text{ kJ/g}) = -1.7 \times 10^8 \text{ kJ/mol}$

21.85 Calculations of decay times make use of Equation 21-5:

$$\ln\left(\frac{N_0}{N}\right) = \frac{t \ln 2}{t_{1/2}} \qquad t = \left(\frac{t_{1/2}}{\ln 2}\right)\ln\left(\frac{N_0}{N}\right)$$

First calculate the amount initially bound to the thyroid gland:

$$N_0 = 0.5 \text{ mg}\left(\frac{10^3 \text{ μg}}{1 \text{ mg}}\right)\left(\frac{45\%}{100\%}\right) = 225 \text{ μg}$$

When 0.1 μg is left, $t = \left(\frac{13.2 \text{ h}}{\ln 2}\right)\ln\left(\frac{225 \text{ μg}}{0.1 \text{ μg}}\right) = 1.5 \times 10^2 \text{ h}.$

21.87 Around $Z = 43$, the $N{:}Z$ ratio for stable isotopes is about 1.27, so the isotope that is most likely to be stable has $(43)(1.27) = 55$ neutrons. This, however, is an odd–odd isotope, so we might expect to find 56 neutrons, $^{99}_{43}\text{Tc}$. This, in fact, is the nuclide used widely in nuclear medicine.

21.89 To determine where the oxygen atom in the water molecule comes from, prepare a sample of the alcohol that is enriched in radioactive ^{18}O and run the reaction using this sample. Separate the products and measure the radioactivity of the ester and the water. If the C–OH bond in the alcohol breaks during the condensation, the ^{18}O will appear in the water, whereas if the C–OH bond in the carboxylic acid breaks during the condensation, the ^{18}O will appear in the ester.

21.91 The precipitate will contain radioactive Na, because once the NaBr is dissolved in solution, its Na^+ cations mix freely with the Na^+ cations of the existing solution. When the solution cools and $NaNO_3$ precipitates, some of the Na^+ cations in the precipitate will be radioactive Na-24.

21.93 First calculate the total energy output of the Sun, and then use mass–energy equivalence to determine the mass loss:

$$\text{Total output} = \frac{3.4 \times 10^{17}\,\text{J/s}}{4.5 \times 10^{-10}} = 7.6 \times 10^{26}\ \text{J/s}$$

(a) $\Delta m = \left(\dfrac{7.6 \times 10^{26}\,\text{J}}{1\,\text{s}}\right)\left(\dfrac{10^{-3}\,\text{kJ}}{1\,\text{J}}\right)\left(\dfrac{1\,\text{g}}{8.988 \times 10^{10}\,\text{kJ}}\right) = 8.5 \times 10^{12}\ \text{g/s}$

(b) The net reaction for the conversion of ^1H to ^4He is

$$4\,^1_1\text{H} + 2\,^{\ 0}_{-1}\text{e} \rightarrow\ ^4_2\text{He} + 6\,\gamma \qquad\qquad \Delta E_{\text{reaction}} = -2.5 \times 10^9\ \text{kJ/mol He}$$

Divide energy output by energy per mole to obtain amounts converted per second:

$$n_{\text{He}} = \frac{7.6 \times 10^{26}\,\text{J/s}}{2.5 \times 10^{12}\,\text{J/mol}} = 3.0 \times 10^{14}\ \text{mol/s } ^4\text{He produced}$$

$$n_{\text{H}} = 4n_{\text{He}} = 1.2 \times 10^{15}\ \text{mol/s } ^1\text{H converted}$$

21.95 (a) Am has $Z = 95$. The n:p ratio is determined from $A = Z + N$:

$$N = 241 - 95 = 146$$

$$\text{n:p} = N{:}Z = \frac{146}{95} = 1.537$$

(b) The amount of a radioactive isotope can be calculated from the decay rate using

Equation 21-4: $\dfrac{dN}{dt} = \dfrac{-N \ln 2}{t_{1/2}}$; $\quad N = \dfrac{-t_{1/2}}{\ln 2}\left(\dfrac{dN}{dt}\right)$. Here, $\dfrac{dN}{dt} = -5.0 \text{ s}^{-1}$.

$$t_{1/2} = 458 \text{ yr}\left(\frac{365 \text{ day}}{1 \text{ yr}}\right)\left(\frac{24 \text{ h}}{1 \text{ day}}\right)\left(\frac{60 \text{ min}}{1 \text{ h}}\right)\left(\frac{60 \text{ s}}{1 \text{ min}}\right) = 1.44 \times 10^{10} \text{ s}$$

$$N = \frac{\left(1.44 \times 10^{10}\text{s}\right)\left(5.0 \text{ s}^{-1}\right)}{0.693} = 1.04 \times 10^{11} \text{ atoms}$$

Convert to moles and then mass:

$$m = 1.04 \times 10^{11} \text{ atoms}\left(\frac{1 \text{ mol}}{6.022 \times 10^{23} \text{ atoms}}\right)\left(\frac{241 \text{ g}}{1 \text{ mol}}\right) = 4.2 \times 10^{-11} \text{ g}$$

(c) To calculate the useful lifetime of the detector, use Equation 21-5:

$$\ln\left(\frac{N_0}{N}\right) = \frac{t \ln 2}{t_{1/2}}; \quad t = \frac{t_{1/2}}{\ln 2}\ln\left(\frac{N_0}{N}\right) = \left(\frac{458 \text{ yr}}{0.693}\right)\ln\left(\frac{5.0}{3.5}\right) = 230 \text{ yr}$$

(d) The α particles would be stopped by 10 cm of air, but you would be exposed to γ rays.

21.97 Advantages of fission include demonstrated effectiveness, relatively plentiful supplies of fuel, and the fact that it is not based on fossil fuels and does not contribute to the greenhouse effect. Disadvantages include risk of serious pollution if there is an accident, and accumulation of harmful long-lived nuclear wastes.

Advantages of fusion include virtually inexhaustible energy source, very high energy output, and very little pollution. Disadvantages include uncertainty about whether a sustained fusion reactor can be built and the extremely high cost of such a reactor.

21.99 To solve this, we must take account of the dilution as well as the half-life of the tracer. Assuming the patient's total blood volume is v mL, the activity will be decreased by a factor of 10.0 mL / v mL due to dilution. The activity will be further decreased due to the decay of the ^{51}Cr between the time of injection and the time of measurement. Rearranging equation 21-5, the fractional decrease in activity will be:

$$\frac{N}{N_0} = \exp\left(\frac{-0.693\,t}{t_{1/2}}\right)$$

$$= \exp\left(\frac{-0.693(5.0 \text{ d})}{27.7 \text{ d}}\right)$$

$$= 0.882$$

The total fractional decrease in activity is therefore 0.882 (10 / v). From the given data, we know this fraction is equal to 640 Bq mL^{-1} / 2.96 x 10^5 Bq mL^{-1}. Thus,

$$\frac{640\,\text{Bq}\,\text{mL}^{-1}}{2.96\times10^{5}\,\text{Bq}\,\text{mL}^{-1}} = 0.882\left(\frac{10\,\text{mL}}{v\,\text{mL}}\right)$$

Solving, v = 4079 mL = 4.1 L